D1554922

CLASSICAL THERMODYNAMICS

CLASSICAL THERMODYNAMICS

ARNOLD MÜNSTER

Professor of Theoretical Physical Chemistry at the University of Frankfurt (Main)

translated by

E. S. HALBERSTADT, B.Sc., Ph.D.

Department of Chemistry, The University, Reading, Berks.

WILEY—INTERSCIENCE

A division of John Wiley & Sons Ltd.

London – New York – Sydney – Toronto

Library of Congress Catalog Card No. 71–122348

ISBN 0 471 62430 6

Made and printed in Great Britain
by John Wright & Sons Ltd.,
at the Stonebridge Press, Bristol

Preface

The present book originates from a course of lectures I have been giving for the last 17 years at the University of Frankfurt (Main). The purpose of this book is to explain, in a concentrated form appropriate to present-day needs, the formal structure of thermodynamics and the technique of handling the subject. The reader should then be capable of applying the theory by himself. The basic plan has been developed from my work in this field over many years and my experience in teaching the subject. This basic plan consists of the purely mathematical derivation of the whole of the structure of thermodynamics from three relationships: the fundamental equation, the equilibrium condition, and the stability condition. These relationships thus play a role similar to that of Maxwell's equations in electrodynamics. This presentation is undoubtedly associated with a certain lack of clarity in the conventional sense; the lack of clarity is, however, amply compensated by a greater depth of understanding and by an enhanced simplicity and confidence in the application of the theory to concrete problems.

The basic plan determines the structure of the book. We start by showing how simple facts of experience lead to the three basic relationships mentioned above (Chapter I). The purpose of this chapter is not (as was erroneously assumed in a review of the German edition) to serve as an introduction to the axioms of thermodynamics in a modern context. Such an introduction would be out of place within the framework of this book and new developments in this field have been largely ignored. Chapter I is designed to lead to a physical understanding of the mathematical formulations which constitute the basis of thermodynamics. Carathéodory's theory has likewise been treated in this way whereas its significance as the basis of modern thermodynamic axiomatics is mentioned only briefly without any detailed discussion.

Chapters II and III comprise an explanation of basic relationships and an investigation of these relationships in terms of their formal properties. Chapter IV (stability conditions) may be regarded as falling within the same field of systematics. The separation of this chapter from the other two is based on certain didactic considerations.

The remaining chapters serve to apply the basic relationships to a number of general problems (heterogeneous equilibria, chemical equilibria, critical phases and higher-order transitions, solids, systems in electric and magnetic fields, electrochemical systems, gravitational and centrifugal fields). This choice of subjects is based not only on their physical relevance but also on the desire to illustrate the power of thermodynamic formalism in as many ways as possible and, furthermore, to show by means of a few examples its usefulness (sometimes underestimated) in problems of practical research.

The design of the book excludes a systematic treatment of various classes of substances (gases, liquids, etc.); it also excludes any discussion of empirically or statistically based formulae for thermodynamic functions as well as numerical calculations. Exceptions to these rules are exemplified by classes of substances whose treatment is connected with the development of particular aspects of thermodynamic formalism (solids, electrochemical systems) and by the definition of molecular weight whose most convenient introduction into thermodynamics is by way of statistical mechanics. I have tried, however, to help the reader to understand the subject by means of numerous examples (often in the form of diagrammatic figures); these examples should also serve to indicate the way in which particular problems may be treated. In this connexion, special care has been taken in the treatment of certain problems which are known to cause particular difficulties to the beginner when he tries to apply the theory to such problems (e.g. the concept of internal parameters, particularly the progress variable; normalization of thermodynamic functions; single electrode potentials). The Nernst heat theorem is discussed at length for the same reason.

The book assumes a knowledge of chemistry, physical chemistry, and physics of about the standard of introductory courses in these subjects. The general mathematical background necessary comprises the basic operations of the differential and integral calculus, particularly partial differentiation of functions of several variables. Some further mathematical aids are developed briefly in the book. Vector and tensor algebra is used in Chapters VIII, IX, and X.

The collection of problems at the end of the book should not be regarded as mere exercises but as a significant supplementation of the material in the text. Certain suggestions in this respect have come from my students and colleagues. Since, however, it is not possible to determine the exact origin of these suggestions, I can merely express my gratitude to these anonymous helpers.

For their valuable help in the construction of the German text I wish to thank Dr. E. Lux, who, in particular, checked all the problems, and Mrs. A. Tüpker. I owe particular gratitude to Dr. E. S. Halberstadt for his excellent translation and for the correction of a number of errors in the original manuscript.

Grateful acknowledgment is made to the following publishers, societies, and authors for permission to reproduce illustrations:

Book or Journal	Publisher	Figure
Ferroelectrics and Antiferroelectrics Vol. 4, p. 44, Fig. 18. W. Känzig	Academic Press, New York	48
J. Am. Chem. Soc.	American Chemical	
62, 331 (1940). K. S. Pitzer	Society	38
51, 1441 (1929). W. F. Giauque and R. Wiebe		39
66, 1397 (1944). C. C. Stephenson and J. G. Hooley		50
Phys. Rev. **A134**, 385 (1964), Fig. 3. N. E. Phillips	American Institute of Physics	58
Rev. Mod. Phys. **10**, 1 (1938). F. C. Nix and W. Shockley	American Institute of Physics	37
Low Temperature Solid State Physics p. 31, Fig. 1.13. p. 150, Fig. 6.3. H. M. Rosenberg	Clarendon Press, Oxford	23
Helium p. 219, Fig. 4.17. W. H. Keesom	Elsevier Publishing Co., Amsterdam	36
Ferroelectrics p. 48, Fig. 2.32. E. Fatuzzo and W. J. Merz	North-Holland Publishing Co., Amsterdam	52
Axiomatic der Thermodynamik (Handbuch der Physik), Vol. 111, p. 169, Fig. 5. G. Falk and H. Jung	Springer-Verlag, Berlin	6
Thermodynamik der Mischphasen p. 106, Figs. 5 and 6. R. Haase	Springer-Verlag, Berlin	9, 10

Finally I wish to thank Messrs. John Wiley & Sons Ltd. for their excellent and understanding co-operation.

A. MÜNSTER

Frankfurt (Main), May 1970

List of the more important symbols

A Affinity
$\mathbf{B}$ Magnetic induction
C_V Molar heat capacity at constant volume
C_P Molar heat capacity at constant pressure
$\mathbf{D}$ Dielectric displacement (electric induction)
$\mathbf{E}$ Electric field strength
F Helmholtz free energy
$\mathfrak{F}$ Faraday constant
H Enthalpy
$\mathbf{H}$ Magnetic field strength
M_i Molecular weight of component i
$\mathbf{M}$ Total magnetization
P Pressure
P_i Intensive parameter
$\mathbf{P}$ Total electric polarization
Q Heat introduced into the system
R Gas constant
S Entropy
$\mathbf{S}$ Strain tensor
S_j Components of the strain tensor represented as the six-dimensional vector $\mathbf{S}$
T Absolute temperature
$\mathbf{T}$ Stress tensor
T_j Components of the stress tensor represented as a six-dimensional vector $\mathbf{T}$
U Internal energy
V Volume
W Work done on the system
X_i Extensive parameter
Y_j Work coefficient, generalized force
Z Function of state
a_i Activity of component i
$c_i = n_i/V$ Concentration of component i in mol cm^{-3}
c_{ij} Isothermal elastic stiffness coefficient
f_i Activity coefficient of component i
g Osmotic coefficient (§ 70)
g Acceleration due to gravity (§ 73)

h_i Partial molar enthalpy of component i
m Number of components
m Mass
$\mathbf{m}$ Magnetic moment per unit volume (magnetization density)
n_i Number of moles (mole number) of component i
$\mathbf{p}$ Electric polarization per unit volume (polarization density)
p_i Partial pressure of component i
s_i Partial molar entropy of component i
v_i Partial molar volume of component i
x_i Mole fraction of component i
y_j Work co-ordinate
z_i Valence of ionic species i
Π Osmotic pressure
Φ Electromotive force (e.m.f)
Φ_k Massieu–Planck function dependent on k intensive parameters
Ψ_k Thermodynamic potential dependent on k intensive parameters
α Coefficient of thermal expansion
α Degree of dissociation
γ_{ij} Piezoelectric coefficient
ε Dielectric constant
κ Isothermal compressibility
κ_s Isentropic compressibility
κ_{ij} Isothermal elastic compliance coefficient
μ Magnetic permeability
μ_i Chemical potential of component i
$\tilde{\mu}_i$ Electrochemical potential of component i
ν_i Stoichiometric coefficient
ξ Progress variable
ρ Density
σ Number of co-existing phases
ϕ Electric potential
χ Electric susceptibility (Chapter IX); magnetic susceptibility (Chapter X)

Molar quantities of pure substances are denoted by small letters (e.g. u, s) or by capital letters with an asterisk (e.g. U^*, S^*).

Contents

Preface v

List of the more important symbols ix

1. **Introduction** 1

CHAPTER I The Laws of thermodynamics . . . 5

2. Definitions 5

A. *Classical formulations of the Laws* . . . 8

3. The First Law. Internal energy . . . 8
4. The Second Law. Entropy and absolute temperature 9
5. Refrigerators and heat pumps . . . 17

B. *Carathéodory's axioms* 19

6. Definitions 19
7. Empirical temperature 20
8. The First Law 23
9. Interlude: Pfaff differentials . . . 25
10. The Second Law applied to quasi-static processes 30
11. Empirical definition of U, S, and T . . 35
12. Measurement of very low temperatures . 39
13. The Second Law applied to non-static processes 41

C. *Generalization of the Second Law to include open systems and chemical reactions* . . . 45

14. The problem 45
15. General formulation of the Second Law . 48

CHAPTER II General conditions for equilibrium and stability 52

16. The concept of equilibrium. Internal parameters 52
17. Gibbs' equilibrium conditions . . . 55
18. Stability conditions 61

CHAPTER III Thermodynamic potentials and Massieu–Planck functions 63
19. Interlude: Legendre transformations. Homogeneous functions and Euler's theorem . . 63
20. The fundamental equation. Extensive and intensive parameters. Equations of state. The Gibbs–Duhem equation 67
21. Thermodynamic potentials 75
22. Massieu–Planck functions 83
23. Transformation of the equilibrium and stability conditions 85
24. Gibbs–Helmholtz equations and Maxwell's relations 90
25. Calculation of partial derivatives. Method of Jacobi determinants. Joule–Thomson effect 94
26. Mean molar and partial molar quantities . 102

CHAPTER IV Heterogeneous equilibria without chemical reactions 109
27. General equilibrium conditions for heterogeneous systems 109
28. Membrane equilibria. Osmotic pressure . 111
29. The phase rule 113
30. Phase reactions 115
31. Invariant and univariant equilibria . . 119
32. Bivariant and multivariant equilibria . . 124

CHAPTER V Chemical equilibrium 128
33. General equilibrium conditions 128
34. Homogeneous reactions. The law of mass action 131
35. Heterogeneous reactions 138
36. The progress variable. Affinity 140
37. The thermodynamic calculation of chemical reactions 141
38. Nernst's heat theorem. The unattainability of the absolute zero. Zero point entropies . 144

CHAPTER VI Stability conditions 159
39. Statement of the problem. Gibbs' criterion 159
40. The stability conditions in the energy representation 162
41. Transformation of the stability conditions . 166

42. Stability conditions for heterogeneous systems 171
43. The Le Châtelier–Braun principle . . . 173
44. The stability of chemical equilibria . . . 176

CHAPTER VII Critical phases. Transitions of higher order . 178
45. Definition and properties of critical phases . 178
46. Gibbs' equations for critical phases . . . 184
47. Transformation of the equations for critical phases. Further properties of critical phases 189
48. Transitions of higher order. The Ehrenfest equations 192
49. Tisza's theory. I: The general basis . . . 200
50. Tisza's theory. II: Formulation of the stability conditions 204
51. Tisza's theory. III: Critical points and higher-order transitions 208

CHAPTER VIII Solids 216
52. The strain tensor 216
53. The stress tensor 222
54. The fundamental equation and thermal equations of state 226
55. Thermodynamic potentials and Maxwell's relations 229
56. Symmetry properties of solids 231

CHAPTER IX Systems in an electric field 238
57. Electrostatic work 238
58. The fundamental equation for a dielectric in an electric field 242
59. Thermodynamic potentials 246
60. Electrostriction 247
61. The electrocaloric effect 251
62. Piezoelectricity 252
63. Ferroelectricity 254

CHAPTER X Systems in a magnetic field 265
64. Magnetostatic work 265
65. The fundamental equation for a system in a magnetic field 267
66. The magnetocaloric effect 270
67. Superconduction 272

CHAPTER XI Electrochemical systems 279
68. Definition and general properties of electrochemical systems 279
69. General conditions for electrochemical equilibrium 282
70. Solutions of strong electrolytes . . . 284
71. Membrane equilibria of electrolyte solutions 292
72. Galvanic cells 295

CHAPTER XII Gravitational field. Centrifugal field. The determination of molecular weights . . 307
73. Systems in a gravitational field . . . 307
74. Systems in a centrifugal field . . . 313
75. The determination of molecular weights . 315

Problems 323

Hints for solving the problems 344

Solutions to problems 350

Bibliography 376

Index 379

Introduction

§1

As a part of theoretical physics thermodynamics is one of the group of phenomenological theories, together with hydrodynamics and electrodynamics. All these theories have the following properties in common:

(a) The atomic structure of matter is ignored.
(b) Accordingly, only measurable quantities defined for macroscopic systems are taken into consideration.
(c) The laws of the theory are derived from experience of macroscopic phenomena. These laws are expressed in suitable mathematical form (Navier–Stokes equations, Maxwell's equations).
(d) Characteristic properties of substances appear as characteristic parameters (viscosity, dielectric constant).

The subject of thermodynamics may be defined, for the time being and using conventional terminology, as being concerned with those physical phenomena which involve heat and temperature. Thermodynamics actually deals with only some of these phenomena. It confines itself to the considerations of equilibrium states and to those changes of state which may be represented by a continuous series of equilibrium states (quasi-static changes of state). Such changes of state can, strictly speaking, only be imagined as being infinitely slow. They can, therefore, not be represented as a function of time. Nonstatic changes of state must sometimes be taken into account but they are not a true part of the theory. The name 'thermostatics' has therefore been thought to be more representative of the subject but has never been generally accepted. In recent times, the thermodynamics of irreversible processes has been developed. Although it is related to classical thermodynamics it has a completely different structure and will not be considered in this book.

It should be clear from what has been said so far that the structure of thermodynamics differs significantly from that of other phenomenological theories. Firstly, time is never a factor in thermodynamic considerations. Secondly, spatial co-ordinates do not, in general, appear since the quantities measured at equilibrium are not functions of spatial co-ordinates. Although the systems considered by thermodynamics are not necessarily homogeneous (e.g. the system liquid–vapour), the spatial distribution of the homogeneous regions is not significant. The situation is somewhat different when external fields (gravitational, electric, and magnetic) or boundaries need to be considered. We shall deal with such cases in Chapters IX, X and XII. Here it is sufficient to say that the basic structure of thermodynamics is not affected by them.

The structure of thermodynamics can be defined, in a negative way, by saying that, unlike hydrodynamics and electrodynamics, it is not a field theory and its geometrical representations are confined to abstract phase space. These considerations determine which aspects of thermodynamics are easy and which are difficult.

It follows from what has just been said that the typical partial differential equations of mathematical physics containing derivatives of time and co-ordinates do not appear in thermodynamics. The mathematics of thermodynamics is, in fact, extremely simple, apart from a few special cases, and consists mainly of the methods of partial differentiation and of ordinary differential equations of simple form. The conceptual aspect of thermodynamics is, in contrast, extraordinarily abstract and it is here that the real difficulties arise. It has long been customary to try to avoid these difficulties by means of spurious analogies. It has, however, become clear that this method makes a deeper understanding leading to mastery of the subject more difficult. The characteristic properties of this field must be accepted and, on the one hand, basic concepts must be developed from concrete experience while, on the other, the mathematical structure must be analysed. These considerations determine the way in which this book is written. Chapter I analyses the basic concepts while subsequent chapters develop the formal structure. Examples of applications are given but the various kinds of substance are not treated systematically. This is justified by the fact that thermodynamic relationships are independent of specific properties which always have to be determined experimentally. The theoretical study of specific properties is the subject of statistical thermodynamics, a complementary part of theoretical physics.

Statistical thermodynamics starts explicitly with the atomic structure of matter and makes possible a deductive foundation for the laws which are introduced into thermodynamics on the basis of experience.

The historical beginnings of thermodynamics lie in the study of the efficiency of heat engines. This explains the name 'thermodynamics'.

1824 Publication of Carnot's thesis 'Réflexions sur la puissance motrice du feu' which includes the theorem which states that the efficiency of a heat engine depends only on the temperature difference and not on the working substance.

1834 Carnot's notes, examined after his death, clearly state the energy principle, i.e. the equivalence of heat and mechanical energy.

1834 Clapeyron applies Carnot's results to vapour–liquid equilibria and arrives at the relationship now called the Clausius–Clapeyron equation which contains an unknown temperature function later identified as the absolute temperature scale by Clausius.

1840–45 Experimental demonstration of the equivalence of heat and work by Joule, published in 1845.

1842 Formulation of the energy principle by Robert Mayer.

1848 W. Thomson (Lord Kelvin) defines an absolute temperature scale (i.e. independent of the thermometric substance) based on Carnot's work.

1850 Clausius publishes the thesis 'Über die bewegende Kraft der Wärme und die Gesetze, welche sich daraus für die Wärmelehre selbst ableiten lassen'. (Concerning the motive force of heat and the thermodynamic laws which can be derived from it.) This, in essence, combines Carnot's theory with the energy principle and thus contains the basis of the Second Law of thermodynamics.

According to Willard Gibbs, the greatest exponent of thermodynamics of a later period, this thesis marks an epoch in the history of physics and the beginning of thermodynamics as a science (Gibbs, *Collected Works*, vol. II, p. 262).

1851 W. Thomson formulates explicitly the two laws of thermodynamics based on the work of Carnot, Joule, and Clausius.

1854 Clausius introduces the concept of entropy leading to a new formulation of the Second Law.

1865 Clausius introduces the term 'entropy'. This work contains the famous quotation: 'The energy of the universe is constant. The entropy of the universe tends towards a maximum.'

1869 Massieu introduces the first 'characteristic functions' from which all thermodynamic quantities can be derived by differentiation.

1873 Horstmann calculates, for the first time, chemical equilibria (dissociation of $CaCO_3$ and PCl_5).

1875 Gibbs publishes his thesis 'On the equilibrium of heterogeneous substances'. Extension of thermodynamics in a general form to heterogeneous systems and chemical reactions. Derivation of equilibrium conditions for various special cases from a completely general formulation. Introduction of characteristic functions. The motto of Gibbs' thesis is the above quotation from Clausius.

1882 Helmholtz, independently of Gibbs, introduces free energy and derives the relationship now known as the Gibbs–Helmholtz equation.

1886 Duhem derives the Gibbs–Duhem equation.

1887 Planck divides changes of state into two classes, reversible and irreversible processes.

1906 Nernst publishes his new heat theorem.

1909 Carathéodory publishes a new axiomatic basis of thermodynamics.

Later work deals with axiomatic problems, extension of the formal framework, and application to special problems.

CHAPTER I

The Laws of thermodynamics

§ 2 Definitions

The so-called Laws of thermodynamics constitute, as has already been explained, its axioms. Starting with certain observed facts they serve to develop the concepts used to build up the formal apparatus of the subject. The formulation of the Laws is, however, the result of a historical process. They do not constitute a complete system of axioms and it must be remembered that thermodynamics uses observed facts which are not contained in the Laws. We shall return to these considerations in due course.

Let us start with some definitions.

A *system* is a part of the physical world, defined in some way by boundaries, which is the subject of an investigation. Usually the system is the sample of material under investigation in the laboratory. It may also be a much more complicated structure (heat engine, electrical network).

The *state* of a system is determined by a set of measurable quantities such that the result of any further measurement applied to the system can be calculated. This formal definition is inherently not particularly useful. Experience shows, however, that the complete set of quantities can be divided into sub-sets which give an approximately complete picture in the above sense, i.e. within each sub-set all other quantities can be calculated from a limited number of measurable quantities. Thus, this limited number of quantities necessary to calculate all others within the sub-set may be taken as the definition of a state. This idealized division corresponds to other branches of theoretical physics: the mechanical state of a chemical system is defined by a set of general co-ordinates and momenta or velocities; the state of a quantum mechanical system is defined by the wave function or Dirac's ket vector; an electromagnetic state is defined by the dependence of electric or magnetic field strength on spatial co-ordinates. A complete division is possible only for ideal systems (incompressible liquid, vacuum). Coupling can be taken into account to a certain extent by the introduction of suitable parameters. Further development leads to the fusion of separate subjects into more comprehensive theories (irreversible thermodynamics). These are, however, usually only of interest when applied to certain problems.

Thermodynamics is a phenomenological theory and therefore deals only with macroscopic measurable quantities, i.e. with quantities which can be defined for a macroscopic system. Such quantities can be defined only for a macroscopic system (a point mass has no temperature) or, at least, the structure of matter does not enter into the definition (in this sense the lattice constant of a crystal is not a macroscopic quantity). The measurable quantities of thermodynamics are defined partly in mechanics,† partly in the Laws of thermodynamics.

A considerable simplification is caused by thermodynamics being limited to equilibrium states and quasi-static processes. Since the existence and properties of *equilibrium* are closely connected with the Second Law, we can, at the moment, define equilibrium merely as that state towards which a system tends spontaneously; alternatively, equilibrium may be defined as that state in which the thermodynamic quantities of the system are no longer time dependent. Experience shows that the number of quantities necessary to describe such a state is smaller than that for any non-equilibrium state. For a given mass of ideal gas, for example, the equilibrium state can be completely described by any two of the three variables—pressure, volume, and temperature—while non-equilibrium states additionally need temperature and density gradients for a complete description.

We shall, from now on, denote as a (thermodynamic) *variable of state* a set of variables which defines the thermodynamic state at equilibrium. We simply call this a (thermodynamic) *state*. The system which has thus been completely described is called a (thermodynamic) *system*. A quantity whose differential is a complete differential of the variables of state is called *a function of state*. The abstract space which corresponds to variables of state is called *co-ordinate* (or *phase*) *space*. Each thermodynamic equilibrium state of the system thus corresponds uniquely to a point in co-ordinate space and vice versa.

A system is called *closed* when it can exchange energy with its surroundings but cannot exchange matter (a liquid under its own

† Certain quantities of thermodynamics are defined by electrodynamics (see Chapters IX–XI).

vapour is an *open system*). An *isolated system* can exchange neither matter nor energy with its surroundings. A substance is called *homogeneous* when temperature, pressure, concentration, and, consequently, all other macroscopic physical properties (crystal structure, refractive index, etc.) within it are independent of location. It should be obvious that this definition makes sense only on a macroscopic scale. If the size and shape of the substance are ignored, we speak of a *phase* (vapour, liquid, crystal). A homogeneous system contains only a single phase. Two *co-existent* phases can occur together separated by a smooth boundary and in an equilibrium which is not determined by 'barriers' (e.g. two different crystals at room temperature pressed against each other are separated by a 'barrier'). A *heterogeneous system* contains two or more co-existent phases. The number (m) of (*independent*) *components* of a system in the thermodynamic sense is equal to the number (c) of substances (or particle species) in the chemical sense less the sum of the number (r) of reaction equations and the number (b) of conditional equations connecting them, i.e.

$$m = c - r - b. \tag{2.1}$$

Example. An aqueous solution of common salt contains the substances H_2O, NaCl, Na^+, Cl^-. There is one reaction equation

$$Na^+ + Cl^- \rightleftharpoons NaCl \tag{2.2}$$

and, if n_A is the mole number of substance A, the conditional equation is

$$n_{Na^+} = n_{Cl^-}. \tag{2.3}$$

According to (2.1), therefore, the system has two (independent) components in the thermodynamic sense (binary system). It is assumed that the reactions corresponding to the reaction equations really occur under the experimental conditions.

Example. The system H_2, O_2, H_2O generally has two components because of the reaction equation

$$2H_2 + O_2 \rightleftharpoons 2H_2O. \tag{2.4}$$

At room temperature and atmospheric pressure, however, this reaction does not take place in the absence of a catalyst. Under these conditions the system is, therefore, a three-component system (ternary system).

The above definitions are important. Insufficient attention to them leads to much misunderstanding and many mistakes.

The following discussion will be restricted to a certain class of systems called *simple systems*. They are characterized by each phase having constant mass and composition and their state is defined by two independent variables of state. We shall see later that at least two variables of state are required for the definition of the thermodynamic state; we shall, therefore, for the time being neglect solid substances, external fields, boundary phenomena, etc. It follows further from the definition that exchange of matter takes place neither between the phases nor with the surroundings. The whole system and each separate phase on its own therefore constitutes a *closed* system. Finally, the definition also excludes chemical reactions. We shall return to these questions in Part C of this chapter.

A. CLASSICAL FORMULATION OF THE LAWS

§ 3 The First Law. Internal energy

The classical formulation of the Laws takes the concepts of temperature and heat from the immediate experience of the senses and does not analyse them further. Their measurability is accepted as innate. We shall therefore defer a detailed discussion of these concepts to Part B of this chapter.

The empirical foundations of the First Law were laid by Joule's experiments (1840–45). He showed that the amount of mechanical work necessary to raise the temperature of a certain amount of water by 1° was always the same. This so-called *equivalence principle* was formulated by Thomson as follows:

When equal amounts of mechanical work are produced from thermal sources or disappear in thermal effects, equivalent amounts of heat disappear or appear.

This theorem can also be put in the form: It is impossible to construct a machine which does mechanical work without consuming an equivalent amount of heat. (Principle of the impossibility of perpetual motion of the first kind.)

Let Q be the heat absorbed by the system and W the work done on the system. If we consider a process which starts from a definite state and returns to the same state via a series of (positive or negative) heat absorptions, ΔQ, and a series of (positive or negative) work performances, ΔW (cyclic process), then according to the equivalence principle

$$\sum \Delta Q = -\sum \Delta W. \tag{3.1}$$

If we now define a quantity

$$\Delta U = \Delta Q + \Delta W \tag{3.2}$$

we have from (3.1), for the limiting case,

$$\oint \mathrm{d}U = 0 \tag{3.3}$$

where the integration is to be performed over a closed curve in co-ordinate space. It follows that

$$\mathrm{d}U = \mathrm{d}'Q + \mathrm{d}'W \tag{3.4}$$

is a complete differential and that U is a function of state. These considerations constitute the appropriate mathematical formulation of the First Law within the framework of the classical theory. Following Clausius, U is called the *internal energy* of the system. According to its definition, U is defined except for an additive constant.

It should be noted that $\mathrm{d}'Q$ and $\mathrm{d}'W$ are not, in general, exact differentials. The values of $\int_1^2 \mathrm{d}'Q$ and $\int_1^2 \mathrm{d}'W$ depend on the path in co-ordinate space over which the integration is performed. This can easily be confirmed by means of simple examples.

For a homogeneous simple system, we have for a quasi-static process

$$\mathrm{d}W = -P\,\mathrm{d}V \tag{3.5}$$

where V is the volume and P the pressure. The internal energy is, therefore, in this case

$$\mathrm{d}U = \mathrm{d}'Q - P\,\mathrm{d}V. \tag{3.6}$$

For a closed system, therefore, the internal energy is constant.

§4 The Second Law. Entropy and absolute temperature

According to Clausius the empirical basis of the Second Law is formulated as follows:

It is impossible for heat to flow spontaneously (i.e. in an isolated system) from a colder to a hotter body.

The formulation used by Thomson and later by Planck is:

It is impossible to construct a cyclic engine which performs mechanical work by merely cooling a heat reservoir. (Principle of the impossibility of perpetual motion of the second kind.)

These two formulations, which at first sight seem to have little connexion, are in fact equivalent.

Proof. (a) Let Thomson's principle be true but not that of Clausius. Part of the heat contained in a heat reservoir R could then be brought to a higher temperature and thus a heat reservoir R′ at a higher temperature could be produced. A part of the heat in R′ could then be changed into work, the rest brought to the temperature of R and returned to R. R′ would then have disappeared and the net effect would be that work had been obtained from the heat contained in R. This process could be repeated indefinitely and would thus constitute perpetual motion of the second kind. This is in conflict with Thomson's principle and, if Thomson's principle is true, Clausius' principle is likewise true.

(b) Let Clausius' principle be true but not that of Thomson. A perpetual motion machine of the second kind could then be constructed. A certain amount of heat could be abstracted from a heat reservoir and this heat could be changed into work. This work can freely be changed into heat at a higher temperature. The heat in question would thus without compensation be brought from a lower to a higher temperature. This denies Clausius' principle and, therefore, if Clausius' principle is true, Thomson's principle is likewise true.

An immediate consequence of Thomson's principle is that a continuously working engine which converts heat into work (heat engine) is possible only if a working substance (e.g. steam) undergoes a cyclic process between two heat reservoirs at different temperatures. The working substance accepts an amount of heat Q_1 (all amounts of heat are expressed as absolute values) from the hotter reservoir and rejects an amount of heat Q_2 into the colder reservoir. The difference $Q_1 - Q_2$ is converted into work. The efficiency of such a cyclic process is defined by the equation

$$\eta = \frac{Q_1 - Q_2}{Q_1}. \tag{4.1}$$

Clausius' principle then leads to the following theorem:

The efficiency of a completely reversible cyclic process K cannot be exceeded by any other cyclic process which uses the same working substance between the same two temperatures.

Proof. If we assume the theorem to be wrong, there exists a cyclic process K′ between the same two temperatures for which

$$W' = W, \quad \eta' > \eta, \quad Q_1' < Q_1, \quad Q_2' < Q_2. \tag{4.2}$$

K′ and K can then be coupled so that K occurs in the reverse direction. According to (4.2) the resulting work would be zero. Since

$Q_1' < Q_1$, however, heat would be added to the hotter reservoir which contradicts Clausius' principle.

An analogous argument is used to prove Carnot's theorem:

The efficiency of a reversible cyclic process depends only on the temperature of the two heat reservoirs and is independent of the nature of the working substance.

Thomson showed that Carnot's theorem can be used to define thermodynamic or absolute temperature.

According to (4.1)

$$\frac{Q_2}{Q_1} = 1 - \eta. \tag{4.3}$$

From Carnot's theorem and (4.3) we have

$$\frac{Q_2}{Q_1} = \mathrm{f}(t_1, t_2) \tag{4.4}$$

where t_1 and t_2 are the temperatures of the two reservoirs on an arbitrary empirical scale. We now imagine a third heat reservoir at a temperature t_3 such that $t_1 > t_3 > t_2$; we also have two reversible cyclic processes K and K′ such that K occurs between t_1 and t_3 while K′ occurs between t_3 and t_2. They are coupled so that K rejects into t_3 the same amount of heat which is absorbed from t_3 by K′. From the above theorem we then have that the resultant efficiency of the coupled processes is the same as the efficiency of a simple cyclic process between t_1 and t_2. In both cases, an amount of heat Q_2 is rejected into the coldest reservoir. But we have

$$\frac{Q_3}{Q_1} = \mathrm{f}(t_1, t_3), \tag{4.5}$$

$$\frac{Q_2}{Q_3} = \mathrm{f}(t_3, t_2). \tag{4.6}$$

From (4.4)–(4.6) we have

$$\frac{Q_2}{Q_3} = \frac{Q_2/Q_1}{Q_3/Q_1} = \frac{\mathrm{f}(t_1, t_2)}{\mathrm{f}(t_1, t_3)} \tag{4.7}$$

and, therefore,

$$\mathrm{f}(t_3, t_2) = \frac{\mathrm{f}(t_1, t_2)}{\mathrm{f}(t_1, t_3)}. \tag{4.8}$$

However, according to the above theorems the efficiency of a cyclic process between the temperatures t_3 and t_2 must be independent of the temperature t_1. Equations (4.7) and (4.8), therefore, give

$$\mathrm{f}(t_3, t_2) = \frac{Q_2}{Q_3} = \frac{\phi(t_2)}{\phi(t_3)} \tag{4.9}$$

where ϕ is a new function. The absolute temperature T is then defined by the equation

$$\phi(t) = \mathrm{C}T \tag{4.10}$$

where C is a universal proportionality factor. According to (4.9) we then have

$$\frac{Q_2}{Q_1} = \frac{T_2}{T_1}. \tag{4.11}$$

Equations (4.10) and (4.11) constitute Thomson's definition of absolute temperature. A reversible cyclic process between the ice point and the boiling point of water at standard pressure (1 atm) gives the absolute temperature scale when taken together with the statement that, over this range, T increases by 100 units. We then have that

$$\frac{Q_1}{Q_2} = \frac{x+100}{x} \tag{4.12}$$

and measurements show that

$$x = 273{\cdot}15\mathrm{K}. \tag{4.13}$$

This defines the absolute or Kelvin scale of temperature.

We now introduce two new definitions. A process which occurs under the condition $T = \text{const}$ is called an *isothermal process.* A process which occurs such that $\mathrm{d}'Q = 0$ is called an *adiabatic process.* For an adiabatic process, therefore, the system must be thermally insulated against the surroundings. Let us consider a homogeneous system and a reversible cyclic process consisting of two adiabatic and two isothermal processes (*Carnot cycle*).

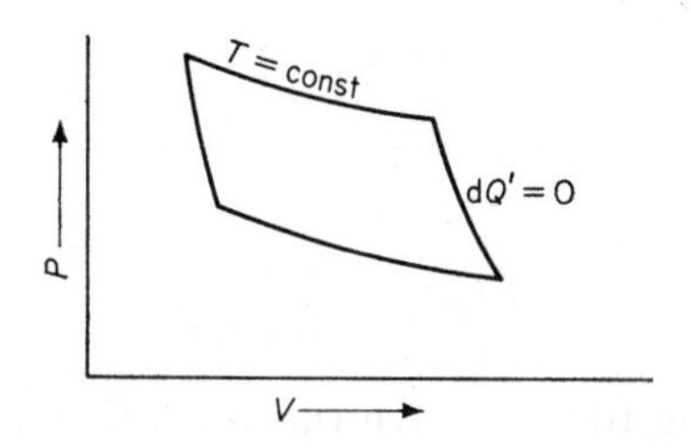

FIG. 1. Carnot cycle

If we again write heats algebraically, (4.11) gives for the Carnot cycle

$$\frac{Q'}{T'}+\frac{Q''}{T''} = 0. \tag{4.14}$$

It can be shown that any reversible cyclic process in the PV-plane may be represented as the sum of infinitesimal Carnot cycles.

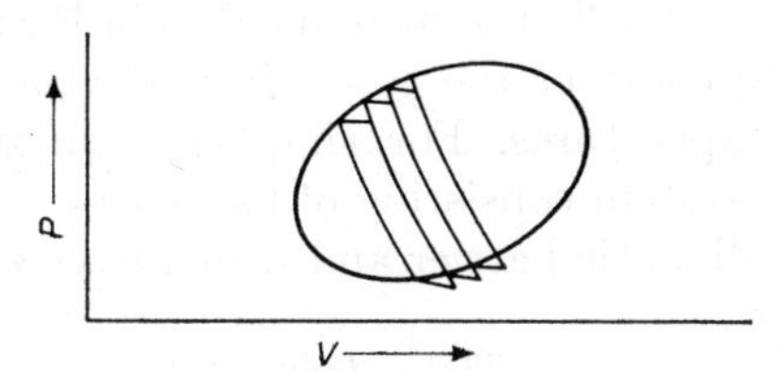

FIG. 2. Arbitrary cyclic process

It must then be generally true that

$$\oint \frac{dQ}{T} = 0. \tag{4.15}$$

If we now define a quantity

$$dS = \frac{d'Q}{T}, \tag{4.16}$$

(4.15) shows that dS is a complete differential and S is therefore a function of state. Clausius (1865) called this function the *entropy* of the system. The reciprocal of the absolute temperature is, therefore, an integrating factor for the incomplete differential $d'Q$.† These considerations constitute the mathematical formulation of the Second Law for reversible processes. According to the definition, entropy is defined except for an additive constant.

From (4.16) we have that for an adiabatic reversible process the entropy remains constant. Such processes are therefore called *isentropic processes*.

From (4.16) and (3.6) we have

$$T\,dS = dU+P\,dV. \tag{4.17}$$

Let us now consider a heterogeneous system whose phases are distinguished by the index α. The definition of a simple system excludes exchange of material with the surroundings or between

† In order to simplify equations we shall, from now on, no longer distinguish incomplete differentials by primes, but write simply dQ and dW.

phases. If there is no heat exchange between phases we have

$$\mathrm{d}S = \sum_{\alpha} \mathrm{d}S^{(\alpha)}. \tag{4.18}$$

The total entropy change of the system is thus equal to the sum of the entropy changes in the parts of the system and $\mathrm{d}S$ is again given by (4.16). If the barriers between the phases allow heat to pass and if we suppose all parts of the system to have the same temperature the same applies. This will be shown in detail in Part B.

We still have to consider the case where the parts of the system have different temperatures. For simplicity we shall consider an adiabatic isolated system consisting of two parts, α and β, which are separated by an adiabatic barrier and which have temperatures $T^{(\alpha)}$ and $T^{(\beta)}$ such that

$$T^{(\alpha)} > T^{(\beta)}. \tag{4.19}$$

The entropy is a function of state and has, in this state (denoted by A) a definite value given by (4.18) as

$$S_{\mathrm{A}} = S_{\mathrm{A}}^{(\alpha)} + S_{\mathrm{A}}^{(\beta)}. \tag{4.20}$$

If the barrier is now made heat conducting, a temperature equilibrium will result. According to Clausius' principle, heat will flow from α to β. In the final state B, the entropy again has a definite value given by

$$S_{\mathrm{B}} = S_{\mathrm{B}}^{(\alpha)} + S_{\mathrm{B}}^{(\beta)}. \tag{4.21}$$

Since the process is adiabatic and irreversible, $S_{\mathrm{A}} \neq S_{\mathrm{B}}$ according to the Second Law. Since, however, the whole system is closed, we have $\mathrm{d}'Q = 0$. Equation (4.16) is, therefore, not applicable to this case. This result was to be expected since we included an irreversible process. This gap in the theory may be filled, on the basis of Clausius' principle, as follows:

Let us suppose that heat exchange between the phases is much slower than heat equilibration within a phase. During the irreversible process each phase is thus in internal equilibrium, i.e. in a thermodynamic state. The two phases, however, are not in equilibrium with each other. During the process, therefore, we have functions of state U and S for each phase. The entropy change for the whole system can then be defined, after the introduction of an adiabatic barrier between the parts of the system, by eq. (3.18). We now consider an infinitesimal step of the process during which an amount of heat

$$-\mathrm{d}Q_i^{(\alpha)} = \mathrm{d}Q_i^{(\beta)} \equiv \mathrm{d}Q_i \tag{4.22}$$

flows from α to β. The temperatures of the two parts of the system may be regarded as constant during this process. If no external work is done ($\mathrm{d}V = 0$) the entropy changes associated with the process are given by the above and eq. (4.17) as

$$T^{(\alpha)}\,\mathrm{d}S^{(\alpha)} = \mathrm{d}U^{(\alpha)}, \quad T^{(\beta)}\,\mathrm{d}S^{(\beta)} = \mathrm{d}U^{(\beta)}. \tag{4.23}$$

Introduction of the general eq. (3.4) (for $\mathrm{d}'W = 0$) gives

$$\mathrm{d}S^{(\alpha)} = \frac{\mathrm{d}Q_i^{(\alpha)}}{T^{(\alpha)}}, \quad \mathrm{d}S^{(\beta)} = \frac{\mathrm{d}Q_i^{(\beta)}}{T^{(\beta)}}. \tag{4.24}$$

This shows that (4.16) applies as long as the phase under consideration is in internal equilibrium, i.e. in a thermodynamic state; it is not, however, necessary for the heat flow to be a reversible process. The total entropy change due to the irreversible process is then

$$\mathrm{d}S_i = \mathrm{d}S^{(\alpha)} + \mathrm{d}S^{(\beta)} \tag{4.25}$$

or, with (4.22) and (4.24),

$$\mathrm{d}S_i = \mathrm{d}Q_i\left(\frac{1}{T^{(\beta)}} - \frac{1}{T^{(\alpha)}}\right). \tag{4.26}$$

Clausius' principle then, because of (4.19), gives the fundamental result

$$\mathrm{d}S_i \geqslant 0. \tag{4.27}$$

The entropy change caused by the irreversible process of heat conduction can only be positive or, in the limiting case, zero.

The preceding argument may be generalized for other exchange processes (exchange of matter, chemical reaction). The equations will, of course, become more complicated. If the whole system is not closed towards the surroundings, dissipative processes may also occur (irreversible change of work into heat, e.g. by friction or an electric current). Even continuous processes may be considered by reducing phases to volume elements such that the differences between adjacent volume elements are infinitesimal. Equation (4.27) is thus shown to be valid for all irreversible processes. For simplicity's sake we shall only consider the case discussed above. The main results are, however, generally valid.

In order to obtain the total entropy change, we still have to consider the exchange of heat with the surroundings, bearing in mind that the two phases have, by definition, different temperatures. The total heat absorbed from the surroundings must, therefore, be divided according to the equation

$$Q = Q^{(\alpha)} + Q^{(\beta)}. \tag{4.28}$$

If the increase in entropy due to absorption of heat from the surroundings is $\mathrm{d}S_{\mathrm{a}}$, we have

$$\mathrm{d}S = \mathrm{d}S_{\mathrm{a}} + \mathrm{d}S_i \tag{4.29}$$

or, introducing (4.16) and (4.26),

$$\mathrm{d}S = \frac{\mathrm{d}Q^{(\alpha)}}{T^{(\alpha)}} + \frac{\mathrm{d}Q^{(\beta)}}{T^{(\beta)}} + \mathrm{d}Q_i\left(\frac{1}{T^{(\beta)}} - \frac{1}{T^{(\alpha)}}\right). \tag{4.30}$$

For simplicity's sake we shall suppose that $\mathrm{d}Q^{(\alpha)} = \mathrm{d}Q^{(\beta)} = \frac{1}{2}\mathrm{d}Q$ and introduce an 'effective temperature'

$$\frac{2}{T} = \frac{T^{(\alpha)} + T^{(\beta)}}{T^{(\alpha)}\,T^{(\beta)}}. \tag{4.31}$$

If we now define $\mathrm{d}Q'$ by

$$\mathrm{d}Q' = 2\mathrm{d}Q_i \frac{T^{(\alpha)} - T^{(\beta)}}{T^{(\alpha)} + T^{(\beta)}} \tag{4.32}$$

then (4.30) becomes

$$\mathrm{d}S = \frac{\mathrm{d}Q}{T} + \frac{\mathrm{d}Q'}{T}. \tag{4.33}$$

This formulation, due to Clausius, is again completely general. If the temperature of the system is uniform (i.e. irreversible processes are confined to transport of matter, chemical reactions and dissipative processes), the effective temperature becomes the same as the actual temperature. The quantity $\mathrm{d}Q'$ is defined, according to Clausius, as 'uncompensated heat'. Its significance for the theory of irreversible processes appears to have been recognized first by Duhem (1911). Comparison of (4.33) with (4.29) shows, together with (4.27), that quite generally

$$\mathrm{d}Q' \geqslant 0 \tag{4.34}$$

must be true. It must be remembered that (4.33) is simply a symbolic way of writing eq. (4.30) for the case of heat conduction. We shall retain this symbolic form to the end of this section in order to obtain a simple formulation. Explicit calculations must, however, obviously be based on (4.30).

From (4.33) and (4.34) we get the completely general relationship

$$\mathrm{d}Q \leqslant T\,\mathrm{d}S. \tag{4.35}$$

This important relationship is called *Clausius' inequality*. In the absence of irreversible processes, it reduces to eq. (4.16).

Let us now consider an adiabatic isolated system. According to (4.29) $dS = dS_i$ and we have from (4.27)

$$dS \geqslant 0 \quad \text{(adiabatic isolated system).} \tag{4.36}$$

The entropy of an isolated system can thus never decrease. In the limiting case when only reversible processes take place, the entropy remains constant. It follows that the entropy of an isolated system in thermodynamic equilibrium (i.e. all possible irreversible processes in the system have gone to completion) has the maximum value compatible with the given conditions. This value is usually, but not necessarily, a stationary value in the mathematical sense.

The preceding considerations amplify the Second Law and constitute the basis for the formulation of the equilibrium conditions (Chapter II). If the universe is regarded as a closed system the theorem of Clausius mentioned in § 1 follows.

§ 5 Refrigerators and heat pumps

The classical development of the Second Law stems from a particular technical problem and can, therefore, serve as a convenient model for the investigation of related problems. We shall look briefly at two such applications. As in § 4, we shall confine our discussion to the idealized case of completely reversible cyclic processes.

The cyclic process considered at the beginning of § 4 constitutes the basic principle of all heat engines and yields the maximum efficiency η, i.e. the greatest efficiency allowed by the Second Law. This maximum efficiency can be reached only in the limiting case of complete reversibility of the process. The equation for the efficiency can be written according to (4.3) and (4.11) as

$$\eta = \frac{T_1 - T_2}{T_1}. \tag{5.1}$$

Since T_2 is in practice usually fixed (temperature of the surroundings), a direct consequence is that the efficiency of a heat engine is increased by increasing the working temperature (use of superheated steam, internal combustion engines).

The process under consideration can also be used in the reverse direction in such a way that an amount of heat Q_2 is taken from the colder reservoir, work W is done on the working substance and heat Q_1 is finally given up to the hotter reservoir. Such an arrangement is called a *refrigerator* when the technical interest is centred on the cooling of the colder reservoir. When the heating of the hotter reservoir is the desired result, the arrangement is called a *heat pump*. Both machines are, therefore, based on the same principle and differ only from a technical point of view.

The simplest example of a refrigerator is the ordinary domestic type. The inside of the 'fridge' constitutes the colder reservoir (temperature T_2) while the surroundings represent the hotter reservoir (temperature T_1). The work is done by the electric current. The definition of the efficiency is naturally based on the technological point of view. For refrigerators, it is therefore defined as the ratio of cooling (i.e. the heat Q_2 extracted from the colder reservoir) to the work done, i.e.

$$\varepsilon_c = \frac{Q_2}{W}. \tag{5.2}$$

Since we have assumed a reversible cyclic process, we again have from (4.11) (if the heats are written as algebraic quantities)

$$\frac{Q_1}{T_1} + \frac{Q_2}{T_2} = 0. \tag{5.3}$$

Now, according to the First Law,

$$W + Q_2 + Q_1 = 0 \tag{5.4}$$

and, therefore,

$$\frac{Q_2}{T_2} - \frac{Q_2 + W}{T_1} = 0. \tag{5.5}$$

We then get the efficiency of the refrigerator from (5.2) and (5.5) to be

$$\varepsilon_c = \frac{T_2}{T_1 - T_2}. \tag{5.6}$$

This shows that the efficiency of a refrigerator becomes greater when the temperature difference $T_1 - T_2$ becomes smaller (the 'fridge' uses more current during the heat of summer). On the other hand, the efficiency is lower (at given T_1) when T_2 is lower. This is one of the reasons why it is difficult to attain very low temperatures.

The heat pump is based on the same principle. In practice the hot reservoir may be represented by the inside of a house while the cold reservoir is represented by the open air. The work is again done by an electric current. According to the technological standpoint the efficiency is here defined as

$$\varepsilon_H = -\frac{Q_1}{W}. \tag{5.7}$$

With eq. (5.4), this gives

$$\varepsilon_{\mathrm{H}} = \frac{T_1}{T_1 - T_2}. \tag{5.8}$$

The efficiency is therefore less at large temperature differences $T_1 - T_2$. If $T_1 - T_2$ is fixed the efficiency increases with increasing temperature T_1 of the hot reservoir.

Such an arrangement can be realized in principle by placing an open 'fridge' by an open window, although the effect is very small. Heat pumps are of considerable technical interest for heating purposes because of their enormous efficiency. We can see, for instance, from eq. (5.8) that for $T_1 = 16\,^\circ\mathrm{C}$ and $T_2 = 0\,^\circ\mathrm{C}$ more than 16 times the electric energy used appears as heat energy (in the ideal case). The heat pump is thus greatly superior to direct electric heating. Practical applications have been almost absent so far owing to the excessive investment and maintenance costs, but now seem promising for smaller installations. The use of the Peltier effect in semiconductors is of special interest in this connexion.

B. CARATHÉODORY'S AXIOMS

§ 6 Definitions

The classical development of the Laws is particularly suitable for an introduction to the basic concepts of thermodynamics. Certain aspects are, however, not satisfactory from a logical standpoint (uncritical introduction of the concepts of temperature and heat, no clear differentiation between purely physical facts of experience and purely mathematical aspects). We shall now develop the Laws, according to Carathéodory (1909), in a strictly systematic way.

We shall first summarize, even though this involves repetition, the most important definitions which we shall need. We shall accept as given only those concepts which are defined in mechanics (mass, volume, pressure, work, etc.). Let us consider a homogeneous isotropic substance in equilibrium in the absence of external fields. According to hydrostatics the volume V is then uniquely determined by the pressure P. We now introduce the first fact of experience which also serves to define a concept.

Theorem based on experience; definition. Processes exist which lead to a change in V even though P is constant. These are called *thermal processes*.

Heating and cooling are therefore thermal processes. The theorem shows immediately that, in thermodynamics, the system we are considering needs two independent variables to define its state. We shall choose, for the time being, the quantities P and V as variables of state. They are defined in mechanics. Each equilibrium state of the substance then corresponds to a point in the P–V-plane.

A wall surrounding the substance is called *adiabatic* when the state of the substance can be changed only by non-thermal (i.e. mechanical, electrical, etc.) processes (movement of parts of the wall, switching on of the current, etc.). Any non-thermal wall is called *diathermic* or *heat conducting*.

A change of state is called *quasi-static* when it occurs via a continuous series of equilibrium states. It must, therefore, occur infinitely slowly and is not a function of time (see § 1). *Example*: Expansion of a gas in a cylinder by slow withdrawal of the piston.

Any other change of state is called *non-static*. Intermediate states in non-static processes need additional variables for their description. Their discussion, apart from certain general considerations, lies beyond the scope of classical thermodynamics. We shall confine our discussion, for the time being, to quasi-static processes.

It should be noted that the concept of quasi-static change of state is pragmatically the same as that of reversible change of state, but the two are different from the point of view of logic. The concepts of reversibility and irreversibility will be introduced much later in connexion with the Second Law.

§ 7 Empirical temperature

Let us consider a homogeneous adiabatic closed substance in the state P_0, V_0. Let there be a quasi-static change of state resulting in the state P, V. Since thermal processes are by definition excluded, all states which are attainable in a quasi-static adiabatic way, starting from the initial state P_0, V_0, must lie on a curve in the P–V-plane. This curve is called the *adiabatic* of the body through P_0, V_0. All the adiabatics of the body constitute a single-parameter family of curves on the P–V-plane. This family of curves completely covers a region of the plane. One, and only one, adiabatic therefore passes through each point in the P–V-plane.

The equation of the adiabatics may be written as

$$s(P, V) = s \tag{7.1}$$

where s denotes the parameter.

Example. The equation of the adiabatic for an ideal gas is

$$PV^{\kappa} = s \tag{7.2}$$

where κ is a constant characteristic of the substance, and

$$s = P_0 V_0^{\kappa}. \tag{7.3}$$

Let us now consider a system which is adiabatically isolated from the surroundings and which consists of the phases $'$ and $''$. The corresponding families of adiabatics are given by

$$s'(P', V') = s', \quad s''(P'', V'') = s''. \tag{7.4}$$

If the two phases are separated by an adiabatic wall, there will be a relationship between P' and V' on the one hand, and one between P'' and V'' on the other. Arbitrary pairs of values can exist separately in equilibrium. If, however, the separating wall is diathermic, experience shows that this is no longer true.

Theorem of experience, and definition. Where two bodies are separated by a diathermic wall and are isolated adiabatically from the surroundings, they are in equilibrium with each other only if one equation of the form

$$\mathrm{F}(P', V'; P'', V'') = 0 \tag{7.5}$$

is obeyed. Such an equilibrium is called a *thermal* equilibrium since, by definition, it can be attained only by means of a thermal process.

This theorem does not tell us whether eq. (7.5) is obeyed because of any special properties of the substances concerned or whether it is obeyed because of a general property possessed, in principle, by all substances. The following example explains what we mean. Let us introduce into our system a third, unprimed, phase in such a way that it can be brought into thermal contact with $'$ and with $''$ (by changing the adiabatic wall into a diathermic wall). We shall suppose that thermal equilibrium exists between the unprimed phase and $'$ and also between the unprimed phase and $''$. The equations

$$\mathrm{F}(P, V; P', V') = 0, \quad \mathrm{F}(P, V; P'', V'') = 0 \tag{7.6}$$

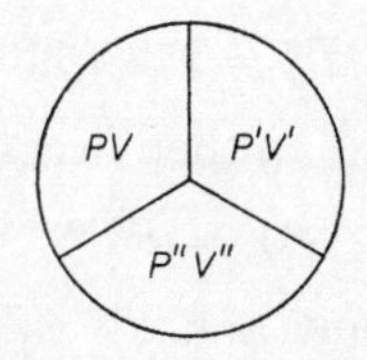

Fig. 3. Definition of empirical temperature

are therefore obeyed. The theorem still does not tell us whether thermal equilibrium exists between ′ and ″. A new fact of experience is therefore required.

Theorem of experience, and definition. If, in a three-phase system, ′ and ″ are in thermal equilibrium with a third phase, ′ and ″ are also in thermal equilibrium with each other. Each phase has, therefore, a measurable property $t(P, V)$ such that $t' = t''$ follows from $t = t'$, $t = t''$.

This property is called the *empirical temperature.* The theorem so far allows us merely to state whether the temperatures of two bodies are the same or different. In order to get a temperature scale which permits comparison of different temperatures, the function

$$t(P, V) = t \tag{7.7}$$

must be defined. This is done by defining two easily reproducible temperatures (thermometric fixed points, e.g. ice point and boiling point of water at standard pressure) and measuring, over this interval, the change in a suitable property of a convenient 'thermometric substance' (e.g. change of volume of mercury or ethanol). If numerical values are assigned to the temperatures of the fixed points (0° and 100° on the Celsius scale), the temperature scale is defined. The most convenient thermometric substance from a theoretical point of view is the ideal gas, i.e. the measurement of PV extrapolated to $PV \to 0$ for a gas with the lowest possible condensation temperature (H_2 or He). For this case, the function $t(P, V)$ is known, apart from scale conversion constants, and only one fixed point is required. The true gas thermometer scale is the one which agrees most accurately with the thermodynamic scale (§ 4).

Equation (7.7) is called the *thermal equation of state* of the body under consideration. For an ideal gas it has the simple form

$$PV = \mathrm{f}(t) \tag{7.8}$$

where $\mathrm{f}(t)$ depends only on the empirical temperature. On the gas thermometer scale $\mathrm{f}(t)$ is identical with the temperature except for a constant factor. One puts

$$T_0^* = 273{\cdot}15° \tag{7.9}$$

for the ice point. Equation (7.8) then becomes (for one mole)

$$PV = RT^*$$

where R is the gas constant and T^* is numerically equal to the temperature on the Kelvin scale (§ 4).

Equation (7.7) represents a further family of single-parameter curves in the P–V-plane, called the *isothermals* of the substance.

§ 8 The First Law

Carathéodory's theory formulates the First Law from a viewpoint different from that of the classical theory. Internal energy is introduced by means of purely mechanical concepts and a basic fact of experience is used to define the concept of heat.

Theorem of experience, and definition. The same amount of mechanical (or electrical, etc.) work is always necessary to change a homogeneous or heterogeneous system from an initial state 1 to a final state 2 by an adiabatic process. This amount of work is independent of the equilibrium or non-equilibrium states traversed during the process. The work done on the system during such a process is called the increase in *internal energy* of the system. The equation

$$U_2 - U_1 = \int_1^2 \mathrm{d}W \quad \text{(adiabatic)} \tag{8.1}$$

therefore applies.

According to this definition, the internal energy is a function of state and is defined except for an additive constant.

The usefulness of this definition (which is also a way of measurement) obviously depends on whether the process is practicable for all pairs of states 1, 2. A somewhat unsatisfactory aspect of Carathéodory's theory is that a consequence of the Second Law must be considered at this point, i.e. that it is not always possible to reach any state 2 from any other state 1 by means of an adiabatic process. For instance, it is impossible to change a homogeneous system at constant volume adiabatically to a state of lower internal energy since such a process would be the reverse of a dissipative process (e.g. production of heat by friction). In such cases, it is, however, always possible to realize the reverse of the desired process adiabatically. This assures the general validity of the definition (8.1).

If the process $1 \to 2$ occurs along an arbitrary (non-adiabatic) path, the work done depends on the path, i.e. on the intermediate states passed through by the system. The change in internal energy must, however, be the same as that for the adiabatic case since it depends only on the initial and final states.

Definition. For any process $1 \to 2$, the difference between the increase in internal energy and the work done on the system is called the *heat absorbed*:

$$\int_1^2 \mathrm{d}'Q = (U_2 - U_1) - \int_1^2 \mathrm{d}'W \quad \text{(any change of state)} \tag{8.2}$$

or, in differential form,

$$\mathrm{d}U = \mathrm{d}'Q + \mathrm{d}'W. \tag{8.3}$$

This definition and eqs. (8.1)–(8.3) together constitute the First Law in the Carathéodory formulation.

We shall consider briefly a few special cases: for homogeneous systems we have for a *quasi-static* process

$$\mathrm{d}W = -P\mathrm{d}V \tag{8.4}$$

and, therefore,

$$\mathrm{d}U = \mathrm{d}Q - P\mathrm{d}V; \tag{8.5}$$

for an *adiabatic* process

$$\mathrm{d}Q = 0, \quad \mathrm{d}U = \mathrm{d}W, \tag{8.6}$$

and for a *quasi-static adiabatic* process

$$\mathrm{d}Q = 0, \quad \mathrm{d}U = -P\mathrm{d}V. \tag{8.7}$$

These equations can easily be generalized for heterogeneous systems since the quantities of eq. (8.3) are obtained by adding together the contributions of the individual phases. For the sake of simplicity, we shall confine ourselves to two phases $'$ and $''$.

We then have for a *quasi-static* process

$$\mathrm{d}U' + \mathrm{d}U'' = \mathrm{d}Q' + \mathrm{d}Q'' - P'\mathrm{d}V' - P''\mathrm{d}V'', \tag{8.8}$$

and for a *quasi-static adiabatic* process

$$\mathrm{d}Q' + \mathrm{d}Q'' = 0, \quad \mathrm{d}U' + \mathrm{d}U'' = -P'\mathrm{d}V' - P''\mathrm{d}V''. \tag{8.9}$$

If we choose t and V as variables of state, the First Law gives for a homogeneous system

$$\mathrm{d}U = \left(\frac{\partial U}{\partial t}\right)_V \mathrm{d}t + \left(\frac{\partial U}{\partial V}\right)_t \mathrm{d}V. \tag{8.10}$$

Combination of (8.7) and (8.10) gives for a *quasi-static adiabatic* process

$$\left[\left(\frac{\partial U}{\partial V}\right)_t + P\right] \mathrm{d}V + \left(\frac{\partial U}{\partial t}\right)_V \mathrm{d}t = 0. \tag{8.11}$$

Accordingly we have for a two-phase system whose phases are in thermal equilibrium

$$\left[\left(\frac{\partial U'}{\partial V'}\right)_t + P'\right] \mathrm{d}V' + \left[\left(\frac{\partial U''}{\partial V''}\right)_t + P''\right] \mathrm{d}V'' + \left[\left(\frac{\partial U'}{\partial t}\right)_{V',V''} + \left(\frac{\partial U''}{\partial t}\right)\right]_{V',V''} \mathrm{d}t = 0. \tag{8.12}$$

The derivative of internal energy with respect to temperature, which occurs in these equations, is called the *heat capacity* at constant volume. If it refers to one mole of a uniform substance, it is called the *molar heat capacity at constant volume*, C_v.

§ 9 Interlude: Pfaff differentials

This section summarizes some mathematical aids necessary for Carathéodory's theory. As was seen in § 8 the First Law leads to equations of the form

$$\mathrm{d}Q = \sum_{i=1}^{n} X_i \,\mathrm{d}x_i \tag{9.1}$$

where the quantities X_i are functions of some or all of the independent variables. Such expressions are called Pfaff differentials. Equation (9.1) can be integrated along a path in n-dimensional space but the value of the integral will, in general, depend on the chosen path. The integral can, therefore, not be expressed in the form $Q(2) - Q(1)$ where 1 and 2 denote the integration limits. $\mathrm{d}Q$ is called an incomplete differential in such a case. We now pose two questions:

(a) Under what conditions is $\mathrm{d}Q$ a complete differential?

(b) Under what conditions is there an integrating factor $1/\tau$ such that $\mathrm{d}Q/\tau$ is a complete differential?

(a) *Condition for* $\mathrm{d}Q$ *to be a complete differential.* The condition for $\mathrm{d}Q$ to be a complete differential can always be written

$$\mathrm{d}Q = \mathrm{d}\mathrm{f}(x_1, \ldots, x_n) \tag{9.2}$$

where $\mathrm{df}(x_1, \ldots, x_n)$ is the differential of a function $\mathrm{f}(x_1, \ldots, x_n)$. It then follows that

$$\int_1^2 \mathrm{d}Q = \mathrm{f}(2) - \mathrm{f}(1). \tag{9.3}$$

By definition,

$$\mathrm{df} = \sum_{i=1}^{n} \frac{\partial \mathrm{f}}{\partial x_i} \mathrm{d}x_i \tag{9.4}$$

and thus, because of (9.1),

$$X_i = \frac{\partial f}{\partial x_i}. \tag{9.5}$$

We have, therefore, that

$$\frac{\partial X_i}{\partial x_j} = \frac{\partial^2 f}{\partial x_i \partial x_j}, \quad \frac{\partial X_j}{\partial x_i} = \frac{\partial^2 f}{\partial x_j \partial x_i}. \tag{9.6}$$

The necessary and sufficient condition for dQ to be a complete differential is, therefore,

$$\frac{\partial X_i}{\partial x_j} = \frac{\partial X_j}{\partial x_i} \quad (i,j = 1,\ldots,n). \tag{9.7}$$

Equation (9.1) in vectorial notation becomes

$$dQ = \boldsymbol{R}.d\boldsymbol{r}. \tag{9.8}$$

Condition (9.7) for $n = 3$ can, therefore, be written in the simple form†

$$\text{rot}\,\boldsymbol{R} = 0. \tag{9.9}$$

The necessary and sufficient condition for the line integral of $\boldsymbol{R}$ to vanish along a closed curve is that $\boldsymbol{R}$ is expressible as the gradient of a scalar function.

(b) *Condition for the existence of an integrating factor*. Let us suppose that dQ is not a complete differential and ask the question: What condition determines whether there is an integrating factor $1/\tau$ such that dQ/τ is a complete differential. If we call

$$dQ = 0 \quad \text{or} \quad \sum X_i\, dx_i = 0 \tag{9.10}$$

the Pfaff differential equation for dQ, we have

Theorem 1. An integrating factor $1/\tau$ exists for the Pfaff differential form of dQ when, and only when, there is a solution of the form

$$\sigma(x_1,\ldots,x_n) = \sigma \tag{9.11}$$

to the Pfaff differential equation for dQ.

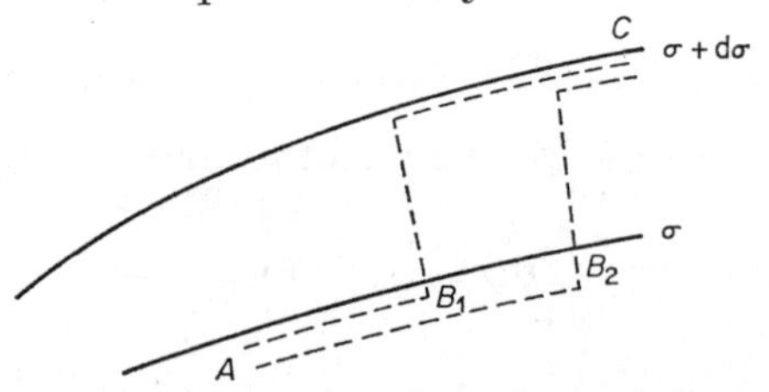

FIG. 4. Existence of an integrating factor

† For $n > 3$ an analogous form in tensor notation is possible.

Proof. The simplest proof is geometric. The solution (9.11) defines a single parameter family of hypersurfaces in n-dimensional space. Let us consider two adjacent infinitesimal surfaces σ and $\sigma+\mathrm{d}\sigma$. Let us now go from A to C, first via B_1 and then via B_2. By definition, $\mathrm{d}Q = 0$ and $\mathrm{d}\sigma = 0$ on both surfaces. However, we have assumed that the change in $\mathrm{d}Q$ depends on the path taken. Since $\mathrm{d}Q$ is zero on both surfaces the change in $\mathrm{d}Q$ can depend only on the crossing point B. We have, therefore, that

$$\mathrm{d}Q = \tau(B)\,\mathrm{d}\sigma. \tag{9.12}$$

Since, however, $\mathrm{d}\sigma$ is by definition a complete differential, we have

$$\mathrm{d}\sigma = \frac{\mathrm{d}Q}{\tau(x_1, \ldots, x_n)} \tag{9.13}$$

which proves the existence of the integrating factor.

Theorem 2. If an integrating factor exists for a Pfaff differential, then an infinite number of integrating factors exists for the differential.

Proof. If $S(\sigma)$ is a single-single valued function of σ, (9.11) becomes

$$S[\sigma(x_1, \ldots, x_n)] = S. \tag{9.14}$$

Using (9.13), we get

$$\mathrm{d}S = \frac{\mathrm{d}S}{\mathrm{d}\sigma}\mathrm{d}\sigma = \frac{\mathrm{d}S}{\mathrm{d}\sigma}\frac{\mathrm{d}Q}{\tau}. \tag{9.15}$$

If we now put

$$T(x_1, \ldots, x_n) = \tau(x_1, \ldots, x_n)\frac{\mathrm{d}\sigma}{\mathrm{d}S} \tag{9.16}$$

we get, from (9.15),

$$\mathrm{d}S(x_1, \ldots, x_n) = \frac{\mathrm{d}Q}{T(x_1, \ldots, x_n)}. \tag{9.17}$$

The reciprocal of $T(x_1, \ldots, x_n)$ is, therefore, also an integrating factor, which proves the theorem.

Theorem 3 (Carathéodory's first theorem). If a Pfaff differential expression has an integrating factor, then there are points P_1 $(x_{1_1}, \ldots, x_{n_1})$ in the vicinity of P_0 $(x_{1_0}, \ldots, x_{n_0})$, and as near as we please to it, which are not accessible by a route which starts at P_0 and traverses only paths characterized by $\mathrm{d}Q = 0$.

Proof. According to Theorem 1, the existence of an integrating factor depends on the existence of the family of surfaces (9.11). According to (9.10) and (9.11), $\mathrm{d}\sigma = 0$ and $\mathrm{d}Q = 0$ in each surface. Let P_0 be a point in such a surface. Then paths $\mathrm{d}Q = 0$ will lead only to points on the same surface, which proves the theorem.

If no integrating factor exists, any point P_1 can be reached from a point P_0 on paths such that $\mathrm{d}Q = 0$. This can easily be verified by means of examples.

Theorem 4. (Carathéodory's second theorem). If a Pfaff differential $\mathrm{d}Q$ has the property that as near as we please to P_0 there are other points P_1 which are not accessible from P_0 via paths $\mathrm{d}Q = 0$, then an integrating factor exists for $\mathrm{d}Q$.

This theorem is the converse of Theorem 3. Its proof is, however, much more complicated and we shall have to omit it here.

Theorem 5. For the existence of a solution (9.11) to a Pfaff differential equation it is necessary and sufficient for the equation

$$X_k\left(\frac{\partial X_j}{\partial x_i}-\frac{\partial X_i}{\partial x_j}\right)+X_j\left(\frac{\partial X_i}{\partial x_k}-\frac{\partial X_k}{\partial x_i}\right)+X_i\left(\frac{\partial X_k}{\partial x_j}-\frac{\partial X_j}{\partial x_k}\right)=0 \tag{9.18}$$

to be obeyed for all triple sets of values i, k, j.

Proof. We shall only show that the condition is necessary and shall suppose, therefore, that the solution in the form (9.11) exists. If we simplify by letting $\tau^{-1} \equiv u$, it follows from (9.13) together with (9.1) that

$$u(x_1, \ldots, x_n)\, X_i = \frac{\mathrm{d}\sigma}{\mathrm{d}x_i}. \tag{9.19}$$

This gives, since $\mathrm{d}\sigma$ is a complete differential,

$$\frac{\partial(uX_j)}{\partial x_i} = \frac{\partial(uX_i)}{\partial x_j}. \tag{9.20}$$

If we carry out the differentiation, we get

$$u\left(\frac{\partial X_j}{\partial x_i}-\frac{\partial X_i}{\partial x_j}\right) = X_i\frac{\partial u}{\partial x_j}-X_j\frac{\partial u}{\partial x_i}$$

and, correspondingly,

$$\left.\begin{aligned} u\left(\frac{\partial X_i}{\partial x_k}-\frac{\partial X_k}{\partial x_i}\right) &= X_k\frac{\partial u}{\partial x_i}-X_i\frac{\partial u}{\partial x_k}, \\ u\left(\frac{\partial X_k}{\partial x_j}-\frac{\partial X_j}{\partial x_k}\right) &= X_j\frac{\partial u}{\partial x_k}-X_k\frac{\partial u}{\partial x_j}. \end{aligned}\right\} \tag{9.21}$$

If the first of the eqs. (9.21) is multiplied by X_k, the second by X_j and the third by X_i and the resulting equations are added together, eq. (9.18) is obtained. This shows that the condition (9.18) is necessary. It can also be proved to be sufficient.

For three independent variables, (9.18) can be written in the vectorial form

$$\boldsymbol{R}.\mathrm{rot}\,\boldsymbol{R} = 0 \quad \text{(three variables).} \tag{9.22}$$

We shall now consider explicitly the special cases involving two or three independent variables since these two cases are of particular importance.

(α) *Two independent variables. Theorem* 6. There is an integrating factor for any Pfaff differential of two independent variables.

Proof. If we put, for instance, $k = j$ in eq. (9.18) it becomes obvious that the condition is always fulfilled for two independent variables. Combination with Theorem 1 then gives the proposition.

Let us consider the Pfaff differential equation

$$\mathrm{d}Q = X\,\mathrm{d}x + Y\,\mathrm{d}y = 0. \tag{9.23}$$

According to Theorem 6, there is a solution

$$\sigma(x, y) = \sigma \tag{9.24}$$

which represents a family of single parameter curves in the x, y-plane.

Along each curve

$$\mathrm{d}Q = 0 \tag{9.25}$$

and

$$\mathrm{d}\sigma = \frac{\partial\sigma}{\partial x}\mathrm{d}x + \frac{\partial\sigma}{\partial y}\mathrm{d}y = 0. \tag{9.26}$$

From (9.26) we get

$$\frac{\mathrm{d}y}{\mathrm{d}x} = -\frac{\partial\sigma/\partial x}{\partial\sigma/\partial y} \tag{9.27}$$

and from (9.23)

$$\frac{\mathrm{d}y}{\mathrm{d}x} = -\frac{X}{Y}. \tag{9.28}$$

From (9.27) and (9.28) we get

$$\frac{X}{\partial\sigma/\partial x} = \frac{Y}{\partial\sigma/\partial y} = \tau(x, y) \tag{9.29}$$

where $\tau(x, y)$ is a function dependent on x and y. But from (9.1) and (9.13) we get

$$\mathrm{d}\sigma = \frac{\mathrm{d}Q}{\tau(x, y)} = \frac{X}{\tau}\mathrm{d}x + \frac{Y}{\tau}\mathrm{d}y. \tag{9.30}$$

The reciprocal, $\tau(x, y)$, of the integrating factor is therefore given by eq. (9.29).

(β) *Three independent variables.* Let us consider the Pfaff differential equation

$$dQ = X\,dx + Y\,dy + Z\,dz = 0. \tag{9.31}$$

In this case, condition (9.18) is not necessarily fulfilled. This can be verified easily by means of examples. Equation (9.31) therefore does not necessarily have a solution of the form

$$\sigma(x, y, z) = \sigma \tag{9.32}$$

and there is thus not necessarily an integrating factor for dQ.

Let us now suppose that the solution (9.32) exists. It defines a family of single-parameter surfaces in three-dimensional space. In these surfaces, we have that

$$d\sigma = \frac{\partial \sigma}{\partial x} dx + \frac{\partial \sigma}{\partial y} dy + \frac{\partial \sigma}{\partial z} dz. \tag{9.33}$$

From this we get

$$\left.\frac{dy}{dx}\right|_z = -\frac{\partial \sigma/\partial x}{\partial \sigma/\partial y}, \quad \left.\frac{dz}{dx}\right|_y = -\frac{\partial \sigma/\partial x}{\partial \sigma/\partial z}, \quad \left.\frac{dz}{dy}\right|_x = -\frac{\partial \sigma/\partial y}{\partial \sigma/\partial z}. \tag{9.34}$$

But from (9.31) we get

$$\left.\frac{dy}{dx}\right|_z = -\frac{X}{Y}, \quad \left.\frac{\partial z}{dx}\right|_y = -\frac{X}{Z}, \quad \left.\frac{dz}{dy}\right|_x = -\frac{Y}{Z}. \tag{9.35}$$

From (9.34) and (9.35) we get

$$\frac{X}{\partial \sigma/\partial x} = \frac{Y}{\partial \sigma/\partial y} = \frac{Z}{\partial \sigma/\partial z} = \tau(x, y, z) \tag{9.36}$$

and we can show, as above, that $\tau(x, y, z)$ is the reciprocal of the integrating factor of the incomplete differential dQ.

This derivation can be generalized quite simply for more than three independent variables.

§ 10 The Second Law applied to quasi-static processes

The facts of experience on which the Second Law is based are introduced by Carathéodory in a form essentially similar to the one we used in the discussion of the First Law (§ 8).

Theorem of experience (*Carathéodory's principle*). As near as we please to any state of a (homogeneous or heterogeneous) system there are neighbouring states which are not accessible via an adiabatic path starting at the original state.

An example has already been mentioned in § 8. We shall confine ourselves, in the present section, to quasi-static processes. It must then be true that the above theorem applies to this special class of changes of state. An example is the adiabatic compression or expansion of an ideal gas which can only occur for values of the P, V pairs which obey eq. (7.2).

Let us now consider the relationship of Carathéodory's principle to Clausius' (or Thomson's) principle (§ 4). We can see clearly that Carathéodory's principle is derived directly from Clausius' principle. The converse is, however, not true: Carathéodory's principle confines itself to the statement that there are adiabatic processes which cannot be carried out, while Clausius' principle states which adiabatic processes are impossible.

Carathéodory's principle would, therefore, be preferable if all the consequences of the Second Law could be derived from it. We shall see that this is not the case and that a further theorem of experience is required (§ 13).

Let us now return to our consideration of quasi-static processes. For a homogeneous system the Pfaff differential $\mathrm{d}Q$ depends on only two independent variables. The existence of an integrating factor and thus of entropy is then, according to Theorem 6 of § 9, a purely mathematical consequence; no further fact of experience is necessary. From this point of view, therefore, the case of three independent variables is the first one of interest. Furthermore, the identification of the integrating factor with reciprocal temperature requires the introduction of a thermal equilibrium; this is impossible if the argument is confined to two independent variables. For this reason we shall start with the analysis of a system consisting of two phases $'$ and $''$ which are separated by a diathermic wall and in thermal equilibrium. We shall choose V', V'' and t as independent variables.

Carathéodory's principle together with Theorem 4 of § 9 leads directly to the existence of an integrating factor for $\mathrm{d}Q$.

We have, therefore, that

$$\mathrm{d}Q = \mathrm{d}Q' + \mathrm{d}Q'' = \tau(V', V'', t)\,\mathrm{d}\sigma(V', V'', t). \tag{10.1}$$

For the two phases, we have

$$\mathrm{d}Q' = \tau(V', t)\,\mathrm{d}\sigma'(V', t), \quad \mathrm{d}Q'' = \tau''(V'', t)\,\mathrm{d}\sigma''(V'', t). \tag{10.2}$$

Equations (10.1) and (10.2) give

$$\tau\,\mathrm{d}\sigma = \tau'\,\mathrm{d}\sigma' + \tau''\,\mathrm{d}\sigma''. \tag{10.3}$$

We now choose σ', σ'', and t as independent variables. Since $\mathrm{d}\sigma$ is a complete differential, (10.3) gives

$$\frac{\partial \sigma}{\partial \sigma'} = \frac{\tau'}{\tau}, \quad \frac{\partial \sigma}{\partial \sigma''} = \frac{\tau''}{\tau}, \quad \frac{\partial \sigma}{\partial t} = 0. \tag{10.4}$$

The function σ is, therefore, independent of t and we have

$$\sigma = \sigma(\sigma', \sigma''). \tag{10.5}$$

Therefore, because of (10.4), we also have

$$\frac{\partial}{\partial t}\left(\frac{\tau'}{\tau}\right) = 0, \quad \frac{\partial}{\partial t}\left(\frac{\tau''}{\tau}\right) = 0. \tag{10.6}$$

Differentiation then leads to

$$\frac{1}{\tau'}\frac{\partial \tau'}{\partial t} = \frac{1}{\tau''}\frac{\partial \tau''}{\partial t} = \frac{1}{\tau}\frac{\partial \tau}{\partial t}. \tag{10.7}$$

Now τ' depends only on t and σ', τ'' only on t and σ''. The first equality sign can, therefore, only be valid when the logarithmic derivatives of τ' and τ'' with respect to t depend on neither σ' nor σ'' but only on t. It follows thus that

$$\frac{\partial \ln \tau'}{\partial t} = \frac{\partial \ln \tau''}{\partial t} = \frac{\partial \ln \tau}{\partial t} = g(t) \tag{10.8}$$

where $g(t)$ is a function of the empirical temperature independent of all special properties of the phases and the same for all phases in thermal equilibrium. Integration of (10.8) for the whole system or for the two phases gives

$$\left.\begin{aligned} \ln \tau &= \int g(t)\,\mathrm{d}t + \ln \phi(\sigma', \sigma''), \\ \ln \tau' &= \int g(t)\,\mathrm{d}t + \ln \phi'(\sigma'), \\ \ln \tau'' &= \int g(t)\,\mathrm{d}t + \ln \phi''(\sigma'') \end{aligned}\right\} \tag{10.9}$$

or

$$\left.\begin{aligned} &\tau = \exp\left[\int g(t)\,\mathrm{d}t\right].\phi(\sigma', \sigma''), \\ &\tau' = \exp\left[\int g(t)\,\mathrm{d}t\right].\phi'(\sigma'), \quad \tau'' = \exp\left[\int g(t)\,\mathrm{d}t\right].\phi''(\sigma''). \end{aligned}\right\} \tag{10.10}$$

Thus, for a system whose phases are in thermal equilibrium, τ the reciprocal of the integrating factor for the whole system and for each of the phases is made up of two factors; one depends only on the common empirical temperature t, the other is a function of the individual variables of state (σ' and σ'' for the system, σ' for $'$ and σ'' for $''$).

The *absolute temperature* is now defined by the equation

$$T = C.\exp\left[\int g(t)\,\mathrm{d}t\right] \tag{10.11}$$

where C is an arbitrary constant which is fixed by the definition of the temperature scale. In other words, T is made to differ by a certain amount (100°) between two fixed points (ice point and normal boiling point of water). Equation (10.11), however, no longer contains an arbitrary constant. The zero point of T is, therefore, defined on a physical basis. Combination of (10.11) with (10.1), (10.2) and (10.10) gives

$$\mathrm{d}Q = \tau\,\mathrm{d}\sigma = T\frac{\phi}{C}\mathrm{d}\sigma, \tag{10.12}$$

$$\mathrm{d}Q' = \tau'\mathrm{d}\sigma' = T\frac{\phi'}{C}\mathrm{d}\sigma', \quad \mathrm{d}Q'' = \tau\,\mathrm{d}\sigma'' = T\frac{\phi''}{C}\mathrm{d}\sigma''. \tag{10.13}$$

Let us now consider a single phase (say $'$) and define the *entropy* of the phase by the equation

$$S' = \frac{1}{C}\int \phi'(\sigma')\,\mathrm{d}\sigma' + \text{const.} \tag{10.14}$$

Entropy, like internal energy, is thus defined except for an arbitrary additive constant. Combination of (10.13) with (10.14) gives

$$\mathrm{d}Q' = T\,\mathrm{d}S', \quad \mathrm{d}Q'' = T\,\mathrm{d}S''. \tag{10.15}$$

We have thus proved the existence of entropy as a function of state for each phase, at least for quasi-static processes, and have also derived the fundamental equation (4.16). Let us now consider the complete system. Equation (10.12) together with (10.3) and (10.10) gives

$$\phi\,\mathrm{d}\sigma = \phi'\mathrm{d}\sigma' + \phi''\,\mathrm{d}\sigma''. \tag{10.16}$$

We want to show that ϕ can be written in the form

$$\phi = \phi(\sigma) \tag{10.17}$$

i.e. that ϕ depends only on σ and thus on σ' and σ''. Differentiation of (10.17) gives

$$\frac{\partial\phi}{\partial\sigma'} = \frac{\partial\phi}{\partial\sigma}\frac{\partial\sigma}{\partial\sigma'}, \quad \frac{\partial\phi}{\partial\sigma''} = \frac{\partial\phi}{\partial\sigma}\frac{\partial\sigma}{\partial\sigma''}. \tag{10.18}$$

Elimination of $\partial\phi/\partial\sigma$ gives

$$\frac{\partial\phi}{\partial\sigma'}\frac{\partial\sigma}{\partial\sigma''} - \frac{\partial\phi}{\partial\sigma''}\frac{\partial\sigma}{\partial\sigma'} = 0 \tag{10.19}$$

or, in the notation usual for Jacobi determinants,

$$\frac{\partial(\phi, \sigma)}{\partial(\sigma', \sigma'')} \equiv \mathrm{J}(\phi, \sigma) = 0. \tag{10.20}$$

The function ϕ, therefore, depends explicitly only on σ if the Jacobi determinant $\mathrm{J}(\phi, \sigma)$ vanishes. This condition is fulfilled since we get from (10.16) that

$$\phi\frac{\partial\sigma}{\partial\sigma'} = \phi', \quad \phi\frac{\partial\sigma}{\partial\sigma''} = \phi''. \tag{10.21}$$

By definition ϕ' depends only on σ', ϕ'' only on σ''. Differentiation of (10.21) with respect to σ'' and σ' respectively, therefore, gives

$$\begin{aligned} \frac{\partial\phi}{\partial\sigma''}\frac{\partial\sigma}{\partial\sigma'} + \phi\frac{\partial^2\sigma}{\partial\sigma'\partial\sigma''} &= 0, \\ \frac{\partial\phi}{\partial\sigma'}\frac{\partial\sigma}{\partial\sigma''} + \phi\frac{\partial^2\sigma}{\partial\sigma'\partial\sigma''} &= 0. \end{aligned} \tag{10.22}$$

Equation (10.20) is a direct consequence of this, and (10.17) is thus proved. The entropy of the complete system can, therefore, be defined by the equation

$$S = \frac{1}{C}\int \phi(\sigma)\, \mathrm{d}\sigma + \text{const.} \tag{10.23}$$

From (10.12), therefore, we have again, for the complete system,

$$\mathrm{d}Q = T\,\mathrm{d}S. \tag{10.24}$$

If the First Law in the form of eq. (8.8) is introduced into eq. (10.15) and the condition of thermal equilibrium, which is used only to define absolute temperature, is ignored we have for each phase that

$$T'\mathrm{d}S' = \mathrm{d}U' + P'\mathrm{d}V' \tag{10.25}$$

and for the complete system

$$\mathrm{d}S = \mathrm{d}S' + \mathrm{d}S''. \tag{10.26}$$

With limitation to quasi-static processes we have deduced the existence of absolute temperature. We have also derived entropy as a function of state defined by (10.15) for each phase and we have shown that the entropy of the complete system is derived additively from the entropies of the separate phases.

One of the important consequences of the Second Law is still missing, i.e. the property of entropy increase which is fundamental for the formulation of equilibrium conditions. We shall return to this subject in § 13.

It should be noted that, recently, σ and S have been called empirical and metrical entropy respectively. In this connexion it should be noted that an empirical variable is defined apart from arbitrary continuous and strictly increasing scale transformations; a metrical variable is defined apart from increasing linear scale transformations (expansion of the scale, change of origin). The existence of empirical entropy depends, in the above argument, on Theorem 6 of § 9 and on Carathéodory's principle. The introduction of thermal coupling is used to deduce metrical entropy and to choose one of the possible pairs of variables, σ, τ (cf. Theorem 2 of § 9).

§ 11 Empirical definition of U, S, and T

The derivation given in § 10 may be regarded essentially as a proof of existence. Since the functions $g(t)$ and $\phi(\sigma)$ are unknown, T and S cannot be derived directly from empirical data.

Let us consider a closed homogeneous system having adiabatics

$$s(P, V) = s \tag{11.1}$$

and isothermals

$$t(P, V) = t. \tag{11.2}$$

We choose s and t as independent variables of state which are measured on arbitrary empirical scales.

We have, therefore, that

$$\mathrm{d}V = \frac{\partial V}{\partial s}\mathrm{d}s + \frac{\partial V}{\partial t}\mathrm{d}t \tag{11.3}$$

and that

$$\mathrm{d}U = \frac{\partial U}{\partial s}\mathrm{d}s + \frac{\partial U}{\partial t}\mathrm{d}t. \tag{11.4}$$

Furthermore,

$$\mathrm{d}U = T(t)\,\mathrm{d}S(s) - P\,\mathrm{d}V. \tag{11.5}$$

It follows that

$$\mathrm{d}U = T(t)\left[\frac{\mathrm{d}S(s)}{\mathrm{d}s} - P\frac{\partial V}{\partial s}\right]\mathrm{d}s - P\frac{\partial V}{\partial t}\mathrm{d}t. \tag{11.6}$$

Since U is a function of state, (11.5) and (11.6) give

$$\frac{\partial}{\partial t}\left[T(t)\frac{\mathrm{d}S(s)}{\mathrm{d}s}-P\frac{\partial V}{\partial s}\right]=\frac{\partial}{\partial s}\left(-P\frac{\partial V}{\partial t}\right) \tag{11.7}$$

and on differentiating we get

$$\frac{\mathrm{d}S(s)}{\mathrm{d}s}\frac{\mathrm{d}T(t)}{\mathrm{d}t}=\frac{\partial V}{\partial s}\frac{\partial P}{\partial t}-\frac{\partial V}{\partial t}\frac{\partial P}{\partial s}\equiv\frac{\partial(V,P)}{\partial(s,t)}. \tag{11.8}$$

The Jacobian $\mathrm{J}(V,P)$ is thus always the product of two functions one of which depends only on s, the other only on t. We can write, therefore,

$$\mathrm{J}(V,P)=\alpha(s)\,\beta(t) \tag{11.9}$$

where the quantities $\alpha(s)$ and $\beta(t)$ can be regarded as empirically known. Equation (11.8) is thus divided into two separate equations which, by integration, give

$$S(s)=\frac{1}{C}\int\alpha(s)\,\mathrm{d}s+S_0, \tag{11.10}$$

$$T(t)=C\int\beta(t)\,\mathrm{d}t+T_0 \tag{11.11}$$

and finally, from (11.6),

$$U(s,t)=\int(s,t)+U_0 \tag{11.12}$$

where C, S_0, T_0, and U_0 are constants. As has been shown, S_0 and U_0 may be chosen arbitrarily (i.e. they have no physical significance) while C is fixed by the choice of temperature scale [cf. eq. (10.11)]. The zero point of T, however, is physically fixed, and the constant T_0 is thus not adjustable but must be found by experiment. Equation (11.11) shows this to be impossible on the basis of quasi-static processes which lead to eqs. (11.1) and (11.2). At this point, therefore, non-static processes must be introduced. Since, however, the absolute temperature is universal, a single experiment on any substance (e.g. Gay-Lussac's streaming experiment using an ideal gas) is sufficient to fix T_0 once and for all.

It can be shown that the values of S, T, and U are independent of the choice of the empirical scales for s and t.

The above argument obviously constitutes the explicit definition of the metrical entropy S using the empirical entropy s. This process assumes the existence of metrical entropy and introduces thermal coupling implicitly by using isotherms (cf. § 7). It should be noted that the existence can be proved using only quasi-static processes but that the explicit definition necessarily requires a non-static experiment.

An example which makes use of these arguments is the ideal gas, for which the equation for the isotherms (cf. § 7) is

$$PV = t \tag{11.13}$$

and the equation for the adiabatics is

$$PV^{\kappa} \equiv PV^{1+\gamma} = s \tag{11.14}$$

where κ and γ are constants. Logarithmic differentiation gives

$$\frac{1}{s}\frac{\partial s}{\partial V} = \frac{1+\gamma}{V}, \quad \frac{1}{s}\frac{\partial s}{\partial P} = \frac{1}{P}, \tag{11.15}$$

$$\frac{1}{t}\frac{\partial t}{\partial V} = \frac{1}{V}, \quad \frac{1}{t}\frac{\partial t}{\partial P} = \frac{1}{P} \tag{11.16}$$

which lead to

$$\frac{\partial(s,t)}{\partial(V,P)} \equiv \frac{\partial s}{\partial V}\frac{\partial t}{\partial P} - \frac{\partial s}{\partial P}\frac{\partial t}{\partial V} = \gamma s. \tag{11.17}$$

Quite generally

$$\frac{\partial(x,y)}{\partial(z,w)}\frac{\partial(z,w)}{\partial(u,v)} = \frac{\partial(x,y)}{\partial(u,v)} \tag{11.18}$$

and thus (11.17) becomes

$$\frac{\partial(V,P)}{\partial(s,t)} = \frac{1}{\gamma s}. \tag{11.19}$$

Equation (11.9), therefore, defines the functions

$$\alpha(s) = \frac{1}{\gamma s}, \quad \beta(t) = 1. \tag{11.20}$$

From (11.10), (11.11) and (11.20) we get

$$S(s) = \frac{1}{C\gamma}\ln s + S_0, \tag{11.21}$$

$$T(t) = Ct + T_0 \tag{11.22}$$

and substitution of these into eq. (11.5) gives

$$dU = (Ct+T_0)\frac{1}{C\gamma}d\ln s - P\,dV \tag{11.23}$$

or, using (11.13) and (11.14),

$$dU = \frac{T_0}{C\gamma} d\ln s + \frac{t}{\gamma} d\ln(sV^{-\gamma})$$

$$= \frac{T_0}{C\gamma} d\ln s + \frac{1}{\gamma} dt. \tag{11.24}$$

On integration, (11.24) gives

$$U = \frac{T_0}{C\gamma} \ln s + \frac{t}{\gamma} + U_0 \tag{11.25}$$

or, again using (11.13) and (11.14),

$$U = \frac{T_0}{C\gamma} \ln(tV^{\gamma}) + \frac{t}{\gamma} + U_0. \tag{11.26}$$

In order to determine T_0 a very dilute gas is allowed to expand into a vacuum (Gay-Lussac's streaming experiment). We find that

$$\left(\frac{\partial U}{\partial V}\right)_t = 0. \tag{11.27}$$

According to (11.26) this is possible only if $T_0 = 0$. We have, therefore, that

$$S = \frac{1}{C\gamma} \ln s + S_0, \tag{11.28}$$

$$T = Ct, \tag{11.29}$$

$$U = \frac{t}{\gamma} + U_0. \tag{11.30}$$

Equation (11.29) shows that absolute temperature and temperature on the ideal gas scale differ only by a factor which may be chosen arbitrarily. They can, therefore, be taken to be identical, which is usual. If we now write (11.13) in the form of (7.10) we have

$$t = RT^* \tag{11.31}$$

and, with $C = 1/R$, this becomes

$$T = T^*. \tag{11.32}$$

For monatomic gases $\gamma = \frac{2}{3}$ and (11.28) with (11.13), (11.14), (11.31), and (11.32) gives the well-known relationships

$$S = \tfrac{3}{2}R\ln T + R\ln V + S_0' \tag{11.33}$$

(where S_0' is a new constant) and

$$U = \tfrac{3}{2}RT + U_0. \tag{11.34}$$

§ 12 Measurement of very low temperatures

In principle the most important processes for the definition of the absolute temperature scale are, as has been shown, reversible cyclic processes between fixed points (§ 4) and the ideal gas thermometer (§ 11). At extremely low temperatures (i.e. $T \leqslant 1$ K) both of these become impracticable. The heat changes in cyclic processes become so small as to be no longer accurately measurable while the gas thermometer becomes useless because all gases condense. The only available scale in this region is, therefore, empirical and the problem of correlating this scale with the absolute scale becomes important for low temperature physics. The solution of this problem arises naturally from the considerations of § 11. We shall follow it through explicitly since the argument differs slightly from the general considerations of § 11.

Let us assume that the absolute scale is known as far down as a temperature T_0. Below T_0, let the temperature be proportional to the volume of a thermometric fluid. Let us choose the proportionality factor such that this empirical scale T^* becomes identical with the absolute scale at T_0, i.e. $T_0^* = T_0$. We shall choose T (or T^*) and P as independent variables of state and investigate the relationship between T and T^* along the straight line $P = 0$.

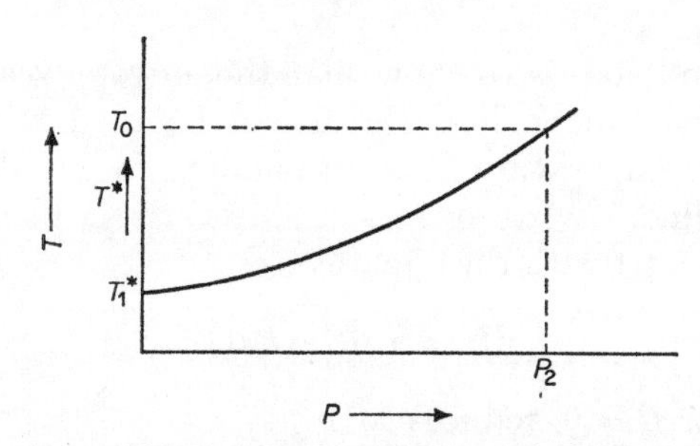

Fig. 5. Measurement of very low temperatures

(a) The first step is to prove that independent entropy values can be determined for all points $(T^*, 0)$ of the empirical scale. We shall assume that the entropy in the usual units is known for the point $(T_0, 0)$ and denote it by S_0. An isothermal compression to a pressure P_2 leads to the entropy at the point (T_0, P_2),

$$S_1 = S_0 + \int_0^{P_2} \left(\frac{\partial S}{\partial P}\right)_{T=T_0} \mathrm{d}P. \tag{12.1}$$

We shall show later (§ 24) that

$$\left(\frac{\partial S}{\partial P}\right)_T = -\left(\frac{\partial V}{\partial T}\right)_P. \tag{12.2}$$

Introduction of (12.2) and of the coefficient of thermal expansion

$$\alpha = \frac{1}{V}\left(\frac{\partial V}{\partial T}\right)_P \tag{12.3}$$

into (12.1) gives

$$S_1 = S_0 - \int_0^{P_2} V\alpha \, \mathrm{d}P. \tag{12.4}$$

S_1 can thus be calculated if the volume and the coefficient of thermal expansion at T_0 in the pressure region $0 \to P_2$ are known.

Starting from the point (T_0, P_2) let us now carry out a quasi-static adiabatic expansion which changes the empirical temperature of the substance to T_1^*. For this process $\mathrm{d}S = 0$, i.e. the entropy has the same value as that at the point (T_0, P_2). This shows that, along the straight line $P = 0$, we can determine an entropy scale independent of the empirical scale T^*.

(b) The second step is to show that the entropy scale determined along the straight line $P = 0$ can be used to calculate the absolute temperature T corresponding to T^* along the same straight line.

Let us start with the equation

$$\mathrm{d}U = T\,\mathrm{d}S - P\,\mathrm{d}V \tag{12.5}$$

which, for $P = 0$, reduces to

$$\mathrm{d}U = \mathrm{d}Q = T\,\mathrm{d}S \tag{12.6}$$

from which follows that

$$\left(\frac{\mathrm{d}U}{\mathrm{d}T^*}\right)_{P=0} = \left(\frac{\mathrm{d}Q}{\mathrm{d}T^*}\right)_{P=0} = T\left(\frac{\mathrm{d}S}{\mathrm{d}T^*}\right)_{P=0}. \tag{12.7}$$

The quantity $(\mathrm{d}Q/\mathrm{d}T^*)_{P=0}$ is the heat capacity at constant pressure on the empirical scale (for simplicity we shall take it to refer to 1 mole and denote it by C_P^*), which is experimentally accessible and we can write

$$T = C_P^*\left(\frac{dS}{dT^*}\right)_{P=0}^{-1}. \tag{12.8}$$

Since the variation of S with T^* is known we can calculate T for any T^*.

In practice we use certain paramagnetic salts as thermometric substances. The adiabatic process used is adiabatic demagnetization (discussed in § 66) and, therefore, the magnetic field strength $H = 0$ corresponds to the pressure $P = 0$. The empirical scale is defined with the aid of susceptibility measurements.

§ 13 The Second Law applied to non-static processes

It has already been mentioned (§ 10) that the increase of entropy under conditions of adiabatic isolation (4.36) cannot be deduced from Carathéodory's principle. This fact was emphasized particularly by Planck (1926). A further fact of experience must, therefore, be used in order to obtain all the consequences of the Second Law formulated in Part A.

As our first step we shall formulate

Theorem 1. Each adiabatic surface s divides the volume of state in co-ordinate space into two half-volumes in such a way that all states z which can be attained adiabatically from a state z_0 lying in the surface are within the same half-volume.

Proof. The theorem follows directly from Carathéodory's principle. Let us suppose that the inaccessible states z' lie on both sides of s. According to Carathéodory's principle these states z' can be as near as we please to z_0 which leads to the conclusion that there is no way of leaving s along an adiabatic path, since accessibility of z' contains implicitly accessibility of all states lying on the adiabatic surface s' through z'. This conclusion, however, contradicts the principle and the theorem is thus proved.

Conditions for an ideal gas are shown diagrammatically in Fig. 6. For simplicity we shall initially confine our discussion to homogeneous systems with two independent variables of state. Carathéodory now introduces, as a definition,

Theorem 2. Adiabatic changes which start from z_0 and end at $V = V'$ cover that half of the V'-line which lies on one side of s.

Since V' can be varied arbitrarily, we immediately get

Theorem 3. The totality of all states adiabatically accessible from s is represented by one of the half-volumes generated by s.

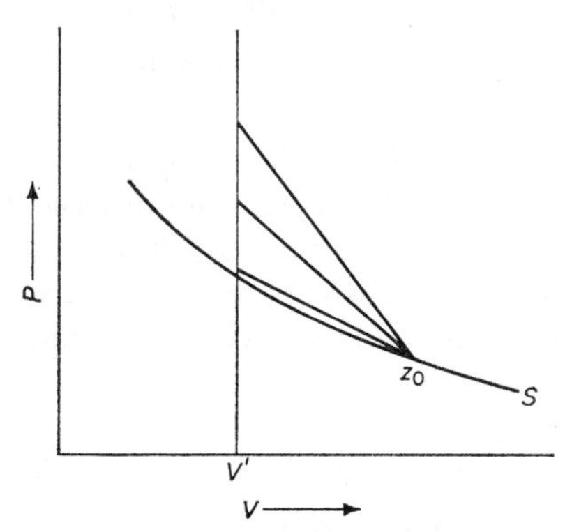

FIG. 6. Non-static processes for an ideal gas

Finally, we have

Theorem 4. An adiabatic s'' lying between s and s' cannot be reached adiabatically from both s and s'.

Proof. The theorem again follows directly from Carathéodory's principle. This is illustrated by Fig. 7 in which two further adiabatics s_1 and s_2 are drawn, one on each side of s''.

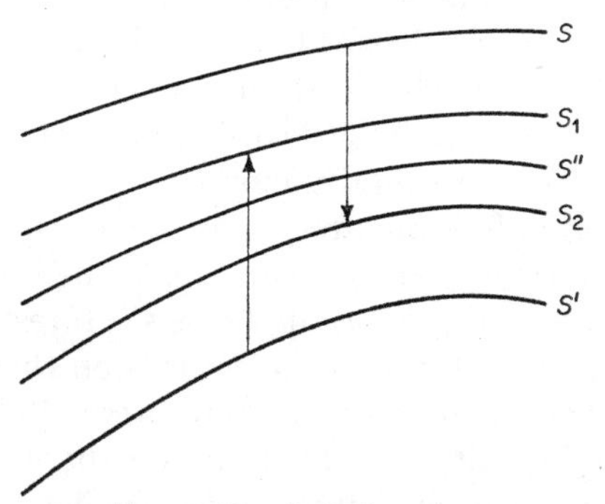

FIG. 7. Proof of Theorem 4

If s'' is adiabatically accessible from s, Theorem 3 tells us that s_2 is similarly accessible. This means that s_2 will be adiabatically accessible from s''. If s'' is adiabatically accessible from s', Theorem 3 tells us that s_1 is similarly accessible; s_1, therefore, would also be adiabatically accessible from s''. This means that there would be no states as near as we please to a state z'' lying on s'' which would be adiabatically inaccessible from z''. This directly contradicts Carathéodory's principle; s'' can, therefore, be reached adiabatically either from s or from s'' but not from both.

Theorem 4 states clearly that the adiabatically accessible half-volumes always lie on the same side of the adiabatic generating the half-volumes.

It follows quite generally from the preceding theorems that irreversible adiabatic processes exist. If, therefore, the adiabatic process $1 \to 2$ is impossible, the adiabatic process $2 \to 1$ must be possible, a conclusion which we have already used in § 8. It follows further that the empirical entropy defined by the family of adiabatics [eqs. (10.1) and (10.2)] can under conditions of adiabatic isolation either only increase or only decrease. Physical experience cannot distinguish between these two possibilities since the direction of increase of the empirical entropy may be chosen arbitrarily. We may say alternatively that the two half-volumes generated by an adiabatic are not labelled physically by the empirical entropy (i.e. are not made distinguishable). The decision between the two possibilities is, therefore, simply conventional and is given by the *definition*:

Under conditions of adiabatic isolation the empirical entropy can never decrease.

We must emphasize again that this definition cannot be verified by experiment and that it could, in principle, be replaced by its opposite.

The characteristic increase in entropy under conditions of adiabatic isolation discussed in § 4 cannot, therefore, be formulated on the basis of empirical entropy. We shall, therefore, have to use metrical entropy and we have from eq. (10.23)

$$\mathrm{d}S = \frac{1}{C}\phi(\sigma)\,\mathrm{d}\sigma. \tag{13.1}$$

On the right-hand side of this equation C must, by convention, be a positive constant since it depends on the definitions of absolute temperature [eqs. (10.11) and (11.11)]. According to the above definition $\mathrm{d}\sigma$ is also a positive quantity while, according to eq. (10.10), the sign of $\phi(\sigma)$ is the same as the sign of the integrating factor τ^{-1}. The derivation of statement (4.36) thus requires, as far as the present argument is concerned, answers to the following questions:

(a) Can the half-volumes generated by an adiabatic be physically distinguished?

(b) If they can be distinguished, do we have a choice of sign for the integrating factor τ^{-1}?

(c) What fact of experience fixes the sign of τ?

Question (a) is answered by

Theorem 5. The half-volumes generated by an adiabatic are physically distinguished in that the internal energy changes monotonically along any line $V = V'$.

Proof. Since U, V is a complete set of variables of state for a simple homogeneous system, U cannot twice take the same value along a line $V = V'$. If it could, Fig. 6 shows that there would be two states having the same value of U and V but different values of P. This contradicts the supposition that U, V is a complete set of variables of state. U must, therefore change monotonically along any line $V = V'$. Any adiabatic process starting from one adiabatic s and proceeding along a line $V = V'$ is, therefore, associated with an increase in internal energy for one half-volume and with a decrease for the other.† The two half-volumes generated by s are thus physically distinguished and the theorem is proved.

For question (b) we have to prove

Theorem 6. If $d\sigma$ is defined as positive (see above) for all possible adiabatic processes, then τ, the reciprocal of the integrating factor, is positive when the adiabatic processes possible along the line $V = V'$ are associated with an increase in internal energy. If they are associated with a decrease in internal energy, then τ is negative.

Proof. According to § 10 the empirical entropy σ is a function of state; therefore σ, V constitutes a complete set of variables of state. $U, V \to \sigma, V$ must, therefore, be a one-one transformation. It follows that, for all states, either $(\partial U/\partial \sigma)_V > 0$ or $(\partial U/\partial \sigma)_V < 0$.

But, according to eq. (10.1)

$$\left(\frac{\partial U}{\partial \sigma}\right)_V = \tau. \tag{13.2}$$

The integrating factor τ^{-1} can, therefore, be either only positive or only negative for all states. The left-hand side represents the change in internal energy during an adiabatic process occurring along $V = V'$, and the change in empirical entropy for a possible process is positive by definition; the original statements, therefore, follow directly.

† It can easily be shown for Fig. 6 that processes which lead into the upper half-volume are associated with an increase in internal energy. This, however, has no significance for the argument which follows.

We have now reduced the problem to the question: Are adiabatic processes occurring along a line $V = V'$ associated with an increase or a decrease in internal energy? The reason why a discussion of such special processes leads to a generally valid result is based on Theorems 1 and 3 which maintain that all states accessible from a given adiabatic lie within the same half-volume. As was mentioned in § 10, the question cannot be answered by using Carathéodory's principle since this does not tell us which of the two half-volumes is adiabatically inaccessible.

We therefore introduce the following new *theorem of experience*:

An adiabatic process which occurs along a line $V = V'$ is always associated with an increase in internal energy.

Examples of such processes are the production of frictional heat by mechanical work or the production of heat by an electric current. It is obvious that, in both these cases, the reverse process is impossible since it would contradict Thomson's principle and, therefore, Clausius' principle (§ 4). The present theorem of experience constitutes simply a new formulation of the older principles in a way more convenient for our present purposes. Furthermore, like the older principles, it implies Carathéodory's principle. For purely logical reasons, therefore, this theorem of experience (or an equivalent statement) should serve as the empirical basis of Carathéodory's theory. Despite this, we have chosen a somewhat different way (which, on the whole, follows the historical development) since it more clearly emphasizes certain aspects.

The consequences of our theorem of experience can now be developed quite simply. The definition and Theorem 6 show immediately that the integrating factor τ^{-1} is always positive. Equation (10.10) then gives

$$\phi'(\sigma') > 0, \quad \phi''(\sigma'') > 0, \quad \phi(\sigma) > 0 \tag{13.3}$$

which, with (13.1) gives for an adiabatic isolated system completely generally

$$\mathrm{d}S \geqslant 0 \quad \text{(adiabatic isolated system).} \tag{13.4}$$

We have now deduced all the results of Part A on the basis of Carathéodory's theory.

C. GENERALIZATION OF THE SECOND LAW FOR OPEN SYSTEMS AND CHEMICAL REACTIONS

§ 14 The problem

We have, so far, confined our discussion to the simple systems defined in § 2. We now have to decide how these limitations can be removed and how the Laws can be formulated in a generally useful way.

The only 'work co-ordinates' we have used so far are the volumes of the phases. It is, however, quite simple to take into account the properties of solids, the action of external fields, surface effects, etc. by the introduction of additional work co-ordinates. These are usually defined so that the differential $\mathrm{d}y_j$ of the work co-ordinate when multiplied by a 'work coefficient' or a 'generalized force' Y_j gives the work done in a closed phase during a quasi-static change of state. These work co-ordinates are, like the volume, adiabatically freely adjustable. Their introduction causes no change in the structure of the theory developed in Part B, it merely leads to an increase in the number of dimensions of the appropriate co-ordinate space. Equation (8.4) is now replaced, for quasi-static changes of state, by

$$\mathrm{d}W = -P\,\mathrm{d}V + \sum_{j=2}^{n} Y_j\,\mathrm{d}y_j = \sum_{j=1}^{n} Y_j\,\mathrm{d}y_j. \tag{14.1}$$

The situation is, however, quite different when we try to include open systems and chemical reactions in the discussion. We note immediately that the definition of the fundamental concepts of work and heat run into difficulties. A more detailed analysis goes beyond the scope of this book and we shall confine our discussion to two simple examples.

(a) Let one phase be surrounded by a semi-permeable membrane and let all work co-ordinates y_j be fixed. Now let some material be forced through the semi-permeable membrane into the phase. Work is obviously done due to compression but the volume remains constant. It is, therefore, not generally possible to define clearly the 'volume work' done on an open phase. This removes, at the same time, the basis for the definition, according to § 8, of the heat absorbed. This state of affairs may be expressed somewhat more precisely by saying that adiabatic work as required by § 8 cannot, by definition, be done on an open phase.

(b) Let us consider a chemical reaction occurring within a closed phase. The reaction may be symbolized by the equation

$$\sum \nu_i X_i = 0. \tag{14.2}$$

ν_i is called the *stoichiometric coefficient* of substance i. Equation (14.2) shows that the changes in mole number of all substances taking part in the reaction are completely determined by a single mole number n_i.

The course of the reaction can, therefore, be described in the form

$$\mathrm{d}n_i = \nu_i \,\mathrm{d}\xi \tag{14.3}$$

where ξ is called the *progress variable.*

The quantity ξ is not, however, a variable of state. Its equilibrium value is completely determined by the two variables of state (e.g. T and V) of the system; ξ is an 'internal parameter' (cf. § 16) which becomes significant only with departure from equilibrium. The formal treatment of Part B, therefore, allows chemical reactions to occur within any (closed) phase since such reactions do not appear explicitly in the treatment. However, it is intrinsically impossible to obtain any information about chemical equilibrium in this way.

We might consider using ξ as a quasi-static adiabatic freely adjustable work co-ordinate by using a (positive or negative) catalyst to inhibit and uninhibit the reaction at will. Let us, in this connexion, consider the following process:†

Let (T_1, V_1, ξ_1) be the initial state in which the reaction is inhibited.

(i) By a quasi-static change in T and V at constant ξ_1 the system is brought to the equilibrium state corresponding to ξ_1.

(ii) The reaction is uninhibited and occurs quasi-statically as far as a value ξ_2. This is possible only if T and V also change quasi-statically.

(iii) The reaction is inhibited at the value ξ_2 and the system brought to the final state (T_2, V_2, ξ_2) by quasi-static changes in T and V at constant ξ_2.

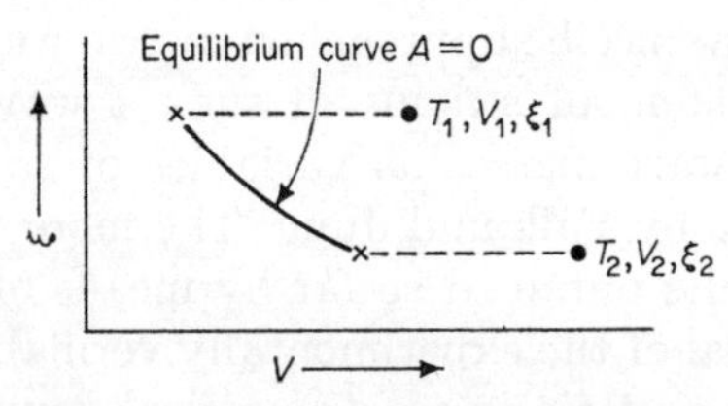

Fig. 8. Quasi-static performance of a chemical reaction

There is, however, a difficulty here: the work done during the quasi-static process does not generally reduce to an infinitesimal expression, even when 1 and 2 are as near to each other as we please. We return, in fact, to the previous result when we construct the Pfaff form

$$\mathrm{d}W = -P\,\mathrm{d}V - A\,\mathrm{d}\xi \tag{14.4}$$

† R. Haase, *Thermodynamik der Mischphasen*, Berlin, 1956.

where A is called the affinity. We shall discuss affinity in § 34. All we need here is that at chemical equilibrium $A = 0$. All states connected by quasi-static adiabatic processes then lie on a surface given by the solution of the Pfaff differential equation

$$\mathrm{d}U + P\,\mathrm{d}V + A\,\mathrm{d}\xi = 0. \tag{14.5}$$

This example shows, however, that in any partial process either $A = 0$ or $\mathrm{d}\xi = 0$ so that the last term vanishes for all quasi-static adiabatic processes and ξ does not, therefore, constitute a work co-ordinate.

§ 15 General formulation of the Second Law

The difficulties outlined in § 14 have their real basis in the fact that we have, so far, neglected to introduce as quantities of state the masses (or mole numbers) of the components. This omission leads to difficulties in the context of Carathéodory's theory since mass is an adiabatically inhibited variable (i.e. it is not adiabatically freely adjustable; the difference can be illustrated by comparing mass and volume). It is, therefore, impossible to show, by the method used in Part B, that entropy exists as a function of mass.

We can now see that we are dealing with a real extension of the theory. This extension, and its consequences, constitute the main part of Gibbs' thermodynamics. It is an innate property of phenomenological theory that such an extension requires new empirical bases. The problem may be approached in two ways. One possibility is to build the system of axioms in such a way as to be able to include from the start masses as variables of state. This has been attempted recently by Falk and Jung. The more usual method is to generalize the results obtained so far by means of analogies, and to regard the sum total of the experimentally verifiable consequences of such a procedure *a posteriori* as the empirical foundation.

We shall choose a middle way. We shall refrain from a rigorous deduction from empirical axioms, but we shall try to make plausible our choice of the particular fact of experience necessary for the extension of the theory and its validity.

Our investigations so far have shown that for any closed phase there is a function of state, entropy, for which we can write (with limitation to one work co-ordinate)

$$T\,\mathrm{d}S = \mathrm{d}U + P\,\mathrm{d}V. \tag{15.1}$$

Each of the quantities of state U and V on the right is related to the attainment of a *contact equilibrium*. For U it is thermal equilibrium made possible by means of a diathermic wall. For V it is pressure equilibrium (mechanical equilibrium) attainable by means of a frictionless movable piston. We again denote the part systems between which contact equilibrium is to be attained by $'$ and $''$. The variables U and V are summarized as *extensive parameters* X_i. We can now formulate the following important properties of contact equilibria:

(a) Within an adiabatically isolated total system any contact equilibrium can be reached and maintained.

(b) The extensive parameters under consideration obey a conservation law

$$X_i' + X_i'' = \text{const.} \tag{15.2}$$

(c) The processes leading to equilibrium are adiabatically irreversible.

For given values X_j', X_j'', therefore, the entropy of the total system as a function of X_i' is a maximum at equilibrium.

(d) Each contact equilibrium defines an *intensive parameter* P^* such that at equilibrium

$$P_i^{*\prime} = P_i^{*\prime\prime} \tag{15.3}$$

and

$$P_i^* = \frac{\partial S}{\partial X_i}. \tag{15.4}$$

From equation (15.1) we have

$$P_1^* = 1/T, \quad P_2^* = P/T. \tag{15.5}$$

The above formulation is descriptive. We shall not investigate whether the statements are independent of each other.

Our results so far show entropy to be a function of state. We have, however, only used certain types of variables of state. If the above conclusions are regarded as necessary and sufficient conditions for the existence of a variable of state X_i we can formulate the following theorem of experience as the basis for the extension of the theory.

Theorem of experience. A contact equilibrium having the properties (a)–(d) above exists for the mass m_k (or the mole number n_k) of every component. Any equilibrium between open phases will serve as an example for this.

We now have directly the *general formulation of the Second Law*:

For every phase α containing m components, there exists a function of state

$$S^{(\alpha)} = S^{(\alpha)}(U^{(\alpha)}, V^{(\alpha)}, y_2^{(\alpha)}, \ldots, y_n^{(\alpha)}, n_1^{(\alpha)}, \ldots, n_m^{(\alpha)}) \tag{15.6}$$

called the entropy of the phase having the following properties:

(a) The differential of the entropy is given by

$$T^{(\alpha)}\,\mathrm{d}S^{(\alpha)} = \mathrm{d}U^{(\alpha)} + P^{(\alpha)}\,\mathrm{d}V^{(\alpha)} - \sum_{j=2}^{n} Y_j^{(\alpha)}\,\mathrm{d}y_j^{(\alpha)} - \sum_{i=1}^{m} \mu_i^{(\alpha)}\,\mathrm{d}n_i^{(\alpha)}. \tag{15.7}$$

(b) The entropy of the total system is given by

$$S = \sum_{\alpha} S^{(\alpha)}. \tag{15.8}$$

(c) When the system is adiabatically isolated we have

$$\mathrm{d}S \geqslant 0 \quad \text{(adiabatic isolated system).} \tag{15.9}$$

(d) The quantity $T^{(\alpha)}$ is a universal function of the empirical temperature of the phase.

Equation (15.7) is called *Gibbs' fundamental equation*. The quantity

$$\mu_i^{(\alpha)} = -T^{(\alpha)}\left(\frac{\partial S^{(\alpha)}}{\partial n_i^{(\alpha)}}\right)_{U^{(\alpha)}, V^{(\alpha)}, y_j, n_{k \neq j},} \tag{15.10}$$

introduced by Gibbs is known as the *chemical potential* of substance i in phase α. It is expressed in terms of the mole as is usual nowadays in phenomenological thermodynamics. In principle it can also be expressed in terms of unit mass (Gibbs) or in terms of the molecule (statistical mechanics). Equation (15.7) is applicable not only to reversible processes although we shall mostly be concerned with such processes. We have here similar considerations to those laid down in connexion with eq. (4.24). The criterion for applicability is that the internal state of the phase at each stage of a change of state is completely described by the indicated variables. Even though this may not be the case, the integrated form of eq. (15.7) is still valid as long as the initial and final states of the process are described by these variables.

We still have to decide how to formulate the First Law within the above extension of the theory. As explained in § 14 the classical point of view (equivalence of heat and work) as well as Carathéodory's point of view (definition of heat) are meaningless for open systems. The usual formulation can be saved by an independent and arbitrary new definition of the heat absorbed. This device is of no interest in connexion with the present discussion. We are thus left with the statement that the internal energy of a phase is a function of state. The simple method of measurement of § 8 is naturally no longer useful and we have to investigate more closely the question of experimental determination.

We consider an adiabatic isolated system consisting of two part-systems $'$ and $''$. If we define reference values, the internal energies U, U', and U'' can be determined by the method of § 8. If we now remove the adiabatic wall between $'$ and $''$ while maintaining the adiabatic isolation of the whole system and exclude the performance of external work, the part-systems $'$ and $''$ will mix, temperatures and pressures will change but the total energy will remain constant. This process may be regarded as a change of state of the open phase $'$ for which the change in internal energy is

$$\Delta U' = U - U'. \tag{15.11}$$

Since all quantities on the right-hand side are measurable the change in internal energy of an open phase is, in principle, a measurable quantity.

CHAPTER II

General conditions for equilibrium and stability

§ 16 Discussion of the equilibrium concept. Internal parameters

We have already discussed the concept of thermodynamic equilibrium in a preliminary form in § 2. We shall now give this concept a precise form by means of the *definition*:

An isolated system is in thermodynamic equilibrium when, in the system, no changes of state are occurring at a measurable rate.

The existence of equilibrium follows directly from Clausius' principle and its generalization for other balancing processes. According to this, changes of state occurring in an isolated system are irreversible, i.e. they can proceed in only one direction in the isolated system (which is not subject to any external influences). It follows, therefore, that such processes must approach a final state asymptotically. This conclusion is reached with even greater clarity if we consider the fact, implicit in Clausius' principle, that these processes are balancing processes whose 'driving force' is the difference in intensive parameters which tends to zero during the process (cf. § 15).

The proviso 'at a measurable rate' implies that we can consider an equilibrium only with respect to specified processes and defined experimental conditions.

Example 1. A certain volume V contains H_2, O_2, and H_2O (vapour), the distribution of both matter and temperature being inhomogeneous. At room temperature and in the absence of a catalyst, the processes of heat conduction and diffusion occur at a measurable rate while the chemical reaction occurs at an immeasurably slow rate. After both temperature and concentration have become uniform, therefore, the system is in thermodynamic equilibrium for any mass ratio of the three substances. In the presence of a catalyst or at a sufficiently high temperature the reaction between oxygen and hydrogen occurs at a measurable rate. The system is now, therefore, in equilibrium only when chemical equilibrium has been established.

Example 2. It is known that the distribution of nucleons in atomic nuclei does not constitute a thermodynamic equilibrium. This can, however, be completely ignored since thermonuclear processes do not occur at a measurable rate under conditions usually considered in thermodynamics.

The concept 'absolute equilibrium' or 'equilibrium with respect to all imaginable processes' has, therefore, no physical significance.

A particular difficulty is encountered in the case of certain solids, e.g. alloys and glasses. Changes of state under the usual experimental conditions occur immeasurably slowly but the properties of the system depend on its previous history. A more thorough investigation into the applicability of thermodynamics then becomes necessary. We shall not discuss such problems here.

The description of equilibrium and non-equilibrium states involving several phases is made possible by the considerations developed in § 15. However, we are still faced with the problem of how to describe the attainment of thermodynamic equilibrium in a liquid one-component system. If such a system is not in thermodynamic equilibrium (apart from any effects due to external fields), it generally contains local inhomogeneities and is, therefore, not a phase according to the general definition of § 2. The conclusions of § 15 must then be applied to the limits of infinitesimally small volume elements regarded as being homogeneous. The quantities of state of § 15 then become field quantities and this takes us into the methods of the thermodynamics of irreversible processes. These methods are not used explicitly in ordinary thermodynamics and we shall, therefore, assume all phases to be macroscopic. Certain possibilities of departure from thermodynamic equilibrium still remain and these are described by means of 'internal parameters' independent of position within the phase.

The most important example of an internal parameter is the progress variable ξ introduced in § 14. The properties of internal parameters important in the present context are:

(a) Internal parameters are not variables of state in the sense of the definition of § 2.

(b) Any internal parameter may be expressed as a combination of the variables appearing in the fundamental equation (15.7) together with certain secondary conditions.

Statement (a) arises directly from a property of the progress variable described in § 14: at equilibrium the intensive parameter conjugate to ξ according to eq. (15.4), i.e. the affinity A, vanishes.

Statement (b) arises directly from eqs. (14.2) and (14.3) for the progress variable. Most internal parameters may be related directly to 'generalized chemical reactions', i.e. they may be introduced by means of equations formally identical with (14.2) and (14.3). In such cases, therefore, internal parameters may again be expressed in terms of mole numbers and suitable secondary conditions. It is also possible to express an internal parameter by means of other variables occurring in eq. (15.7) together with corresponding secondary conditions. An example of this can be seen in § 60.

According to (a), therefore, internal parameters are independent variables only as long as there is no equilibrium. This simply corresponds to the statement in § 2 that the number of measurable quantities necessary completely to describe a state is smaller at equilibrium than in any non-equilibrium state. According to (b) the set of variables of the fundamental equation (15.7) may be retained here as long as secondary conditions may be included. Since, however, we have defined entropy as a function in co-ordinate space, i.e. with the assumption of internal equilibrium, we still have to consider whether entropy is definable in cases of departure from equilibrium as described above.

Let us illustrate the problem by means of a simple example such as a homogeneous dissociation reaction

$$\mathrm{A} \rightarrow \mathrm{B} + \mathrm{C}. \tag{16.1}$$

The system contains only the components A, B and C, is adiabatically isolated and the volume is chosen as the only work co-ordinate. Let us suppose that reaction (16.1) is inhibited by a negative catalyst. Entropy will then certainly exist in the form

$$S = S(U, V, n_\mathrm{A}, n_\mathrm{B}, n_\mathrm{C}). \tag{16.2}$$

Now let the inhibitor be removed for a short time so that an infinitesimal amount of reaction can take place. The inhibitor is then rapidly reintroduced. According to the Second Law the entropy will increase during the reaction. After the inhibitor has been replaced the system is again in internal equilibrium. The increase in entropy is, therefore, according to (15.7) and (16.1),

$$T\,\mathrm{d}S = \mu_\mathrm{A}\,\mathrm{d}n_\mathrm{A} - \mu_\mathrm{B}\,\mathrm{d}n_\mathrm{B} - \mu_\mathrm{C}\,\mathrm{d}n_\mathrm{C} \quad (U = \text{const},\ V = \text{const}). \tag{16.3}$$

If we now introduce the progress variable ξ and the affinity A, as in § 14, eq. (16.3) becomes

$$\mathrm{T}\,\mathrm{d}S = A\,\mathrm{d}\xi \quad (U = \text{const},\ V = \text{const}). \tag{16.4}$$

We have thus defined changes in entropy for a homogeneous system for conditions other than those of chemical equilibrium.

These considerations may be extended quite simply to other internal parameters. The considerations are based on the fact that equilibrium is expressible only with respect to definite processes and, furthermore, that the introduction of an inhibitor increases the number of variables of state.

It is, of course, absurd to suppose that, in the foregoing considerations, entropy is 'created' by the introduction of the inhibitor. The introduction of an inhibitor is simply an intellectual aid which makes possible the definition of entropy at non-equilibrium using our previous conclusions. The concept is similar to the drawing of a tangent to the distance–time curve in the definition of non-uniform velocity. The question as to whether suitable inhibitors are actually available in any particular case is, therefore, of no significance.

The introduction of internal parameters is not confined to considerations of homogeneous phases. It is easy to see that the concept of a progress variable is similarly applicable to heterogeneous reactions; but one or more internal parameters may also be introduced into the discussion of other heterogeneous systems when changes in variables of state are limited by secondary conditions. A simple example is discussed in § 17.

We have now also defined entropy for all non-equilibrium states which we have to consider. We can thus use the principle of entropy increase to formulate conditions for thermodynamic equilibrium. This problem will be discussed in the following section.

§ 17 Gibbs' equilibrium conditions

One of the main problems of thermodynamics is the explicit formulation of conditions for thermodynamic equilibrium for various special cases (e.g. vapour pressure, osmotic and chemical equilibria). The older method of solving this problem (used in Europe fairly generally until about 1930) was to construct a reversible cyclic process for every special case.

This procedure has yielded important results. It is, however, not only complicated but also apt to errors. Nowadays, therefore, the method in general use is that due to Gibbs. This starts by formulating completely general conditions for equilibrium and deduces all special cases from these by purely mathematical considerations.

From a purely formal point of view Gibbs' formulation of the equilibrium conditions is analogous to the use of extrema in mechanics. It can, therefore, like the latter, be considered from an axiomatic standpoint. We shall not, however, concern ourselves further with this standpoint although it explains why we shall deduce from Gibbs' formulation several points which were introduced as facts of experience in Chapter I. We shall first deal with the mathematical formulation, followed by some explanations and proofs in so far as they are not already given in Chapter I.

Gibbs' equilibrium conditions are:

For a closed system, whose work co-ordinates are fixed, to be in equilibrium it is necessary and sufficient that either

$$(\delta S)_U \leqslant 0 \tag{17.1}$$

or

$$(\delta U)_S \geqslant 0. \tag{17.2}$$

Explanations.

(a) The symbol δ denotes a virtual displacement in the sense used in analytical mechanics. It thus represents an imaginary change of state which is infinitesimally of the first order and obeys the following conditions:
 (α) The change of state must be possible, i.e. consonant with the general conditions governing the system [cf. (b)].
 (β) The change of state is not a function of time.
 (γ) The entropy is defined for the changed state as well as for the initial state.

(b) The conditions governing the system are initially given by the definition 'closed system with fixed work co-ordinates'. Furthermore, the subscript U means that the internal energy remains constant for the change. This implies that condition (17.1) for the entropy is applied to an isolated system. Condition (a α), furthermore, excludes in particular
 (α) changes of state prevented by some inhibition such as the oxygen–hydrogen reaction at room temperature in the absence of a catalyst,
 (β) changes of state for which the entropy of adiabatically isolated part systems decreases.

If the equilibrium condition is (17.2), however, the system is not isolated since the secondary condition demands constant entropy and this does not imply adiabatic isolation. In any case, (17.2) would not make sense for an isolated system for which variation of internal energy is impossible.

(c) The equality sign in (17.1) means that entropy as a function of the variables under consideration possesses a stationary value at equilibrium. Since (17.1) contains only the first derivative, it does not state whether the stationary value is a maximum or a minimum. We shall return to this problem in § 18.

The inequality sign (which is frequently misunderstood) refers to a state of affairs when virtual displacements are possible in only one direction. It is, therefore, impossible to decide whether the equilibrium value of the entropy is a stationary point in the mathematical sense.

Let us illustrate the above by means of an example† and consider the distribution of a substance i between two phases $'$ and $''$. A particular case would be the distribution of a solute between two immiscible solvents (e.g. benzene and water would constitute a good approximation). The distribution may be described by the parameter

$$\xi = \frac{n_i'}{n_i''}. \tag{17.3}$$

We must distinguish between two cases:

(α) The substance i at equilibrium occurs in both phases. The equilibrium value ξ_{eq} is then positive and finite (e.g. the distribution of benzoic acid between water and benzene). The function $S(\xi)$ is, therefore, physically meaningful for both $\xi < \xi_{eq}$ and $\xi > \xi_{eq}$. Virtual displacements are thus possible on both sides of ξ_{eq}. Statement (17.1) shows that, for this case, ξ_{eq} is a stationary point of the function $S(\xi)$. (Fig. 9.)

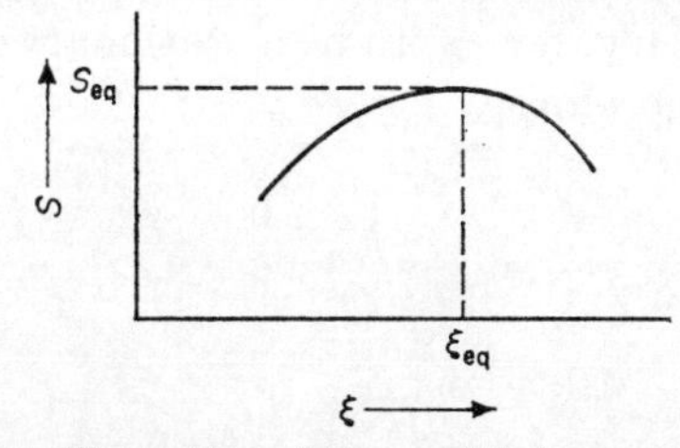

FIG. 9. Illustration of the equilibrium condition

For a stationary point we have

$$\left(\frac{\partial S}{\partial \xi}\right)_{eq} = 0. \tag{17.4}$$

† R. Haase, *Thermodynamik der Mischphasen*, Berlin, 1956.

The first-order variation is, therefore, with all other variables constant,

$$\delta S = \left(\frac{\partial S}{\partial \xi}\right)_{\text{eq}} \delta\xi = 0. \tag{17.5}$$

(β) Substance i at equilibrium occurs in only one phase, say ″. The equilibrium value ξ_{eq} is zero (e.g. approximately for the distribution of naphthalene between water and benzene). The $S(\xi)$ curve becomes physically meaningless for $\xi < \xi_{\text{eq}}$ and virtual displacements are possible only for $\delta\xi > 0$. (Fig. 10.)

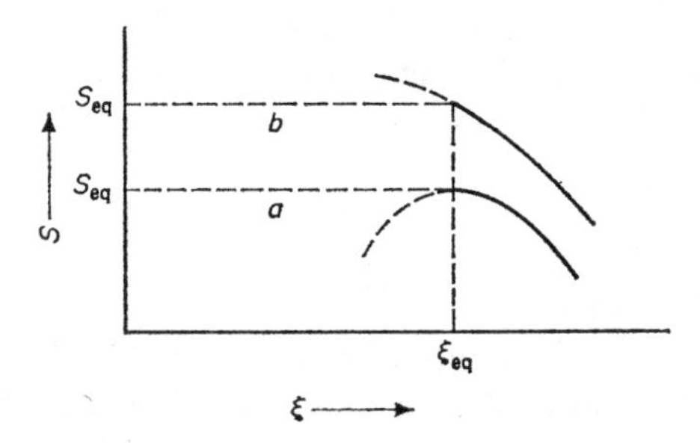

Fig. 10. Illustration of the equilibrium condition

Equation (17.1) now states that δS is zero or negative for a virtual displacement. Figure 10 shows that the derivative at the point ξ_{eq} may be zero or negative. It is therefore possible, but not necessary, for ξ_{eq} to be a stationary point.

We have, therefore,

$$\left(\frac{\partial S}{\partial \xi}\right)_{\text{eq}} \leqslant 0 \tag{17.6}$$

and thus

$$\delta S = \left(\frac{\partial S}{\partial \xi}\right)_{\text{eq}} \delta\xi \leqslant 0. \tag{17.7}$$

The quantity used here is an internal parameter, as described in § 16, for the complete system. This simple example shows clearly how this internal parameter is related to the variables of state of § 15. We therefore choose as variables the mole numbers n_i' and n_i''. Since the system is closed the secondary condition that the sum $n_i' + n_i''$ must remain constant applies to all virtual changes. Case (α) thus now becomes

(α')

$$\delta S = \delta S' + \delta S'' = \left(\frac{\partial S'}{\partial n_i'}\right)_{\text{eq}} \delta n_i' + \left(\frac{\partial S''}{\partial n_i''}\right)_{\text{eq}} \delta n_i'' \tag{17.8}$$

together with the secondary condition

$$\delta n_i' + \delta n_i'' = 0. \tag{17.9}$$

For case (β) we have, in this notation,

(β')

$$\delta S = \delta S' + \delta S'' = \left(\frac{\partial S'}{\partial n_i'}\right)_{\text{eq}} \delta n_i' + \left(\frac{\partial S''}{\partial n_i''}\right)_{\text{eq}} \delta n_i'' \tag{17.10}$$

with secondary conditions

$$\delta n_i' + \delta n_i'' = 0, \quad \delta n_i' > 0, \quad \delta n_i'' < 0. \tag{17.11}$$

These formulations show explicitly that the general equilibrium condition is in the form of an extremum expression with secondary conditions.

We chose as our example the problem of the distribution of a substance between phases because of its simplicity. Case (β) can in practice be ignored in this connexion. We shall generally not consider it for explicit calculations although it was discussed in detail by Gibbs. It is, however, of interest in connexion with chemical reactions (§ 36) and also occurs with certain internal parameters which are introduced in statistical thermodynamics, e.g. long-range order in binary mixed crystals.† Its discussion is, therefore, justified not only on historical grounds.

Proofs. The equilibrium condition (17.1) follows directly from the Second Law, statement (15.9), and the definition of equilibrium in § 16. We can, therefore, here dispense with a repeated proof of the necessity and sufficiency of the condition. We must, however, remember that, in the case of the equality sign, (17.1) contains only a part of the statement (15.9) since (17.1) simply states that the entropy must have a stationary value at equilibrium. The possibility of such a division (first carried out by Gibbs) is based on the formulation in terms of the extremum principle. It has proved extraordinarily fruitful for thermodynamics.

† Cf. § 49.

However, the equilibrium condition (17.2) does not follow directly from (15.9). We shall prove, therefore, that it constitutes a formulation equivalent to the condition (17.1). The proof rests on the following two facts:

(α) The system is, by definition, not isolated. When [as in the formulation (17.1)] no additional conditions are imposed heat may be gained or lost by the system without restriction.

(β) According to (α) it is always possible to increase or decrease the energy and entropy of the system concurrently by the introduction or abstraction of heat.

Proof of the equivalence of (17.1) *and* (17.2).

(a) Suppose that the condition (17.1) is not valid. There must then be a virtual change for which

$$\delta S > 0, \quad \delta U = 0. \tag{17.12}$$

Starting from this changed state II a state III can be attained by a simultaneous decrease of the energy and the entropy. If state III is regarded as a change from the initial state I, then

$$\delta S = 0, \quad \delta U < 0. \tag{17.13}$$

This, however, also contradicts the condition (17.2) (Fig. 11a).

(b) Suppose that the condition (17.2) is not valid. A change must then be possible for which

$$\delta U < 0, \quad \delta S = 0. \tag{17.14}$$

Starting from this changed state II, a state III will be accessible by a simultaneous increase of the energy and the entropy. If state III is regarded as a change from the initial state I, then

$$\delta U = 0, \quad \delta S > 0. \tag{17.15}$$

The condition (17.1) is, therefore, likewise contradicted (Fig. 11b). The equivalence of (17.1) and (17.2) is, therefore, established.

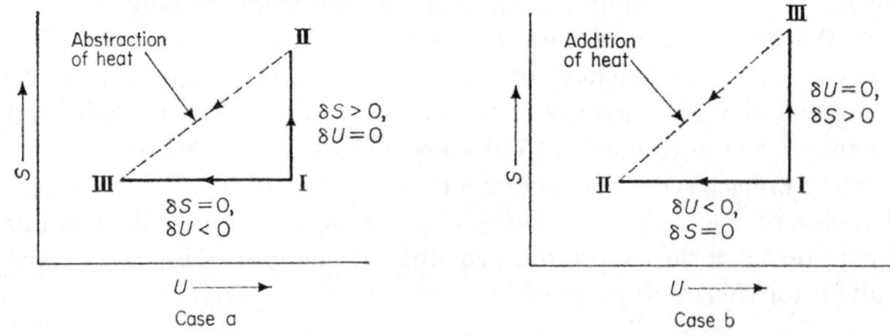

FIG. 11. Proof of the equivalence of (17.1) and (17.2)

The fact that consequences of the Second Law can also be represented on the basis of internal energy if entropy is chosen as an independent variable of state was discovered by Gibbs. It shows the division into the First and Second Law to be fundamentally arbitrary and is to be understood merely on historical grounds (cf. the discussion of the First Law in § 8). The use of internal energy as a function of state and of entropy as an independent variable of state furthermore leads to a simplification of the formal representation; this will become apparent in Chapter III.

§ 18 The stability conditions

We have already mentioned several times that the equilibrium conditions (17.1) and (17.2) (for the case of the equality sign) merely contain the statement that, at equilibrium, under the given conditions, the entropy or, as the case may be, the internal energy has a stationary value. This situation is entirely analogous to the state of affairs in mechanics where the definition of equilibrium leaves open the question whether the equilibrium is stable, unstable, neutral, or metastable.

This limitation derives formally from the fact that (17.1) and (17.2) contain only first-order differentials, i.e. that the Taylor series expansions (17.5) and (17.8) end with the linear term. Higher-order differentials must, therefore, be considered if all the consequences of the Second Law are to be used for the formulation of equilibrium conditions. We find that no differentials of order greater than two are significant in thermodynamics; this will become apparent later, in Chapter IV. We shall ignore this for the time being and shall use the symbol Δ to denote virtual displacements defined by including higher-order differentials.

As in mechanics we then distinguish between the following cases:

(a) *Stable equilibrium.* For all virtual displacements, we have

$$(\Delta S)_U < 0 \tag{18.1}$$

or

$$(\Delta U)_S > 0. \tag{18.2}$$

At stable equilibrium under given conditions, therefore, the entropy is a maximum (highest possible value) or the internal energy a minimum (lowest possible value). This statement is derived from the Second Law.

(b) *Unstable equilibrium.* There are virtual displacements for which

$$(\Delta S)_U > 0 \tag{18.3}$$

or

$$(\Delta U)_S < 0. \tag{18.4}$$

In this case the stationary value of the entropy is a minimum, and that of the internal energy a maximum.

Unstable equilibria are not physically realizable. This statement which has occasionally been challenged on the basis of phenomenological arguments can be proved within the framework of statistical thermodynamics.

(c) *Neutral equilibrium.* There are virtual displacements for which

$$(\Delta S)_U = 0 \tag{18.5}$$

or

$$(\Delta U)_S = 0. \tag{18.6}$$

This case is possible in principle but will not be discussed until Chapter VI.

(d) *Metastable equilibrium.* We understand by this the case where an equilibrium is stable with respect to infinitesimally near neighbouring states but is unstable with respect to states with finite differences. A well-known example is a supercooled liquid. Within the framework of thermodynamics this case has no innate significance. The basic state of affairs can best be represented by means of the concept of inhibition.

CHAPTER III

Thermodynamic potentials and Massieu–Planck functions

§ 19 Interlude: Legendre transformations. Homogeneous functions and Euler's theorem

We shall need some further mathematical aids to build the formal framework of thermodynamics. We shall develop these aids in the present section.

(a) *Legendre transformations.* Consider the following problem: Let there be a function

$$y = y(x_1, \ldots, x_n). \tag{19.1}$$

We now look for a transformation such that

(α) one or more of the differentials

$$P_i \equiv \frac{\partial y}{\partial x_i} \tag{19.2}$$

are introduced as independent variables;

(β) it is a one–one transformation.

Requirement (β) means that, starting with the transformed function, the original function (19.1) must be capable of unique reconstruction. In other words, it is essential that the transformation conserves the mathematical content (or the corresponding physical information) of (19.1).

Let us illustrate the problem and its solution by using the example of the single variable. Our original function is now

$$y = y(x). \tag{19.3}$$

This equation represents a curve in the x, y-plane. The gradient for each x-value is given by

$$p = \frac{\mathrm{d}y}{\mathrm{d}x} \equiv y'(x). \tag{19.4}$$

It might be thought best to solve the problem by eliminating x from eqs. (19.3) and (19.4); y is indeed obtained as a function of p, viz.

$$y = y(p). \tag{19.5}$$

If, however, we attempt to reconstruct the original function (19.3) from (19.5), we find that no unique solution is obtainable and that condition (β) is thus not satisfied.

The formal reason for this is simply that (19.5) is a first-order differential equation whose solution represents a family of curves in the x, y-plane. The change from (19.3) to (19.5) therefore involves the loss of part of the mathematical content of (19.3). The solution of the problem which satisfies the two conditions (α) and (β) is based on the fact that a curve in the x, y-plane may be regarded either as the geometrical locus of the points (x, y) which obey equation (19.3), or as the envelope of the family of tangents to (19.3) (Fig. 12).

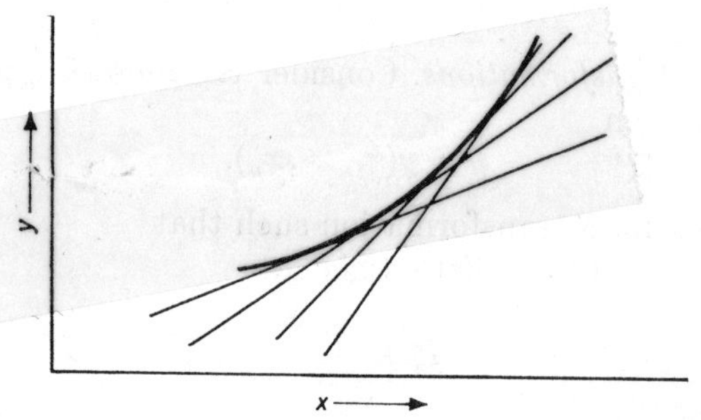

FIG. 12. A curve as the envelope of its family of tangents

An equation which defines the family of tangents is thus mathematically equivalent to eq. (19.3). If, therefore, we denote by p the gradient of the tangent and by ψ its intercept on the ordinate, the equation

$$\psi = \psi(p), \tag{19.6}$$

which assigns a value of ψ to each value of p, constitutes the formal solution of the problem. Since (19.6) represents the family of tangents to the curve (19.3), p is defined by eq. (19.4) and (α) is satisfied. Furthermore, the representations (19.3) and (19.6) bear a unique and reversible relation each to the other and thus (β) is also satisfied. We still have to calculate explicitly the function (19.6) from (19.3). For this purpose we consider a tangent to the curve (19.3) at the point (x, y). This tangent has a gradient p and an intercept ψ on the ordinate. By definition we then have

$$p = \frac{y - \psi}{x} \tag{19.7}$$

or

$$\psi = y - px. \tag{19.8}$$

If the quantities x and y are now eliminated from eqs. (19.3), (19.4), and (19.8), the required relation between ψ and p is obtained. It is easy to see that the process can be reversed to recover (19.3) from (19.6). By differentiating (19.8) and introducing (19.4) we get

$$\mathrm{d}\psi = -p\,\mathrm{d}x - x\,\mathrm{d}p + \mathrm{d}y = -x\,\mathrm{d}p \tag{19.9}$$

or

$$-x = \frac{\mathrm{d}\psi}{\mathrm{d}p}. \tag{19.10}$$

If the variables ψ and p are eliminated from eqs. (19.6), (19.8), and (19.10), then (19.3) is recovered. The elimination of the variables is possible only if p is a function of x. If (19.4) is expanded into a Taylor series according to

$$\Delta p = \frac{\mathrm{d}p}{\mathrm{d}x}\Delta x + \ldots, \tag{19.11}$$

the sufficient condition

$$\frac{\mathrm{d}p}{\mathrm{d}x} \neq 0 \tag{19.12}$$

or

$$\frac{\mathrm{d}^2 y}{\mathrm{d}x^2} \neq 0, \quad \frac{\mathrm{d}^2 \psi}{\mathrm{d}p^2} \neq 0 \tag{19.13}$$

is obtained. These conditions are adequate for most applications. The few special cases in which they are not satisfied require separate discussion.

Let us now represent the above procedure once more, this time schematically:

$$\left.\begin{aligned} y &= y(x), & \psi &= \psi(p), \\ p &= \frac{\mathrm{d}y}{\mathrm{d}x}, & -x &= \frac{\mathrm{d}\psi}{\mathrm{d}p}, \\ \psi &= -px + y, & y &= xp + \psi. \end{aligned}\right\} \tag{19.14}$$

Elimination of x and y gives $\psi = \psi(p)$.
Elimination of p and ψ gives $y = y(x)$.

The transformation $y(x) \rightleftharpoons \psi(p)$ defined by the above equations is called a Legendre transformation; $\psi(p)$ is called the Legendre transform of $y(x)$. The Legendre transformation is a special case of a contact transformation. It occurs in classical mechanics on changing from a Lagrangian to a Hamiltonian formulation. The property which is important for our purposes is as follows:

The Legendre transformation does *not* result in assigning to each point in the x, y-plane a corresponding point in the ψ, p-plane. It does, however, assign reversibly and uniquely a point on the curve $\psi(p)$ to each point on the curve $y(x)$.

These considerations can be generalized for the function (19.1) with n independent variables by substituting $(n+1)$-dimensional space for the plane. This causes no difficulty at all. We shall, therefore, not carry out the process in detail but simply quote the necessary formulae. We shall also consider the important possibility that only a sub-group $(x_1, \ldots, x_k)$ of the complete group $(x_1, \ldots, x_n)$ needs to be transformed. In geometric terms this means that the transformation is performed into a $(k+1)$-dimensional sub-space of the $(n+1)$-dimensional space. The sub-space must obviously contain the y-co-ordinate. The variables $x_{k+1}, \ldots, x_n$ are to be regarded as parameters for the k-fold Legendre transformation. We then have the following scheme analogous to (19.14):

$$\left.\begin{aligned}
& y = y(x_1, \ldots, x_k, \ldots, x_n), && \psi = \psi(p_1, \ldots, p_k, x_{k+1}, \ldots, x_n), \\
& p_i = \frac{\partial y}{\partial x_i}, && -x_i = \frac{\partial \psi}{\partial p_i} \quad (i \leqslant k), \\
& \mathrm{d}y = \sum_{i=1}^{n} p_i \, \mathrm{d}x_i, && p_j = \frac{\partial \psi}{\partial x_j} \quad (j > k), \\
& \psi_k = y - \sum_{i=1}^{k} p_i x_i, && \mathrm{d}\psi = -\sum_{i=1}^{k} x_i \, \mathrm{d}p_i + \sum_{j=k+1}^{n} p_j \, \mathrm{d}x_j, \\
& && y = \psi + \sum_{i=1}^{k} x_i p_i.
\end{aligned}\right\} \tag{19.15}$$

Elimination of y and $x_1, \ldots, x_k$ gives

$$\psi = \psi(p_1, \ldots, p_k, x_{k+1}, \ldots, x_n).$$

Elimination of ψ and $p_1, \ldots, p_k$ gives $\quad y = y(x_1, \ldots, x_k, \ldots, x_n)$.

For this general case the sufficient condition for the existence of the transformation is

$$\frac{\partial(p_1, \ldots, p_k)}{\partial(x_1, \ldots, x_k)} = \left| \frac{\partial^2 y}{\partial x_i \, \partial x_j} \right| \neq 0 \quad (i, j \leqslant k), \tag{19.16}$$

i.e. the Jacobi determinant for the p_i and x_i must differ from zero.

(b) *Homogeneous functions and Euler's theorem.*

Definition. If a function

$$\phi = \phi(x_1, \ldots, x_n) \tag{19.17}$$

obeys the equation

$$\phi(\alpha x_1, \ldots, \alpha x_n) = \alpha^l \phi(x_1, \ldots, x_n) \tag{19.18}$$

where l is a positive integer, it is said to be a homogeneous function of the lth degree.

Euler's theorem. For homogeneous functions of the lth degree it is true that

$$l\phi(x_1, \ldots, x_n) = \sum_{i=1}^{n} \frac{\partial \phi}{\partial x_i} x_i. \tag{19.19}$$

Proof. If the eq. (19.18) in the definition is differentiated with respect to α, we have

$$\sum_{i=1}^{n} \frac{\partial \phi}{\partial(\alpha x_i)} \frac{\partial(\alpha x_i)}{\partial \alpha} = l\alpha^{l-1} \phi(x_1, \ldots, x_n). \tag{19.20}$$

This relationship must be true for all values of α. If we put α equal to unity, eq. (19.19) follows directly.

As far as thermodynamics is concerned, homogeneous functions of the first degree ($l = 1$) are the only ones required.

§ 20 The fundamental equation. Extensive and intensive parameters. Equations of state. The Gibbs–Duhem equation

We shall now discuss Gibbs' fundamental equation (19.7) in somewhat greater detail. For simplicity's sake we shall confine ourselves to a consideration of volume as the only work co-ordinate. We shall discontinue using the phase index which is of no significance for our present purpose. We therefore write the fundamental equation in the form

$$\mathrm{d}S = \frac{1}{T}\mathrm{d}U + \frac{P}{T}\mathrm{d}V - \sum_{i=1}^{m} \frac{\mu_i}{T}\mathrm{d}n_i \tag{20.1}$$

or, by integration,

$$S = S(U, V, n_1, \ldots, n_m). \tag{20.2}$$

In §§ 17 and 18 we have already shown that, for the equilibrium and stability conditions, an equivalent representation is possible by considering the internal energy. Let us, therefore, change the fundamental equation into the corresponding form.

Since, by definition (§ 10),

$$\left(\frac{\partial S}{\partial U}\right)_{V,n_1,\ldots,n_m} > 0 \tag{20.3}$$

the entropy constitutes an unique, continuous, and differentiable function of the internal energy if all other quantities of state are constant. Equation (20.2), therefore, can be solved uniquely for U and we obtain

$$U = U(S, V, n_1, \ldots, n_m) \tag{20.4}$$

or, in differential form,

$$\mathrm{d}U = T\,\mathrm{d}S - P\,\mathrm{d}V + \sum_{i=1}^{m} \mu_i\,\mathrm{d}n_i. \tag{20.5}$$

This equation, too, is called Gibbs' fundamental equation. Equation (20.1) is called the *entropy representation*; eq. (20.5) is the *energy representation.* In thermodynamics the energy representation is nowadays used almost exclusively; the entropy representation is of significance mainly for the thermodynamics of irreversible processes and for statistical mechanics. We shall, therefore, from now on mention the entropy representation only in connexion with a few general considerations. The variables of state which occur in the fundamental equation fall into two classes according to their properties. The variables of the first class have the property that, when two part systems ′ and ″ are combined into an (unprimed) total system, the relationship

$$X_i' + X_i'' = X_i \tag{20.6}$$

applies. These variables of state are called *extensive parameters.* Volume and mole numbers obviously belong to this class.

Variables of the second class are defined by means of contact equilibria in such a way that, at equilibrium,

$$P_i' = P_i''. \tag{20.7}$$

This was discussed in greater detail in § 15. Such variables, e.g. temperature, pressure, and chemical potential, are called *intensive parameters.* If we now consider the Laws [cf. eqs. (8.8) and (10.26)], it follows that internal energy and entropy are extensive parameters.

We are now in a position to formulate the next important properties of the fundamental equation. We shall first give a summary of these properties, followed by some explanations.

Properties of the fundamental equation.

(a) The fundamental equation is, in both the entropy and the energy representation, a function which depends only on extensive parameters.

(b) The fundamental equation is, in both the entropy and the energy representation, a homogeneous function of the first degree in all independent variables.

(c) The fundamental equation is, in both the entropy and the energy representation, a *characteristic function*, i.e. one which contains every statement that thermodynamics can make about the system.

(d) The definition of intensive parameters is, in the entropy representation,

$$P_i^* = \frac{\partial S}{\partial X_i} \tag{20.8}$$

and in the energy representation

$$P_i = \frac{\partial U}{\partial X_i}. \tag{20.9}$$

(e) Any intensive parameter can be represented as a function of the same independent variables which occur in the corresponding form of the fundamental equation. Such equations which have the general form

$$P_i = P_i(X_1, \ldots, X_r) \tag{20.10}$$

are called *equations of state*. The functions (20.10) are homogeneous and of the zeroth degree.

(f) The equations of state are not independent of each other since there is an additional relationship between these intensive parameters. This relationship in the differential form is called the *Gibbs–Duhem equation*. It is, in the entropy representation,

$$U\,\mathrm{d}\left(\frac{1}{T}\right) + V\,\mathrm{d}\left(\frac{P}{T}\right) - \sum_{i=1}^{m} n_i\,\mathrm{d}\left(\frac{\mu_i}{T}\right) = 0 \tag{20.11}$$

and in the energy representation

$$S\,\mathrm{d}T - V\,\mathrm{d}P + \sum_{i=1}^{m} n_i\,\mathrm{d}\mu_i = 0. \tag{20.12}$$

Explanations and proofs.

Re (a): This statement follows from eqs. (20.1) and (20.5) together with the Laws and the definition of extensive parameters.

Re (b): If we assume that the part systems ′ and ″ in eq. (20.6) are equal, the statement follows from (a) and the definition (19.18).

Re (c): This statement summarizes the following properties:

(α) If the fundamental equation is known for all homogeneous regions of a system, the thermodynamic equilibrium of the total system can be calculated explicitly by substitution into (17.1) or (17.2).

(β) Every statement concerning the stability of the equilibrium is deducible from (18.1) to (18.6).

(γ) The partial first derivatives of the fundamental equation give the intensive parameters. In the energy representation we have in particular that

$$\left(\frac{\partial U}{\partial S}\right)_{V,n_1,\dots,n_m} = T, \tag{20.13}$$

$$\left(\frac{\partial U}{\partial V}\right)_{S,n_1,\dots,n_m} = -P, \tag{20.14}$$

$$\left(\frac{\partial U}{\partial n_i}\right)_{\substack{S,V,n_j\\ i\neq j}} = \mu_i. \tag{20.15}$$

(δ) The partial second derivatives yield other measurable quantities. On the basis of the general relationship

$$\frac{\partial^2 \mathrm{f}}{\partial x\,\partial y} = \frac{\partial^2 \mathrm{f}}{\partial y\,\partial x} \tag{20.16}$$

for a complete differential df, equations which interconnect these measurable quantities are also obtained. We shall give only one example for the energy representation since we shall discuss the second derivatives in detail in §§ 24 and 25. According to (20.5) and (20.13) we have

$$\frac{\partial^2 U}{\partial V\,\partial S} = \frac{\partial}{\partial V}\left(\frac{\partial U}{\partial S}\right) = \left(\frac{\partial T}{\partial V}\right)_{S,n}. \tag{20.17}$$

This quantity represents the temperature increase per unit volume for a quasi-static adiabatic expansion. From (20.5) and (20.14) we have that

$$\frac{\partial^2 U}{\partial S\,\partial V} = \frac{\partial}{\partial S}\left(\frac{\partial U}{\partial V}\right) = -\left(\frac{\partial P}{\partial S}\right)_{V,n}. \tag{20.18}$$

The quantity on the right may also be written in the form $T(\partial P/\partial Q)_V$. It therefore denotes the increase in pressure when one unit of heat is introduced at constant volume. From (20.16) we get

$$\left(\frac{\partial T}{\partial V}\right)_{S,n} = -\left(\frac{\partial P}{\partial S}\right)_{V,n}. \tag{20.19}$$

Relationships of this type are known as *Maxwell relations*. It is important to remember that the properties of characteristic functions do not belong to the entropy or internal energy as such; these properties depend entirely on the set of variables chosen for the fundamental equation. This can be shown easily by means of a counter-example. The internal energy may be represented as a function of the variables T, V, n (for a one-component system). We then have

$$\mathrm{d}U = \left(\frac{\partial U}{\partial T}\right)_{V,n} \mathrm{d}T + \left(\frac{\partial U}{\partial V}\right)_{T,n} \mathrm{d}V + \left(\frac{\partial U}{\partial n}\right)_{V,T} \mathrm{d}n. \tag{20.20}$$

This equation is perfectly correct and, according to the First Law, $\mathrm{d}U$ is a complete differential in this representation as well. The function

$$U = U(T, V, n) \tag{20.21}$$

is, however, *not* a characteristic function. It can neither be used to calculate thermodynamic equilibrium nor can the intensive parameters P and μ be obtained from it by differentiation. The reason is that the entropy is not uniquely defined in eq. (20.21). Comparison with eq. (20.13) shows that (20.21), when considered from the point of view of the fundamental equation, represents a partial differential equation of the first order whose solution includes arbitrary functions. Equation (20.21) can, therefore, be obtained from the fundamental equation (20.4) but the reverse process is not possible. In other words, the representations (20.4) and (20.21) are *not equivalent* and (20.21) contains less information than (20.4). The representation (20.20) is, in principle, already known from § 8. We shall see that (20.20) is of considerable importance and that it leads directly to the definition of the molar heat capacity at constant volume, C_v. For $u = U/n$, we have that

$$C_v = \left(\frac{\partial u}{\partial T}\right)_{V,n}. \tag{20.21a}$$

Re (d): Extensive parameters are the same whether the representation is in the entropy or energy form. Entropy and energy simply exchange roles. The explicit definition of intensive parameters is, however, different in the two representations as can be seen by comparing (20.1) and (20.5). The great advantage of the energy representation is that it uses the directly measurable intensive parameters $(T, -P, \mu_i)$ while the intensive parameters $(1/T, P/T, -\mu_i/T)$ of the entropy representation are not directly measurable.

Re (e): The equations of state (for a one-component system) are, in the entropy representation, expressed explicitly as

$$\frac{1}{T} = \frac{1}{T}(U, V, n), \tag{20.22}$$

$$\frac{P}{T} = \frac{P}{T}(U, V, n), \tag{20.23}$$

$$\frac{\mu}{T} = \frac{\mu}{T}(U, V, n), \tag{20.24}$$

and in the energy representation as

$$T = T(S, V, n), \tag{20.25}$$

$$P = P(S, V, n), \tag{20.26}$$

$$\mu = \mu(S, V, n). \tag{20.27}$$

An older terminology which is, however, still used quite frequently refers to the equation

$$P = P(T, V, n) \tag{20.28}$$

as the *thermal equation of state*, while the equation

$$U = U(T, V, n) \tag{20.29}$$

is called the *caloric equation of state*. Comparison with (20.22) and (20.23) shows that (20.28) and (20.29) are part of the entropy representation and simply represent a transformation of the clearer expressions (20.22) and (20.23). The thermal equation of state is identical with eq. (7.7) introduced in § 7 as the equation for the isotherms. In eq. (20.28) the empirical temperature has simply been replaced by the absolute temperature.

That the equations of state are homogeneous functions of the zeroth degree follows from (b) together with the definitions (19.18) and (20.8) or (20.9). According to these we have that

$$\frac{\partial(\alpha S)}{\partial(\alpha X_i)} = \frac{\partial S}{\partial X_i} = P_i^* \tag{20.30}$$

or

$$\frac{\partial(\alpha U)}{\partial(\alpha X_i)} = \frac{\partial U}{\partial X_i} = P_i. \tag{20.31}$$

Therefore,

$$P_i(\alpha X_1, \ldots, \alpha X_r) = P(X_1, \ldots, X_r). \tag{20.32}$$

The system is thermodynamically not completely defined by a single equation of state. Knowledge of all the equations of state is, on the other hand, equivalent to knowledge of the fundamental equation. The property (b) shows, however, that (for a one-component system) a knowledge of two equations of state is enough to define the thermodynamic system completely. If, in eq. (20.32) (with $r = 3$), we put $\alpha = X_3^{-1}$, we obtain three equations of state of the form

$$P_i = P_i\left(\frac{X_1}{X_3}, \frac{X_2}{X_3}, 1\right). \tag{20.33}$$

The variables X_1/X_3 and X_2/X_3 can be eliminated from these three equations. We then get an equation involving the three intensive parameters P_i. If, therefore, we know two equations of state, we can calculate the third intensive parameter from them.

The variables which appear in (20.33) are, in the entropy representation, the *molar quantities*

$$u = \frac{U}{n}, \quad v = \frac{V}{n} \tag{20.34}$$

or the *densities*

$$\frac{U}{V}, \quad \rho = \frac{n}{V}. \tag{20.35}$$

In the energy representation we have the *molar quantities*

$$s = \frac{S}{n}, \quad v = \frac{V}{n} \tag{20.36}$$

or the *densities*

$$\frac{S}{V}, \quad \rho = \frac{n}{V}. \tag{20.37}$$

Molar quantities and densities are independent of the size of the system, i.e. they do not possess the properties of extensive parameters defined by eq. (20.6). They are, therefore, often called intensive variables of state in the literature. This terminology is misleading and should be avoided. Molar quantities and densities do not possess the fundamental property of intensive parameters as given by eq. (20.7). This is obvious from the simple example of the molar density for e.g. a liquid–vapour equilibrium.

These ideas may be looked at in a slightly different way. According to (b) and (19.18) we have for the fundamental equation (20.4) (for one-component systems)

$$\alpha U = U(\alpha S, \alpha V, \alpha n). \qquad (20.38)$$

If we again let $\alpha = 1/n$ and use the definitions (20.36), we have that

$$u = u(s, v, 1) \qquad (20.39)$$

or, in differential form,

$$\mathrm{d}u = \frac{\partial u}{\partial s}\,\mathrm{d}s + \frac{\partial u}{\partial v}\,\mathrm{d}v. \qquad (20.40)$$

According to (20.31), however, we have that

$$\frac{\partial u}{\partial s} = \frac{\partial U}{\partial S} = T, \quad \frac{\partial u}{\partial v} = \frac{\partial U}{\partial V} = -P. \qquad (20.41)$$

Equation (20.40) thus becomes

$$\mathrm{d}u = T\,\mathrm{d}s - P\,\mathrm{d}v. \qquad (20.42)$$

Equations (20.39) and (20.42) are referred to as the fundamental equation per mole. If the equations of state for T and P are given, the explicit fundamental equation per mole is obtained by substitution into (20.42) and integration. Corresponding considerations apply to the entropy representation.

Re (f): The additional relationship between the intensive parameters which can be deduced from (b) may be expressed explicitly in differential form. From (b) together with Euler's theorem and eqs. (20.13)–(20.15) we find that the fundamental equation (20.4) may be written as

$$U = TS - PV + \sum_{i=1}^{m} \mu_i n_i. \qquad (20.43)$$

This gives on differentiation

$$\mathrm{d}U = T\,\mathrm{d}S + S\,\mathrm{d}T - P\,\mathrm{d}V - V\,\mathrm{d}P + \sum_{i=1}^{m} \mu_i\,\mathrm{d}n_i + \sum_{i=1}^{m} n_i\,\mathrm{d}\mu_i. \qquad (20.44)$$

Comparison of (20.44) with (20.5) gives

$$S\,\mathrm{d}T - V\,\mathrm{d}P + \sum_{i=1}^{m} n_i\,\mathrm{d}\mu_1 = 0. \tag{20.45}$$

This extremely important relationship is called the *Gibbs–Duhem equation.*

The Gibbs–Duhem equation in the entropy representation is

$$U\,\mathrm{d}\left(\frac{1}{T}\right) + V\,\mathrm{d}\left(\frac{P}{T}\right) - \sum_{i=1}^{m} n_i\,\mathrm{d}\left(\frac{\mu_i}{T}\right) = 0. \tag{20.46}$$

The Gibbs–Duhem equation may be used in place of the method described under re (e) to obtain the fundamental equation from two equations of state. The various methods may easily be carried out explicitly by using, for example, the case of an ideal gas.

§ 21 Thermodynamic potentials

We have shown that the fundamental equation in either the entropy or the energy representation contains the maximum possible amount of information about a system. The explicit development of this information within the formal framework discussed in § 20, however, often encounters extraordinarily great difficulties. The basic reason for these difficulties lies in the fact that only extensive parameters occur as independent variables in the fundamental equation. Extensive parameters are, however, usually measurable or controllable only with difficulty, or not at all. To give only two examples: there is no instrument which measures entropy directly nor is there any device to keep the entropy constant; it is practically impossible to keep the volume of a condensed phase constant. On the other hand, as has been previously mentioned, the intensive parameters in the energy representation are directly measurable and experimentally easily controlled. This is the natural consequence of the definition of intensive parameters by means of contact equilibria. We are, therefore, now faced with the problem of extending the formal apparatus of thermodynamics. The problem may be formulated as follows:

The fundamental equation is to be transformed by the introduction of one or more intensive parameters as independent variables while the complete information contained in the fundamental equation is preserved.

The latter requirement merely states that the transformed function must retain the properties of a characteristic function. We are now obviously faced with the same question from which we formally started in § 19. The problem is, therefore, solved by a Legendre transformation of the fundamental equation.

In order to make the formal development as clear as possible we shall now make considerable use of the previously used 'generalized quantities of state' X_i and P_i. The advantage of this notation derives from the fact that the only relevant factor for many thermodynamic relationships is the difference between extensive and intensive parameters. Many similar relationships can, therefore, be summarized by means of a single equation in this way. We must, however, bear in mind that energy and entropy possess, apart from the general properties of extensive parameters, additional individual properties derived from the Laws of thermodynamics. Whenever convenient we shall, therefore, introduce energy in the entropy representation and entropy in the energy representation by means of explicit notation.

Analogously, chemical potential occupies a special position among intensive parameters as can be seen from the way in which chemical potential was introduced (§ 15). The definition of neither T nor P contains an arbitrary constant. An examination of (21.40), however, shows immediately that chemical potential is defined apart from a term $a+bT$ where a and b are arbitrary constants.

The Legendre transformation may be applied to both the entropy and the energy representation leading to two series of characteristic functions. We shall limit the discussion in this section to the energy representation which has by far the greater significance in thermodynamics.

The fundamental equation in terms of generalized quantities of state is

$$U = U(X_i, \ldots, X_r) \quad (r \geqslant m+2). \tag{21.1}$$

The intensive parameters are given by

$$P_i = \left(\frac{\partial U}{\partial X_i}\right)_{X_j} \quad (i \neq j). \tag{21.2}$$

The quantities X_i and P_i as related by eq. (21.2) are called *conjugate parameters*. The fundamental equation in differential form now becomes

$$\mathrm{d}U = \sum_{i=1}^{r} P_i \,\mathrm{d}X_i. \tag{21.3}$$

The Legendre transforms of the fundamental equation in the energy representation are called *thermodynamic potentials*. The general definition of thermodynamic potentials is, therefore,

$$\Psi_k = U - \sum_{i=1}^{k} P_i X_i. \tag{21.4}$$

The sufficient condition for the existence of these transformations is

$$|U_{ij}| \equiv \frac{\partial(P_1, \ldots, P_k)}{\partial(X_1, \ldots, X_k)} \neq 0. \tag{21.5}$$

Differentiation of (21.4) and introduction of (21.3) gives

$$d\Psi_k = -\sum_{i=1}^{k} X_i \, dP_i + \sum_{j=k+1}^{r} P_j \, dX_j. \tag{21.6}$$

For $k < r$ all thermodynamic potentials are, by definition, characteristic functions. Furthermore, they are homogeneous first-order functions of the extensive parameters. Equation (20.43) in terms of generalized quantities of state is

$$U = \sum_{i=1}^{r} P_i X_i. \tag{21.7}$$

Substitution into (21.4) gives

$$\Psi_k = \sum_{j=k+1}^{r} P_j X_j \tag{21.8}$$

which, with Euler's theorem, proves the proposition. The partial derivatives of a thermodynamic potential with respect to the intensive parameters P_i give (with a negative sign) the conjugate extensive parameters X_i. We have, therefore, that

$$\frac{\partial \Psi_k}{\partial P_i} = -X_i. \tag{21.9}$$

Partial derivatives with respect to extensive parameters X_j give, as in the fundamental equation, the conjugate intensive parameter P_j. We have, therefore, that

$$\frac{\partial \Psi_k}{\partial X_j} = P_j. \tag{21.10}$$

If $k = r$, we have a special case, for which eqs. (21.4) and (21.7) give

$$\Psi_r = \sum_{i=1}^{r} P_i X_i - \sum_{i=1}^{r} P_i X_i = 0 \tag{21.11}$$

and, therefore, according to (21.6)

$$d\Psi_r = \sum_{i=1}^{r} X_i \, dP_i = 0. \tag{21.12}$$

This is simply the Gibbs–Duhem equation in terms of generalized quantities of state. The initially surprising fact that the complete (r-fold) Legendre transform of the fundamental equation, i.e. the thermodynamic potential, is identically equal to zero is mathematically based, as shown by the derivation, on the fundamental equation being a homogeneous function of the first degree. Physically it is explained by the need to have at least one extensive parameter for the definition of the system.

Let us now consider those special cases of eq. (21.4) that are important in thermodynamics. We shall use the explicit notation for quantities of state.† The thermodynamic potential

$$U = U(S, V, n_1, \dots, n_m) \tag{21.13}$$

is merely the fundamental equation in the energy representation which has already been discussed in detail in § 20. We shall use the function again in the course of certain general considerations. It is, however, inconvenient for the treatment of particular problems for the reasons given at the beginning of this section.

The thermodynamic potential

$$H = H(S, P, n_1, \dots, n_m) \tag{21.14}$$

is called the *enthalpy* or *heat content*.

Equation (21.4) in this connexion becomes

$$H = U + PV \tag{21.15}$$

and eq. (21.6) becomes

$$\mathrm{d}H = T\,\mathrm{d}S + V\,\mathrm{d}P + \sum_{i=1}^{m} \mu_i\,\mathrm{d}n_i. \tag{21.16}$$

The partial derivatives are, therefore,

$$\left(\frac{\partial H}{\partial S}\right)_{P,n_i} = T, \quad \left(\frac{\partial H}{\partial P}\right)_{S,n_i} = V, \quad \left(\frac{\partial H}{\partial n_i}\right)_{S,P,n_j} = \mu_i. \tag{21.17}$$

Let us now discuss briefly the physical significance of enthalpy. Any concrete physical interpretation inherently seizes on only one aspect and can, therefore, never represent the total nature of a thermodynamic potential. Any such interpretation must, therefore, never be identified with the function itself.

† The symbols for entropy and the important thermodynamic potentials are fixed by international convention.

Let us now recall two results which were discussed in detail in Chapter I. If the reversible work done by a system is denoted by $W^* = -W$, we have

$$-(\mathrm{d}U)_{S,n} = \mathrm{d}W^*. \tag{21.18}$$

The work done by a closed system at constant entropy, i.e. under quasi-static adiabatic conditions, is equal to the decrease in internal energy. Also,

$$(\mathrm{d}U)_{V,n} = T\,\mathrm{d}S = \mathrm{d}Q. \tag{21.19}$$

The heat introduced into a closed system at constant volume is equal to the increase in internal energy.

In order to arrive at an analogous result for the enthalpy, we imagine the system to be coupled to a reservoir which maintains the pressure in the system constant. These conditions allow work due to volume changes to be done, but this work cannot appear as useful work since it is done only in order to keep the pressure constant. If, therefore, a result analogous to (21.18) is wanted at least *one further work co-ordinate* must be introduced. We shall now introduce the extra work co-ordinate; for the sake of simplicity, however, we shall not introduce the extra co-ordinate generally and shall suppose the system and reservoir to be adiabatically isolated. The reservoir will be denoted by the subscript R. The reversible work done by the system and reservoir is then, according to (21.18),

$$\mathrm{d}W^* = -\mathrm{d}(U+U_{\mathrm{R}}). \tag{21.20}$$

Since the system and the reservoir are in equilibrium with respect to pressure and, furthermore, $P = \text{const}$, we have

$$-\mathrm{d}U_{\mathrm{R}} = P\,\mathrm{d}V_{\mathrm{R}} = -P\,\mathrm{d}V = -\mathrm{d}(PV). \tag{21.21}$$

We now get from (21.15), (21.20), and (21.21)

$$-(\mathrm{d}H)_{S,P,n} = \mathrm{d}W^*. \tag{21.22}$$

Thus, the reversible work done by a closed system at constant entropy and constant pressure is equal to the decrease in its enthalpy. Analogously to (21.19) we have

$$(\mathrm{d}H)_{P,n} = \mathrm{d}Q, \tag{21.23}$$

i.e. the heat introduced into a closed system at constant pressure is equal to the increase in enthalpy. Since the heats of various processes (mixing, chemical reactions) are usually measured at constant pressure, (21.23) gives directly the significance of enthalpy for the theoretical interpretation of such measurements.

An important application of enthalpy is found in the theory of the Joule–Thomson effect since this effect occurs at constant enthalpy. Theoretical consideration of the Joule–Thomson effect will have to be deferred to § 24 since certain formal aids must be developed first.

The use of enthalpy as a thermodynamic potential is difficult since enthalpy contains entropy as an independent variable. Enthalpy can be represented as a function of state by expressing it as a function of any complete set of variables of state; enthalpy is then, however, no longer a thermodynamic potential. The representation

$$H = H(T, P, n_1, \ldots, n_m) \tag{21.24}$$

which is analogous to (20.20) and (20.21) and where

$$\mathrm{d}H = \left(\frac{\partial H}{\partial T}\right)_{P,n} \mathrm{d}T + \left(\frac{\partial H}{\partial P}\right)_{T,n} \mathrm{d}P + \sum_{i=1}^{m} \left(\frac{\partial H}{\partial n_i}\right)_{T,P,n_j} \mathrm{d}n_i \tag{21.25}$$

is of particular importance. The reason for the special significance of this representation will be explained in § 24. It is sufficient to note here that (21.25) directly yields the definition of the molar heat at constant pressure, C_p. Putting $h = H/n$ (for one-component systems) we have, because of (21.23), that

$$C_p = \left(\frac{\partial h}{\partial T}\right)_{P,n}. \tag{21.26}$$

The derivative $(\partial H/\partial n_i)_{T,P,n_j}$ is, however, *not* a chemical potential.

The thermodynamic potential

$$F = F(T, V, n_1, \ldots, n_m) \tag{21.27}$$

is called the *Helmholtz free energy*. Equation (21.4) now becomes

$$F = U - TS \tag{21.28}$$

and eq. (21.6) becomes

$$\mathrm{d}F = -S\,\mathrm{d}T - P\,\mathrm{d}V + \sum_{i=1}^{m} \mu_i\,\mathrm{d}n_i. \tag{21.29}$$

The partial derivatives are, therefore,

$$\left(\frac{\partial F}{\partial T}\right)_{V,n} = -S, \quad \left(\frac{\partial F}{\partial V}\right)_{T,n} = -P, \quad \left(\frac{\partial F}{\partial n_i}\right)_{T,V,n_j} = \mu_i. \tag{21.30}$$

Under certain conditions the Helmholtz free energy can acquire a real interpretation. Let us consider a system in thermal equilibrium with a reservoir (thermostat) which keeps the temperature constant. Let the system and the reservoir together be adiabatically isolated. The diathermal wall between system and reservoir is assumed to be rigid so that the reservoir cannot do work. The reversible work done by the system is, therefore, according to (21.18),

$$\mathrm{d}W^* = -\mathrm{d}(U+U_{\mathrm{R}}). \tag{21.31}$$

Since the reservoir does no work, the decrease in its internal energy is determined solely by the transfer of heat to the system necessary to keep the temperature constant. Since T = const we have

$$-\mathrm{d}U_{\mathrm{R}} = -\mathrm{d}Q_{\mathrm{R}} = \mathrm{d}Q = T\,\mathrm{d}S = \mathrm{d}(TS). \tag{21.32}$$

From (21.28), (21.31), and (21.32) we get

$$-(\mathrm{d}F)_{T,n} = \mathrm{d}W^*. \tag{21.33}$$

The isothermal reversible work done by a closed system is equal to the decrease in Helmholtz free energy. This is the basis of the misleading terms 'work function' and 'maximum work' (for $-F$) and the symbol A which is still frequently used in the American literature.†

The Helmholtz free energy is a useful thermodynamic potential when the volume can be kept constant without undue difficulty, i.e. particularly for gaseous phases. In this connexion, it is of interest to note that the second one of eq. (21.30) represents the previously mentioned thermal equation of state (20.28). In the earlier literature, the Helmholtz free energy is frequently used in problems involving condensed phases since these problems were primarily concerned with temperature dependence; the volume changed very little with change in pressure within the usually accessible pressure ranges and the difference between constant volume and constant pressure was neglected. The present-day significance of the Helmholtz free energy is due to the fact that this thermodynamic potential can be most conveniently calculated explicitly by the methods of statistical thermodynamics.

The thermodynamic potential

$$G = G(T, P, n_1, \dots, n_m) \tag{21.34}$$

is called the *Gibbs free energy*. Equation (21.4) now appears as

$$G = U - TS + PV = H - TS = F + PV \tag{21.35}$$

and eq. (21.6) as

$$\mathrm{d}G = -S\,\mathrm{d}T + V\,\mathrm{d}P + \sum_{i=1}^{m} \mu_i\,\mathrm{d}n_i. \tag{21.36}$$

† A is now the internationally recommended symbol for the Helmholtz free energy. (*Translator.*)

The corresponding partial derivatives are, therefore,

$$\left(\frac{\partial G}{\partial T}\right)_{P,n} = -S, \quad \left(\frac{\partial G}{\partial P}\right)_{T,n} = V, \quad \left(\frac{\partial G}{\partial n_i}\right)_{T,P,n_j} = \mu_i. \qquad (21.37)$$

A real interpretation can be given in a way analogous to that used in the previous discussion. We consider a system coupled with two reservoirs R and R′. One is a constant-temperature reservoir (thermostat), the other a constant-pressure reservoir (manostat). We shall have to introduce an additional work co-ordinate as we did in the case of enthalpy. The reversible work done by the system is then

$$\mathrm{d}W^* = -\mathrm{d}(U + U_{\mathrm{R}} + U_{\mathrm{R}'}). \qquad (21.38)$$

$\mathrm{d}U_{\mathrm{R}}$ is given by eq. (21.32), $\mathrm{d}U_{\mathrm{R}'}$ by eq. (21.21). Using (21.35) we therefore get

$$-(\mathrm{d}G)_{T,P,n} = dW^*. \qquad (21.39)$$

The isothermal–isobaric reversible work done by a system is equal to the decrease in Gibbs free energy.

The Gibbs free energy is the most useful of all the thermodynamic potentials. This special importance is due to the fact that mole numbers are here the only extensive parameters and are easily measurable and controllable. Furthermore, the Gibbs free energy can thus be computed completely from the mole numbers and the similarly easily measurable chemical potentials, since eq. (21.8) now becomes

$$G = \sum_{i=1}^{m} \mu_i n_i. \qquad (21.40)$$

This equation also shows that, for one-component systems, the chemical potential is identical with the molar Gibbs free energy. This coincidence, however, must not be allowed to obscure the fact that chemical potential has a much more general significance.

The thermodynamic potential

$$\Omega = \Omega(T, V, \mu_1, \ldots, \mu_m) \qquad (21.41)$$

has occasionally been called the *grand potential*. However, neither the name nor the symbol is universally accepted. Equation (21.4) here appears as

$$\Omega = U - TS - \sum_{i=1}^{m} \mu_i n_i \qquad (21.42)$$

and eq. (21.6) as

$$\mathrm{d}\Omega = -S\,\mathrm{d}T - P\,\mathrm{d}V - \sum_{i=1}^{m} n_i\,\mathrm{d}\mu_i. \tag{21.43}$$

The partial derivatives are

$$\left(\frac{\partial\Omega}{\partial T}\right)_{V,\mu} = -S, \quad \left(\frac{\partial\Omega}{\partial V}\right)_{T,\mu} = -P, \quad \left(\frac{\partial\Omega}{\partial \mu_i}\right)_{T,V,\mu_j} = -n_i. \tag{21.44}$$

Equations (21.28), (21.35), (21.40), and (21.42) give

$$\Omega = -PV. \tag{21.45}$$

The quantity PV is, therefore, often used instead of the grand potential.

The grand potential is hardly ever used within the framework of thermodynamics, but has outstanding significance for statistical thermodynamics since it is readily amenable to explicit calculations in connexion with the grand canonical ensemble. The name grand potential is explained by this connexion.

§ 22 Massieu–Planck functions

We shall now briefly look at the characteristic functions derived from the entropy representation of the fundamental equation. The fundamental equation in terms of generalized quantities of state is

$$S = S(X_1, \ldots, X_r) \quad (r \geqslant m+2). \tag{22.1}$$

The intensive parameters are now

$$P_i^* = \left(\frac{\partial S}{\partial X_i}\right)_{\substack{X_j \\ j\neq i}}. \tag{22.2}$$

The differential form of the fundamental equation is, therefore,

$$\mathrm{d}S = \sum_{i=1}^{r} P_i^*\,\mathrm{d}X_i. \tag{22.3}$$

The Legendre transforms of the entropy representation of the fundamental equation are called *Massieu–Planck functions*. The general definition of Massieu–Planck functions is, therefore, given by

$$\Phi_k = S - \sum_{i=1}^{k} P_i^*\,X_i \tag{22.4}$$

and the sufficient condition for the existence of these transforms is

$$|S_{ij}| \equiv \frac{\partial(P_1^*, \ldots, P_k^*)}{\partial(X_1, \ldots, X_k)} \neq 0. \tag{22.5}$$

Equation (22.4) with (22.3) gives

$$\mathrm{d}\Phi_k = -\sum_{i=1}^{k} X_i \, \mathrm{d}P_i^* + \sum_{j=k+1}^{r} P_j^* \, \mathrm{d}X_j. \tag{22.6}$$

The general formal properties of Massieu–Planck functions are completely analogous to those of thermodynamic potentials. The reader is, therefore, referred to the discussion in § 21. We shall here deal only with explicit formulae for two special cases which are of interest in thermodynamics.

The Massieu–Planck function

$$\Phi_1 = \Phi_1\left(\frac{1}{T}, V, n_1, \ldots, n_m\right) \tag{22.7}$$

obviously corresponds directly to the Helmholtz free energy. Guggenheim proposes the name Massieu function for it. Equation (22.4) here becomes

$$\Phi_1 = S - \frac{1}{T} U \tag{22.8}$$

and eq. (22.6)

$$\mathrm{d}\Phi_1 = -U \, \mathrm{d}\left(\frac{1}{T}\right) + \frac{P}{T} \mathrm{d}V - \sum_{i=1}^{m} \frac{\mu_i}{T} \mathrm{d}n_i. \tag{22.9}$$

The partial derivatives are, therefore,

$$\left(\frac{\partial \Phi_1}{\partial(1/T)}\right)_{V,n} = -U, \quad \left(\frac{\partial \Phi_1}{\partial V}\right)_{1/T,n} = \frac{P}{T}, \quad \left(\frac{\partial \Phi_1}{\partial n_i}\right)_{1/T,V,n_j} = -\frac{\mu_i}{T}. \tag{22.10}$$

Similarly, the Massieu–Planck function

$$\Phi_2 = \Phi_2\left(\frac{1}{T}, \frac{P}{T}, n_1, \ldots, n_m\right) \tag{22.11}$$

corresponds to the Gibbs free energy. Guggenheim calls this the Planck function.

Equation (22.4) now becomes

$$\Phi_2 = S - \frac{1}{T} U - \frac{P}{T} V \tag{22.12}$$

and eq. (22.6)

$$\mathrm{d}\Phi_2 = -U\,\mathrm{d}\left(\frac{1}{T}\right) - V\,\mathrm{d}\left(\frac{P}{T}\right) - \sum_{i=1}^{m} \frac{\mu_i}{T}\,\mathrm{d}n_i. \tag{22.13}$$

The partial derivatives are, therefore,

$$\left(\frac{\partial\Phi_2}{\partial(1/T)}\right)_{P/T,n} = -U, \quad \left(\frac{\partial\Phi_2}{\partial(P/T)}\right)_{1/T,n} = -V, \quad \left(\frac{\partial\Phi_2}{\partial n_i}\right)_{1/T,P/T,n_j} = -\frac{\mu_i}{T}. \tag{22.14}$$

These two functions are of interest on historical grounds since they were the first characteristic functions introduced into thermodynamics (Massieu, 1865) and also because Planck frequently used the function Φ_2 in his investigations. Their present practical significance relates to two peculiarities: firstly, the calorific quantities U and H appear explicitly as variables in this representation (we shall discuss this in detail in § 24) and, secondly, Φ_1 and Φ_2 are connected very simply with the corresponding thermodynamic potentials (which is not generally true for Massieu–Planck functions). Comparison of (22.8) with (21.28) and of (22.12) with (21.35) shows that

$$\Phi_1 = -\frac{F}{T}, \quad \Phi_2 = -\frac{G}{T}. \tag{22.15}$$

It is, therefore, simple to change from thermodynamic potentials to Massieu–Planck functions without introducing the latter explicitly. The advantages of both representations are thus readily available. We shall discuss this further in § 24.

As has been mentioned earlier, the Massieu–Planck functions have a very general significance in connexion with the thermodynamics of irreversible processes and with statistical thermodynamics. Discussion of this is outside the scope of this book.

§ 23 Transformation of the equilibrium and stability conditions

As we have seen already, physical information is completely conserved in the Legendre transformation. It must, therefore, be possible to formulate the general equilibrium and stability conditions by means of the Legendre transform of the fundamental equation as well as by means of the fundamental equation itself, i.e. either by thermodynamic potentials or by Massieu–Planck functions. We shall now perform this transformation for the thermodynamic potentials but we shall give only the result for the Massieu–Planck functions since the proof is obtained by a completely analogous method.

(a) *Equilibrium conditions.* In the formulation (17.2) the secondary conditions belonging to the extremum principle are expressed by means of extensive parameters of the whole system; these extensive parameters are part of the energy representation. It is therefore to be expected that one or more secondary conditions will, in the formulation of the equilibrium condition by means of the Legendre transform of the internal energy, be expressed by means of intensive parameters of the whole system. This expectation (which will be proved to be correct) shows immediately the nature of the problem which we meet here: for a heterogeneous system each extensive parameter is equal to the sum of the extensive parameters of the individual phases; according to § 15, however, intensive parameters have so far been defined only for each phase, not for the whole system. The definition of extensive parameters for the whole system is based on the fundamental property (20.6). Analogously, the definition of intensive parameters is based on the fundamental property (20.7) and thus implicitly contains the assumption that the intensive parameter has the same value for all co-existing phases. Any formulation of the equilibrium condition with the aid of the thermodynamic potential Ψ_k must, therefore, assume that each one of the k intensive parameters which are independent variables for Ψ_k has the same value for all phases of the system, e.g. the use of the Helmholtz free energy assumes that all phases are at the same temperature, i.e. in thermal equilibrium. Any formulation of the equilibrium condition by means of the Legendre transform of the fundamental equation thus already assumes certain assertions concerning the equilibrium. We shall be returning later to this problem.

The derivation of the transformed equilibrium condition consists of two steps. The condition (17.2) is first put into an equivalent form which no longer explicitly contains the secondary conditions but includes them in the extremum formulation. In the second step the equilibrium conditions for the thermodynamic potentials are derived from this formulation by means of the Legendre transformation (21.4). We shall, for the present, assume that no additional relationships exist between the parameters X_j $(j > k)$.

Theorem 1. For a system, in which the intensive parameters $P_1, \ldots, P_k$ are fixed and constant throughout the system, to be at equilibrium it is necessary and sufficient that

$$\delta U - \sum_{i=1}^{k} P_i \, \delta X_i \geqslant 0 \quad (X_j = \text{const for } j > k). \tag{23.1}$$

Proof. For the case of the equality sign, (23.1) is a direct consequence of the fundamental equation. The general proof is obtained by showing the equivalence of (17.2) and (23.1) for the assumptions made. For this we can use the fact, derived from the fundamental equation, that changes of state for which

$$\delta U - \sum_{i=1}^{k} P_i \delta X_i = 0 \quad (X_j = \text{const for } j > k) \tag{23.2}$$

can always be brought about. For the sake of simplicity we shall confine ourselves to the case $k = 1$ and $X_1 = S$. Equation (23.2) shows that changes of state for which

$$\delta U - T\delta S = 0 \quad (V, n = \text{const}) \tag{23.3}$$

can always be made to occur (by introduction or abstraction of heat). We shall use this fact to show that the condition

$$\delta U - T\delta S \geqslant 0 \tag{23.4}$$

is equivalent to (17.2) for systems of uniform temperature. We distinguish between the following cases:

(α) $\delta S = 0$.
Equation (23.4) transforms directly into (17.2).

(β) $\delta U = 0$.
Equation (23.4) transforms into (17.1) whose equivalence to (17.2) was proved in § 17.

(γ) $\delta U > 0$, $\delta S > 0$.
According to (23.4) we now have $\delta U \geqslant T\delta S$. We can, therefore, always abstract heat and attain a state which may be regarded as a variation of the initial state with $\delta U \geqslant 0$, $\delta S = 0$. Condition (17.2) is thus satisfied.

(δ) $\delta U < 0$, $\delta S < 0$.
Equation (23.4) tells us that now $|\delta U| \leqslant |T\delta S|$. We can, therefore, always introduce heat and attain a state which may be regarded as a variation of the initial state with $\delta U = 0$, $\delta S \leqslant 0$. Equation (17.1) is thus satisfied and, therefore, also (17.2).

(ε) $\delta U > 0$, $\delta S < 0$.
Heat may be abstracted and a state reached for which, if it is regarded as a variation of the initial state, we have $\delta U = 0$, $\delta S < 0$. Condition (17.1) is satisfied and, therefore, also (17.2).

(ζ) $\delta U < 0$, $\delta S > 0$.
In this case (23.4) cannot be satisfied. A state can always be reached by abstraction of heat for which $\delta U < 0$, $\delta S = 0$ when

this state is regarded as a variation of the initial state. In this case, therefore, (17.2) is likewise not satisfied.

In an analogous way it can be shown that the possibility of a variation which contradicts (23.4) necessarily implies the possibility of a variation which contradicts (17.2). The equivalence of (17.2) and (23.4) is thus completely proved. The proof for the general case of the condition (23.1) is obtained in an analogous manner.

Although the condition (17.2) is formally obtained by means of a specialization [cases (α) and (β)] from (23.1) or (23.4), it does not constitute a special case but (as shown by the proof) an equivalent formulation.

Theorem 2. For a system whose intensive parameters $P_1, \ldots, P_k$ are fixed and constant throughout the system to be in equilibrium it is necessary and sufficient that

$$(\delta \Psi_k)_{P_1,\ldots,P_k,X_{k+1},\ldots,X_r} \geqslant 0. \tag{23.5}$$

Proof. Since $X_{k+1}, \ldots, X_r$ are constant, the condition (23.1) can here be written as

$$(\delta U)_{P_i,X_j} - \sum_{i=1}^{k} P_i(\delta X_i)_{P_i,X_j} \geqslant 0 \quad (i \leqslant k, j > k). \tag{23.6}$$

From eq. (21.4), however, we have that

$$\delta U - \sum_{i=1}^{k} P_i \, \delta X_i = \delta \Psi_k + \sum_{i=1}^{k} X_i \, \delta P_i \tag{23.7}$$

and, furthermore, for $P_1, \ldots, P_k = \text{const}$

$$(\delta U)_{P_i,X_j} - \sum_{i=1}^{k} P(\delta X_i)_{P_i,X_j} = (\delta \Psi_k)_{P_i,X} \,. \tag{23.8}$$

Substitution of (23.8) into (23.6) gives the required statement.

The most important special cases of (23.5) are: The equilibrium condition for constant entropy, volume, and mole number is

$$(\delta H)_{S,P,n} \geqslant 0. \tag{23.9}$$

The equilibrium condition for constant temperature, volume, and mole number is

$$(\delta F)_{T,V,n} \geqslant 0. \tag{23.10}$$

The equilibrium condition for constant temperature, pressure, and mole number is

$$(\delta G)_{T,P,n} \geqslant 0. \tag{23.11}$$

Condition (23.5) is less general than conditions (17.1), (17.2), and (23.1) since (23.5) does not yield any information about internal equilibria which are described by the internal parameters $P_1, \ldots, P_k$. The condition for thermal equilibrium within a system, for example, cannot be obtained from (23.10). The physical explanation for this is that the formulation of (23.5) assumes internal equilibrium with respect to the parameters $P_1, \ldots, P_k$. Comparison of (23.7) with (23.8) shows formally that some of the variations possible for (23.7) have been excluded in the derivation of (23.5). In an analogous manner we obtain for the Massieu–Planck functions:

Theorem 3. For a system in which the intensive parameters $P_1^*, \ldots, P_k^*$ are fixed and constant throughout the system to be in equilibrium, it is necessary and sufficient that

$$(\delta\Phi_k)_{P_1^*,\ldots,P_k^*,X_{k+1},\ldots,X_r} \leqslant 0. \tag{23.12}$$

We have assumed so far that no additional conditions connect the parameters X_j $(j > k)$. The most important case for which this assumption is not true is a chemical reaction where the stoichiometric relationships of the reaction equation apply to variations in mole numbers. Formally analogous relationships may also occur between other variables of state. An example is given in § 60. Two methods are available for the treatment of all such cases as has been shown in §§ 16 and 17. The first method is the introduction of a suitable internal parameter. The extensive parameters which are connected by additional relationships then no longer appear in the secondary conditions and we have an extremum formulation in which variations of the internal parameters are limited only by the remaining secondary conditions (e.g. for a chemical reaction, constant temperature and pressure, and constant mole numbers of those components which do not take part in the reaction). In the second method the number of variables is not reduced (i.e. the mole numbers of all the species concerned in the reaction are retained in the fundamental equation). The new secondary conditions arising from the additional relationships are, however, introduced into the extremum formulation. The above derivations remain unchanged by the generalization. Since, however, a general formulation is inconvenient here, chemical reactions are discussed in §§ 33 and 36 and an example for other variables is given in § 60.

(b) *Stability conditions.* Conditions (23.5) and (23.12) for the case (the most important in practice) of the equality sign only state that at equilibrium the thermodynamic potential Ψ_k, or the Massieu–Planck function Φ_k for the appropriate secondary conditions, assumes a stationary value. Information about the nature of the stationary value requires consideration of second-order variations. As has been mentioned earlier, only variations of order up to 2 are of importance in thermodynamics. A detailed analysis of the problem will be given in Chapter VI. The following statement is sufficient for our present purpose (a similar method was adopted in § 18). For a *stable equilibrium* the stationary value for *thermodynamic potentials* is a *minimum* while that for *Massieu–Planck functions* is a *maximum*.

Formally, this becomes

$$(\Delta\Psi_k)_{P,X} > 0, \tag{23.13}$$

$$(\Delta\Phi_k)_{P^*,X} < 0. \tag{23.14}$$

§ 24 Gibbs–Helmholtz equations and Maxwell's relations

We have so far regarded thermodynamic potentials as Legendre transforms of internal energy. It is, of course, possible to obtain a thermodynamic potential Ψ_k $(k > l)$ from another thermodynamic potential Ψ_l by means of a Legendre transformation. We then have that

$$\Psi_k = \Psi_l - \sum_{i=l+1}^{k} P_i X_i \quad (k > l). \tag{24.1}$$

But, according to (21.9), we have that

$$\frac{\partial \Psi_k}{\partial P_i} = -X_i \tag{24.2}$$

and so we get

$$\Psi_k = \Psi_l + \sum_{i=l+1}^{k} P_i \frac{\partial \Psi_k}{\partial P_i}. \tag{24.3}$$

The only case useful in practice is $k - l = 1$, where

$$\Psi_k = \Psi_l + P_k \frac{\partial \Psi_k}{\partial P_k}. \tag{24.4}$$

In the form given Ψ_k depends explicitly not on P_k but on X_k. From Ψ_l, however, we obtain the equation of state

$$P_k = \frac{\partial \Psi_l}{\partial X_k} = P_k(P_1, \ldots, P_l, X_k, \ldots, X_r) \tag{24.5}$$

which can be used to eliminate X_k.

All quantities appearing in (24.4) are now functions of P_k and we have a linear first-order differential equation which allows the calculation of $\Psi_k(P_k)$ from given $\Psi_l(P_k)$. Equation (24.4) for the free energy was derived first by Gibbs and later independently by Helmholtz. It is, therefore, known as the *Gibbs–Helmholtz equation.*

The following special cases are the most important in thermodynamics:

(a)
$$H = U+P\left(\frac{\partial H}{\partial P}\right)_{S,n}. \tag{24.6}$$

This equation makes it possible to calculate the enthalpy when the internal energy is known as a function of pressure.

(b)
$$F = U+T\left(\frac{\partial F}{\partial T}\right)_{V,n}. \tag{24.7}$$

This equation permits the calculation of the Helmholtz free energy when the internal energy is known as a function of temperature.

(c)
$$G = H+T\left(\frac{\partial G}{\partial T}\right)_{P,n}. \tag{24.8}$$

This equation permits the calculation of the Gibbs free energy when the enthalpy is known as a function of temperature.

(d)
$$G = F+P\left(\frac{\partial G}{\partial P}\right)_{T,n}. \tag{24.8a}$$

This equation may be used to calculate the Gibbs free energy from the Helmholtz free energy when the latter is known as a function of pressure.

These equations illustrate the remark made previously that the representation of U and H in sets of variables in which they are not thermodynamic potentials may be important (cf. §§ 20 and 21). We shall confine further discussion mainly to case (c) which is by far the most important.

The special importance of (24.8) is due to the following properties:

(α) The equation is applicable to the most frequent experimental condition P = const.

(β) According to eq. (21.26) the function $H(T)$ can be found by integration of the molar heat capacity at constant pressure. This is easily determined by experiment.

(γ) According to (23.11) the function G which can be calculated from (24.8) completely determines all material equilibria at T = const, P = const, including chemical equilibria.

The formal solution of eq. (24.8) is obtained directly by writing this equation in the form

$$\left[\frac{\partial(G/T)}{\partial(1/T)}\right]_{P,n} = H. \tag{24.9}$$

This gives

$$G = T\int H\,\mathrm{d}\left(\frac{1}{T}\right) \tag{24.10}$$

for which the constant of integration remains to be found. It should be easy to see that this formulation is basically a change to Massieu–Planck functions. The secondary conditions of the partial differentiation and the function H are, however, derived from the formulation of thermodynamic potentials. An analogous treatment of eq. (24.7), however, takes us completely into the formulation of Massieu–Planck functions and simply gives the first of the equations (22.10).

We end this section by a brief discussion of a problem which appears in connexion with statistical thermodynamics. An explicit statistical calculation of the Gibbs free energy is possible in principle; in practice, however, such calculations generally meet unsurmountable difficulties. What is usually done, therefore, is to calculate the Helmholtz free energy statistically and then to calculate the Gibbs free energy from it purely by thermodynamic methods. This can be done by using eq. (24.8a). In order to calculate the integral

$$G = P\int F\,\mathrm{d}\left(\frac{1}{P}\right) \tag{24.10a}$$

which is analogous to (24.10), it is necessary to eliminate V from the expression for the Helmholtz free energy; this is done by using the thermal equation of state.

The integration can be avoided by using a system of equations which parallels the Gibbs–Helmholtz equations. We shall confine our discussion to the case $k-l = 1$. We then have from (24.1) and (24.5) that

$$\Psi_k = \Psi_l - X_k\left(\frac{\partial \Psi_l}{\partial X_k}\right) \tag{24.10b}$$

whose most important special case is

$$G = F - V\left(\frac{\partial F}{\partial V}\right)_{T,n}. \tag{24.10c}$$

If F has been calculated statistically the derivative on the right-hand side can be stated directly and the only remaining operation is the elimination of the volume.

We have stated previously (cf. § 20) that the fact that characteristic functions are necessarily functions of state (i.e. complete differentials of quantities of state) allows us to derive relationships between their second derivatives. We shall now do this quite generally for the thermodynamic potentials. Equation (21.6) leads directly to

$$\frac{\partial^2 \Psi_k}{\partial P_i \partial X_j} = -\frac{\partial X_i}{\partial X_j} = \frac{\partial P_j}{\partial P_i}, \tag{24.11}$$

$$\frac{\partial^2 \Psi_k}{\partial P_i \partial P_m} = -\frac{\partial X_i}{\partial P_m} = -\frac{\partial X_m}{\partial P_i}. \tag{24.12}$$

$$\frac{\partial^2 \Psi_k}{\partial X_j \partial X_n} = \frac{\partial P_j}{\partial X_n} = \frac{\partial P_n}{\partial X_j} \quad (i, m \leqslant k; j, n > k). \tag{24.13}$$

These equations are called *Maxwell's relations*. The most important special cases are (for one-component systems):

(a) *Internal energy* [only type (24.13) occurs].

$$\left(\frac{\partial T}{\partial V}\right)_{S,n} = -\left(\frac{\partial P}{\partial S}\right)_{V,n}, \tag{24.14}$$

$$\left(\frac{\partial T}{\partial n}\right)_{S,V} = \left(\frac{\partial \mu}{\partial S}\right)_{V,n}, \tag{24.15}$$

$$-\left(\frac{\partial P}{\partial n}\right)_{S,V} = \left(\frac{\partial \mu}{\partial V}\right)_{S,n}. \tag{24.16}$$

(b) *Enthalpy* [only types (24.11) and (24.13) occur].

$$\left(\frac{\partial V}{\partial S}\right)_{P,n} = \left(\frac{\partial T}{\partial P}\right)_{S,n}, \tag{24.17}$$

$$\left(\frac{\partial V}{\partial n}\right)_{S,P} = \left(\frac{\partial \mu}{\partial P}\right)_{S,n}, \tag{24.18}$$

$$\left(\frac{\partial T}{\partial n}\right)_{S,P} = \left(\frac{\partial \mu}{\partial S}\right)_{P,n}. \tag{24.19}$$

(c) *Helmholtz free energy* [only types (24.11) and (24.13) occur].

$$\left(\frac{\partial S}{\partial V}\right)_{T,n} = \left(\frac{\partial P}{\partial T}\right)_{V,n}, \tag{24.20}$$

$$-\left(\frac{\partial S}{\partial n}\right)_{T,V} = \left(\frac{\partial \mu}{\partial T}\right)_{V,n}, \tag{24.21}$$

$$-\left(\frac{\partial P}{\partial n}\right)_{T,V} = \left(\frac{\partial \mu}{\partial V}\right)_{T,n}. \tag{24.22}$$

(d) *Gibbs free energy* [only types (24.11) and (24.12) occur]

$$\left(\frac{\partial V}{\partial n}\right)_{T,P} = \left(\frac{\partial \mu}{\partial P}\right)_{T,n}, \tag{24.23}$$

$$-\left(\frac{\partial S}{\partial P}\right)_{T,n} = \left(\frac{\partial V}{\partial T}\right)_{P,n}, \tag{24.24}$$

$$-\left(\frac{\partial S}{\partial n}\right)_{T,P} = \left(\frac{\partial \mu}{\partial T}\right)_{P,n}. \tag{24.25}$$

§ 25 Conversion of partial derivatives. Method of Jacobi determinants. The Joule–Thomson effect

The thermodynamic treatment of special problems often involves partial derivatives which are not directly measurable. The conversion of such derivatives to experimentally easily accessible quantities therefore plays an extraordinarily important part in thermodynamic applications. We shall, therefore, now discuss several methods which are useful in this way while *confining our discussion to one-component systems*.

The generalization for multi-component systems involves no formal difficulties but uses concepts which will not be introduced until § 28. Since the mole number is trivial as a variable in one-component systems, as has been shown in § 20, we shall keep it constant throughout. All functions of state therefore involve only two independent variables.

We shall start by looking a little more closely at the problem. If a function of state

$$z = z(x, y) \tag{25.1}$$

is given, we have that

$$dz = \left(\frac{\partial z}{\partial x}\right)_y dx + \left(\frac{\partial z}{\partial y}\right)_x dy \tag{25.2}$$

and thus, if w is a further variable of state, we get

$$\left(\frac{\partial z}{\partial x}\right)_w = \left(\frac{\partial z}{\partial x}\right)_y + \left(\frac{\partial z}{\partial y}\right)_x \left(\frac{\partial y}{\partial x}\right)_w. \tag{25.3}$$

By combining equations of the type (25.3) it is possible to express any partial derivative by at most three other arbitrarily chosen partial derivatives. For 10 variables of state there are 720 partial derivatives and the number of relationships of type (25.3) is of the order of 10^{10}. These figures emphasize the need for rational methods in calculations of this kind.

The first basic simplification is obtained by expressing all the partial derivatives in terms of the same standard set of three independent partial derivatives. Following a proposal by Bridgeman we choose as our standard set the three experimentally easily accessible quantities:

molar heat capacity at constant pressure:

$$C_P = T\left(\frac{\partial s}{\partial T}\right)_P, \tag{25.4}$$

coefficient of thermal expansion:

$$\alpha = \frac{1}{V}\left(\frac{\partial V}{\partial T}\right)_P, \tag{25.5}$$

isothermal compressibility:

$$\kappa = -\frac{1}{V}\left(\frac{\partial V}{\partial P}\right)_T. \tag{25.6}$$

This reduced set is sufficient for most practical purposes. If other relationships are required they can easily be deduced from the standard set (25.4) to (25.6).

(a) *Bridgeman's method* (*method of stepwise elimination*). Bridgeman's method uses the following relationships which are familiar from the theory of partial differentiation:

$$\left(\frac{\partial x}{\partial y}\right)_z = \left(\frac{\partial y}{\partial x}\right)_z^{-1} \tag{25.7}$$

$$\left(\frac{\partial x}{\partial y}\right)_z = \left(\frac{(\partial x/\partial w)_z}{(\partial y/\partial w)_z}\right), \tag{25.8}$$

$$\left(\frac{\partial x}{\partial y}\right)_z = -\left(\frac{(\partial z/\partial y)_x}{(\partial z/\partial x)_y}\right). \tag{25.9}$$

Bridgeman now divides the 720 partial differentials into ten classes depending on the constant parameter z. Equation (25.8) tells us that, for each class, a knowledge of the nine quantities $(\partial x/\partial w)_z$ is sufficient for the construction of all the partial derivatives of the class. Our task has then been reduced to the determination of 90 equations. The use of Maxwell's relations (§ 24) further reduces this number by half so that 45 basic equations remain. These have been tabulated by Bridgeman.

We shall now show briefly how this method may be used to achieve the reduction without the use of Bridgeman's table. The individual steps are given in the form of a recipe† and each step is illustrated by means of a simple example. A complete calculation is not attempted.

(α) If the derivative under consideration contains thermodynamic potentials eq. (25.7) or (25.9) is used to bring them into the numerator. They are then eliminated by using equation (21.6).

Example:

$$\left(\frac{\partial P}{\partial U}\right)_G = \left[\left(\frac{\partial U}{\partial P}\right)_G\right]^{-1} \quad \text{[using (25.7)]}$$

$$= \left[T\left(\frac{\partial S}{\partial P}\right)_G - P\left(\frac{\partial V}{\partial P}\right)_G\right]^{-1} \quad \text{[using (20.5)]}, \tag{25.10}$$

etc.

The resulting expression no longer contains any thermodynamic potentials. The derivatives which occur are reduced one by one according to the following rules:

(β) If a derivative contains the chemical potential this is brought into the numerator and eliminated by means of the Gibbs–Duhem eq. (20.45).

† H. B. Callen, *Thermodynamics*, New York, 1960.

Example:

$$\left(\frac{\partial \mu}{\partial V}\right)_S = -s\left(\frac{\partial T}{\partial V}\right)_S + v\left(\frac{\partial P}{\partial V}\right)_S. \tag{25.11}$$

(γ) If a derivative contains the entropy, this is brought into the numerator and either eliminated by using Maxwell's relations or reduced to a standard derivative by means of (25.8), (for $w = T$).

Example 1:

$$\left(\frac{\partial T}{\partial P}\right)_S = -\frac{(\partial S/\partial P)_T}{(\partial S/\partial T)_P} \quad \text{[using (25.9)]}$$

$$= \frac{(\partial V/\partial T)_P}{(n/T)\,C_P} \quad \text{[using (24.24)].} \tag{25.12}$$

Example 2:

$$\left(\frac{\partial S}{\partial V}\right)_P = \frac{(\partial S/\partial T)_P}{(\partial V/\partial T)_P} = \frac{(n/T)C_P}{(\partial V/\partial T)_P} \quad \text{[using (25.8)].} \tag{25.13}$$

The entropy cannot here be eliminated by using Maxwell's relations [cf. eq. (24.17)].

(δ) The volume is brought into the numerator.

Example:

$$\left(\frac{\partial T}{\partial P}\right)_V = -\frac{(\partial V/\partial P)_T}{(\partial V/\partial T)_P} = \frac{\kappa}{\alpha} \quad \text{[using (25.9)].} \tag{25.14}$$

The molar heat capacity at constant volume is eliminated by means of the equation (whose derivation will be given later)

$$C_P - C_V = T\frac{v\alpha^2}{\kappa}. \tag{25.15}$$

The original derivative has thus been reduced to derivatives of the standard set.

(b) *The method using Jacobi determinants.* A more general and more elegant method than that of Bridgeman is the method using Jacobi determinants, developed by Shaw. We shall now summarize, without proof, those properties of Jacobi determinants (confined to two independent variables) which are necessary for our present purpose.

The Jacobi determinant in x and y for the independent variables u and v is defined as

$$\mathrm{J}(x,y) \equiv \frac{\partial(x,y)}{\partial(u,v)} \equiv \begin{vmatrix} \left(\frac{\partial x}{\partial u}\right)_v & \left(\frac{\partial x}{\partial v}\right)_u \\ \left(\frac{\partial y}{\partial u}\right)_v & \left(\frac{\partial y}{\partial v}\right)_u \end{vmatrix} = \left(\frac{\partial x}{\partial u}\right)_v \left(\frac{\partial y}{\partial v}\right)_u - \left(\frac{\partial x}{\partial v}\right)_u \left(\frac{\partial y}{\partial u}\right)_v. \tag{25.16}$$

This definition gives

$$\left.\begin{aligned} &\mathrm{J}(u,v) = -\mathrm{J}(v,u) = 1, \\ &\mathrm{J}(x,x) = 0, \quad \mathrm{J}(k,x) = 0 \quad (k = \text{const}) \\ &\mathrm{J}(x,y) = \mathrm{J}(y,-x) = \mathrm{J}(-y,x) = -\mathrm{J}(y,x). \end{aligned}\right\} \tag{25.17}$$

We also have that

$$\frac{\partial(x,y)}{\partial(z,w)}\frac{\partial(z,w)}{\partial(u,v)} = \frac{\partial(x,y)}{\partial(u,v)}. \tag{25.18}$$

Equation (25.16) therefore gives for $y = v$

$$\frac{\partial(x,y)}{\partial(u,y)} = \left(\frac{\partial x}{\partial u}\right)_y. \tag{25.19}$$

Any partial derivatives can, therefore, be written as a Jacobi determinant.

The method is based on the following consideration. According to (25.19), eq. (25.2) may be written as

$$\mathrm{d}z = \frac{\partial(z,y)}{\partial(x,y)}\mathrm{d}x + \frac{\partial(z,x)}{\partial(y,x)}\mathrm{d}y. \tag{25.20}$$

If we now introduce the new independent variables r and s and the abbreviation $\mathrm{J}(z,y) = \partial(z,y)/\partial(r,s)$, and use eq. (25.18), eq. (25.20) becomes

$$\mathrm{d}z = \frac{\mathrm{J}(z,y)}{\mathrm{J}(x,y)}\mathrm{d}x + \frac{\mathrm{J}(z,x)}{\mathrm{J}(y,x)}\mathrm{d}y. \tag{25.21}$$

Multiplication by $\mathrm{J}(x,y)$ and introduction of (25.17) now gives

$$\mathrm{J}(z,y)\,\mathrm{d}x + \mathrm{J}(x,z)\,\mathrm{d}y + \mathrm{J}(y,x)\,\mathrm{d}z = 0. \tag{25.22}$$

This relationship represents the basic equation of the method. We shall not discuss the general formulation (which may be expressed in the form of tables), but shall confine our discussion to a fairly simple special case illustrated by means of examples.

If, in (25.22), we put $\mathrm{d}y = 0$, we have that

$$\mathrm{J}(z,y)+\mathrm{J}(y,x)\left(\frac{\partial z}{\partial x}\right)_y = 0 \tag{25.23}$$

or, explicitly by using (25.17),

$$\left(\frac{\partial z}{\partial x}\right)_y = \frac{\partial(y,z)/\partial(r,s)}{\partial(y,x)/\partial(r,s)}. \tag{25.24}$$

This equation represents a transformation from the independent variables x, y to the variables r, s. The representation of partial derivatives by the standard derivatives (25.4)–(25.6) always implies a transformation into the independent variables of the Gibbs free energy, i.e. into T and P. In eq. (25.24) we always have, therefore, that $r = T$, $s = P$.

Example 1. This is the example (25.10) treated by the method of Jacobi determinants. We have that

$$\left(\frac{\partial P}{\partial U}\right)_G = \frac{\partial(G,P)/\partial(T,P)}{\partial(G,U)/\partial(T,P)} = \frac{(\partial G/\partial T)_P}{(\partial G/\partial T)_P(\partial U/\partial P)_T-(\partial G/\partial P)_T(\partial U/\partial T)_P} \tag{25.25}$$

and, therefore,

$$\left(\frac{\partial P}{\partial U}\right)_G = \frac{S}{S(\partial U/\partial P)_T+V(\partial U/\partial T)_P}. \tag{25.26}$$

The expression on the right-hand side can easily be reduced to standard derivatives by using (25.3), (25.7)–(25.9), and Maxwell's relation (24.20).

Example 2. The molar heat capacity at constant volume is to be expressed in terms of standard derivatives. We have that

$$C_V = T\left(\frac{\partial s}{\partial T}\right)_V = T\frac{\partial(v,s)/\partial(T,P)}{\partial(v,T)/\partial(T,P)} \tag{25.27}$$

or, using (25.19) and (25.6),

$$C_V = -\frac{T}{v\kappa}\frac{\partial(s,v)}{\partial(T,P)}. \tag{25.28}$$

Expansion of the determinant gives

$$C_V = -\frac{T}{v\kappa}\left[\left(\frac{\partial s}{\partial T}\right)_P\left(\frac{\partial v}{\partial P}\right)_T-\left(\frac{\partial s}{\partial P}\right)_T\left(\frac{\partial v}{\partial T}\right)_P\right]$$

$$= -\frac{T}{v\kappa}\left\{-\frac{C_P}{T}v\kappa+\left[\left(\frac{\partial v}{T}\right)_P\right]^2\right\} \tag{25.29}$$

where we have again used Maxwell's relation (24.24). We have, therefore, that

$$C_P - C_V = T\frac{v\alpha^2}{\kappa}. \tag{25.30}$$

This formula is of great importance and is often used in thermodynamic applications.

(c) *The Joule–Thomson effect.* We are now able to develop the theory of the Joule–Thomson effect. The effect has been mentioned previously in the discussion of enthalpy (§ 21). The effect is based on the experimental arrangement shown in Fig. 13. A cylindrical

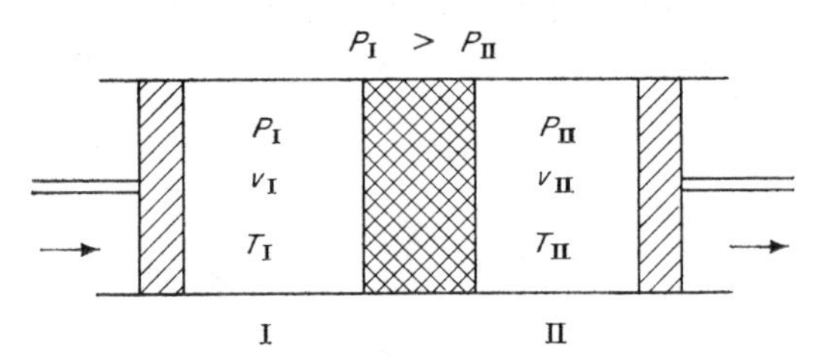

FIG. 13. Diagrams of the Joule–Thomson experiment

tube is closed at both ends by movable pistons and divided into spaces I and II by a porous plug. The plug must have properties such that it will prevent spontaneous pressure equalization between I and II by convection or diffusion but will allow movement of gas when movement of the pistons results in a higher pressure in space I (P_I) than in space II (P_{II}). Suitable materials for such a plug are, for example, cotton-wool or glass-wool. The whole apparatus must be adiabatically insulated. Let us assume that I contains 1 mole of gas while the piston which closes II is touching the plug, and consider the transfer of the gas from I to II while the pressures P_I and P_{II} remain constant. Since the process is an adiabatic one we have, according to the First Law [equation (8.1)], that

$$u_{II} = u_I + P_I v_I - P_{II} v_{II} \tag{25.31}$$

or, with (21.15),

$$h_I = h_{II}. \tag{25.32}$$

The enthalpy therefore remains constant during the process which is accordingly called an isenthalpic expansion. The gas is thus at equilibrium in I and in II but undergoes non-equilibrium changes within the plug. As has been shown at the end of § 15, this fact has no significance when the discussion is confined to initial and final states described by means of thermodynamic variables of state.

In the experiment described above, we find that generally $T_{\mathrm{I}} \neq T_{\mathrm{II}}$ and that T_{I} may be greater or less than T_{II}. This is the Joule–Thomson effect. The aim of our theory is now to calculate this temperature change from a given change in pressure.

If this change in pressure is assumed to be infinitesimal we get, from (25.32),

$$\mathrm{d}T = \left(\frac{\partial T}{\partial P}\right)_H \mathrm{d}P \tag{25.33}$$

which is our basic equation for the Joule–Thomson effect. We now use (25.24) to transform the partial derivative from the variables P, H into the standard variables T, P. We thus have

$$\mathrm{d}T = -\frac{(\partial H/\partial P)_T}{(\partial H/\partial T)_P}\,\mathrm{d}P. \tag{25.34}$$

Using (21.16) this becomes (since the number of moles remains constant)

$$\mathrm{d}T = -\frac{T(\partial s/\partial P)_T + V}{T(\partial s/\partial T)_P}\,\mathrm{d}P. \tag{25.35}$$

This, together with Maxwell's relation (24.24) and the definitions (25.4) and (25.5), finally gives

$$\mathrm{d}T = \frac{v}{C_P}\,(T\alpha - 1)\,\mathrm{d}P. \tag{25.36†}$$

This equation shows immediately that $\mathrm{d}T = 0$ for an ideal gas (since $\alpha = 1/T$). The Joule–Thomson effect therefore depends on deviations from ideality. Measurements of the Joule–Thomson effect can, therefore, be used to calculate absolute temperatures from empirical temperatures measured by means of the real gas thermometer (§ 7). A more detailed examination (which we shall not undertake here) shows that $\mathrm{d}T$ is positive at high temperatures and negative at low temperatures. The inversion temperature is determined by the equation

$$T\alpha = 1 \tag{25.37}$$

whose solution for T requires a knowledge of the thermal equation of state of the gas.

† Equation (25.36) describes the so-called differential Joule–Thomson effect. The quantity $\delta \equiv (\partial T/\partial P)_H$ is called the Joule–Thomson coefficient.

The Joule–Thomson effect at low temperatures is one of the most important aspects of low-temperature technology. A technical exploitation of the effect requires first of all that the experiment described above should be transformed into a continuous process and secondly that, if necessary, the gas to be used is precooled to a temperature lower than the inversion temperature of the Joule–Thomson effect.† The Joule–Thomson effect may then be used down to the liquefaction temperature of the gas.

§ 26 Mean molar and partial molar quantities

As already mentioned in § 21 the Gibbs free energy is the most important characteristic function as far as applications of thermodynamics are concerned. We shall now develop several concepts and relationships which are useful in the application of the Gibbs function to multi-component systems.

We shall develop the present discussion from § 20(e). The definition of density may be directly extended to include multi-component systems but is little used in equilibrium thermodynamics. The concept of molar quantities cannot, however, be directly generalized since it would obviously be futile to relate the function of state to 1 mole of one component.

We must, therefore, introduce a new concept in the following way. According to eq. (21.40) and Euler's theorem we have

$$\alpha G = G(T, P, \alpha n_1, \ldots, \alpha n_m). \tag{26.1}$$

We now let $\alpha = 1 \Big/ \sum_{i=1}^{m} n_i$ and write

$$G^* = G^*(T, P, x_1, \ldots, x_{m-1}). \tag{26.2}$$

The quantities

$$x_i = \frac{n_i}{\sum_{i=1}^{m} n_i} \tag{26.3}$$

are called *mole fractions* and we have

$$\sum_{i=1}^{m} x_i = 1 \tag{26.4}$$

so that only $(m-1)$ mole fractions are independent variables. The function G^* defined by eq. (26.2) is called the *mean molar Gibbs free energy*.

† Precooling is not necessary for most gases since their inversion temperature T_i is far above room temperature (e.g. T_i for N_2 = 621 K). H_2 (T_i = 195 K) and He (T_i = 23·6 K) which are particularly important in low-temperature physics do, however, need to be precooled.

Equations (21.37) and (24.8) show that, within the representational framework based on the Gibbs free energy, the extensive quantities of state S, V, and H also appear as functions of T, P, and the mole numbers n_i. It seems reasonable, therefore, to generalize the above concept. Let Z be an extensive function of state of the independent variables T, P, n_1, ..., n_m. We then have

$$Z^* = \frac{Z}{\sum_{i=1}^{m} n_i} \tag{26.5}$$

which is the general definition of a *mean molar quantity of state*. The quantities

$$z_i = \left(\frac{\partial Z}{\partial n_i}\right)_{T,P,n_j}, \tag{26.6}$$

first introduced by Lewis, are called *partial molar quantities* of component i. They are important in the thermodynamics of multicomponent systems.† Comparison of (21.37) with the definition (26.6) shows that the chemical potential may be formally regarded as the partial molar Gibbs free energy. This inherently correct statement has often been regarded as the definition of chemical potential and has thus obscured its far more general significance based on the fundamental equation and its Legendre transforms.

From (21.37) and the derivatives (26.2) (26.3), and (26.5) we find

$$\left(\frac{\partial G^*}{\partial T}\right)_{P,x} = -S^*, \quad \left(\frac{\partial G^*}{\partial P}\right)_{T,x} = V^*. \tag{26.7}$$

We thus get from (26.2)

$$\mathrm{d}G^* = -S^*\,\mathrm{d}T + V^*\,\mathrm{d}P + \sum_{i=1}^{m-1}\left(\frac{\partial G^*}{\partial x_i}\right)_{T,P,x_j}\mathrm{d}x_i. \tag{26.8}$$

The derivatives of G^* with respect to the mole fractions are, therefore, *not* chemical potentials. We therefore have to ask the following questions:

(a) What is the significance of the derivatives $(\partial Z^*/\partial x_i)_{T,P,x_j}$?

(b) What is the relationship between mean molar and partial molar quantities?

† For one-component systems both mean molar and partial molar quantities reduce to molar quantities which we have previously denoted by lower-case letters [eqs. (20.34) and (20.36)]. From now on we shall, when convenient, also use the symbol Z^* for molar quantities.

The two questions are closely related and we shall now examine them briefly.

By definition we have

$$\mathrm{d}Z = \left(\frac{\partial Z}{\partial T}\right)_{P,n} \mathrm{d}T + \left(\frac{\partial Z}{\partial P}\right)_{T,n} \mathrm{d}P + \sum_{i=1}^{m} \left(\frac{\partial Z}{\partial n_i}\right)_{T,P,n_j} \mathrm{d}n_i. \tag{26.9}$$

According to the definition of extensive parameters, Z is homogeneous of the first degree in the mole numbers. Euler's theorem together with the definition (26.6) therefore gives

$$Z = \sum_{i=1}^{m} z_i n_i. \tag{26.10}$$

By differentiation of (26.10) and comparison with (26.9) we get

$$\sum_{i=1}^{m} n_i \, \mathrm{d}z_i - \left(\frac{\partial Z}{\partial T}\right)_{P,n} \mathrm{d}T - \left(\frac{\partial Z}{\partial P}\right)_{T,n} \mathrm{d}P = 0. \tag{26.11}$$

This relationship is formally identical with (20.43) and is called the *generalized Gibbs–Duhem equation*. The generalization is based on the fact that the Gibbs–Duhem equation merely derives from the fact that the fundamental equation is a homogeneous function of the first degree.

By using (26.3) and (26.5), eq. (26.11) may be written

$$\sum_{i=1}^{m} x_i \, \mathrm{d}z_i - \left(\frac{\partial Z^*}{\partial T}\right)_{P,x} \mathrm{d}T - \left(\frac{\partial Z^*}{\partial P}\right)_{T,x} \mathrm{d}P = 0. \tag{26.12}$$

Isothermal–isobaric changes of state ($\mathrm{d}T = 0, \mathrm{d}P = 0$) constitute a particularly important special case of this equation, for which we have

$$\sum_{i=1}^{m} x_i \, \mathrm{d}z_i = 0 \quad (\mathrm{d}T = 0, \mathrm{d}P = 0). \tag{26.13}$$

We also have, for $\mathrm{d}T = \mathrm{d}P = 0$, that

$$\mathrm{d}z_i = \sum_{j=1}^{m-1} \left(\frac{\partial z_i}{\partial x_j}\right)_{T,P,x_k} \mathrm{d}x_j, \tag{26.14}$$

and thus

$$\sum_{i=1}^{m} x_i \, \mathrm{d}z_i = \sum_{i=1}^{m} x_i \left[\sum_{j=1}^{m-1} \left(\frac{\partial z_i}{\partial x_j}\right)_{T,P,x_k} \mathrm{d}x_j\right] = \sum_{j=1}^{m-1} \left[\sum_{i=1}^{m} x_i \left(\frac{\partial z_i}{\partial x_j}\right)_{T,P,x_k}\right] \mathrm{d}x_j = 0. \tag{26.15}$$

According to eq. (26.4), however, the $(m-1)$ dx_j are independent of each other. Equation (26.15) can, therefore, only be satisfied if

$$\sum_{i=1}^{m} x_i \left(\frac{\partial z_i}{\partial x_j}\right)_{T,P,x_k} = 0. \tag{26.16}$$

This form of the Gibbs–Duhem equation is often used in the investigation of isothermal–isobaric problems.

If we write eq. (26.10) for mean molar quantities we get

$$Z^* = \sum_{i=1}^{m} z_i x_i. \tag{26.17}$$

According to eq. (26.4) we have

$$\frac{\partial x_m}{\partial x_j} = -1. \tag{26.18}$$

Partial differentiation of (26.17) with respect to x_j therefore gives

$$\frac{\partial Z^*}{\partial x_j} = \sum_{i=1}^{m} x_i \frac{\partial z_i}{\partial x_j} + z_j - z_m \tag{26.19}$$

or, because of (26.16),

$$\frac{\partial Z^*}{\partial x_j} = z_j - z_m \quad (j = 1, \ldots, m-1). \tag{26.20}$$

This equation answers the above question (a) about the meaning of the derivatives $(\partial Z^*/\partial x_i)_{T,P,x_j}$. It also partially answers question (b). We shall, however, derive one further relationship which expresses partial molar quantities by the derivatives $(\partial Z^*/\partial x_i)_{T,P,x_j}$.

We first solve (26.20) for z_j and substitute this value into eq. (26.17). We then get

$$Z^* = \sum_{i=1}^{m-1} x_i \frac{\partial Z^*}{\partial x_i} + \sum_{i=1}^{m-1} x_i z_m + x_m z_m, \tag{26.21}$$

which, with (26.4), gives

$$z_m = Z^* - \sum_{i=1}^{m-1} x_i \frac{\partial Z^*}{\partial x_i}. \tag{26.22}$$

Question (b) has thus also been completely answered. The most important special case of eq. (26.22) is $m = 2$ (binary systems), where

$$z_1 = Z^* - x_2 \frac{\partial Z^*}{\partial x_2}, \tag{26.23}$$

$$z_2 = Z^* - x_1 \frac{\partial Z^*}{\partial x_1}. \tag{26.24}$$

These equations make it possible to calculate partial molar quantities from mean molar quantities and this is important in a number of applications since mean molar quantities (e.g. volume, enthalpy) may be experimentally more easily accessible than partial molar ones. The usual method is called the *tangent method* (Fig. 14).

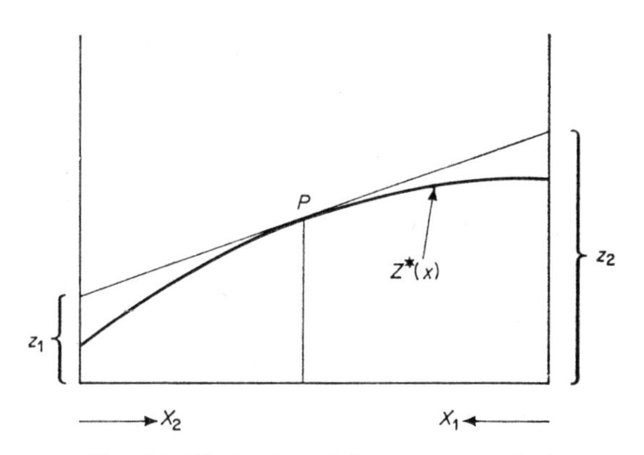

FIG. 14. Illustration of the tangent method

The tangent to a curve at a point x, y where the gradient is $\mathrm{d}y/\mathrm{d}x$ is given by the equation

$$\eta = y + (\xi - x)\frac{\mathrm{d}y}{\mathrm{d}x}. \tag{26.25}$$

The intercept on the ordinate for this is obtained by putting $\xi = 0$. Comparison with (26.23) and (26.24) shows that the intercept on the ordinate in this construction gives the partial molar quantities for the point P.

We now summarize certain special relationships involving the Gibbs free energy. These relationships are of importance for many applications. According to (26.6) we have the following definitions of partial molar quantities:

$$\left(\frac{\partial G}{\partial n_i}\right)_{T,P,n_j} = \mu_i, \quad \left(\frac{\partial S}{\partial n_i}\right)_{T,P,n_j} = s_i, \quad \left(\frac{\partial H}{\partial n_i}\right)_{T,P,n_j} = h_i, \quad \left(\frac{\partial V}{\partial n_i}\right)_{T,P,n_j} = v_i. \tag{26.26}$$

By means of (21.36) we then have

$$\left(\frac{\partial \mu_i}{\partial T}\right)_{P,n} = -s_i, \quad \left(\frac{\partial \mu_i}{\partial P}\right)_{T,n} = v_i, \tag{26.27}$$

and, according to (21.35) and (26.26), we have

$$\mu_i = h_i - Ts_i. \tag{26.28}$$

From (26.27) and (26.28) we finally arrive at the analogue of the Gibbs–Helmholtz equation

$$\mu_i = h_i + T\left(\frac{\partial \mu_i}{\partial T}\right)_{P,n}. \tag{26.29}$$

One of the components in a multi-component system may be present in excess. If this component is a liquid in the pure state it is often called the *solvent*. It is usual to denote quantities relating to the solvent by the subscript 1. The quantity

$$\Delta\mu_1 \equiv \mu_1 - \mu_{10} \tag{26.30}$$

where μ_{10} is the chemical potential of the pure solvent at the same temperature and pressure is called the *free energy of dilution*. The *heat of dilution* Δh_1 and the *entropy of dilution* Δs_1 are defined in a similar way. Equations (26.27)–(26.29) lead directly to the commonly used relationships

$$\Delta\mu_1 = \Delta h_1 - T\Delta s_1, \tag{26.31}$$

$$\left(\frac{\partial \Delta\mu_1}{\partial T}\right)_{P,n} = -\Delta s_1, \quad \frac{\partial(\Delta\mu_1/T)}{\partial(1/T)} = \Delta h_1, \tag{26.32}$$

$$\Delta\mu_1 = \Delta h_1 + T\left(\frac{\partial \Delta\mu_1}{\partial T}\right)_{P,n}. \tag{26.33}$$

The great significance of the free energy of dilution for the thermodynamics of solutions is based on the easy experimental accessibility of this quantity by various methods. Δs_1 and Δh_1 are thus also easily obtained. In particular, the whole thermodynamics of the system is completely defined for binary solutions if $\Delta\mu_1$ is known as a function of the mole fraction x_2 over the whole concentration range. Thus we have from the isothermal–isobaric form of the Gibbs–Duhem equation (26.16) that

$$(1-x_2)\frac{\partial \mu_1}{\partial x_2} + x_2\frac{\partial \mu_2}{\partial x_2} = 0 \quad (T = \text{const}, P = \text{const}), \tag{26.34}$$

and, therefore, that

$$\Delta\mu_2 = \int_{x_2}^{1} \frac{1-x_2}{x_2}\,\mathrm{d}(\Delta\mu_1), \tag{26.35}$$

or, after partial integration,

$$\Delta\mu_2 = \int_0^{x_1} \frac{\Delta\mu_1}{1-x_1}\,\mathrm{d}x_1 - \frac{x_1\,\Delta\mu_1}{1-x_1}. \tag{26.36}$$

The mean molar Gibbs free energy with respect to the pure components (also called the free energy of mixing) is then, according to (26.17), given by

$$\Delta G^* = x_1\,\Delta\mu_1 + x_2\,\Delta\mu_2. \tag{26.37}$$

CHAPTER IV

Heterogeneous equilibria without chemical reactions

§ 27 General equilibrium conditions for heterogeneous systems

Let us consider a closed system with fixed work co-ordinates and consisting of m components and σ phases. The phases are denoted by small Greek letters, α being used as the general index. External fields, boundary effects, electrolyte solutions, and chemical reactions are excluded. If the system contains solid phases we shall assume that a uniform hydrostatic pressure is acting on them from all directions, and that there are no uni-axial normal or tangential stresses (cf. §§ 53 and 54). The volume is, therefore, the only work co-ordinate. The phases are assumed to be in internal equilibrium and completely open to each other so that entropy, volume, and all mole numbers can be varied for each phase. Following the discussion in § 17 we shall assume that equilibrium conditions and secondary conditions will be given in the form of equations.

The application of the fundamental equation (20.5) to each phase accordingly gives

$$\left.\begin{aligned} \delta U^{(\alpha)} &= T^{(\alpha)}\,\delta S^{(\alpha)} - P^{(\alpha)}\,\delta V^{(\alpha)} + \sum_{i=1}^{m} \mu_i^{(\alpha)}\,\delta n_i^{(\alpha)}, \\ &\cdot \quad \cdot \quad \cdot \quad \cdot \quad \cdot \quad \cdot \\ \delta U^{(\sigma)} &= T^{(\sigma)}\,\delta S^{(\sigma)} - P^{(\sigma)}\,\delta V^{(\sigma)} + \sum_{i=1}^{m} \mu_i^{(\sigma)}\,\delta n_i^{(\sigma)}. \end{aligned}\right\} \tag{27.1}$$

According to the general equilibrium condition (17.2) we have

$$\sum_{\alpha=1}^{\sigma} \delta U^{(\alpha)} = 0 \tag{27.2}$$

together with the secondary conditions

$$\sum_{\alpha=1}^{\sigma} \delta S^{(\alpha)} = 0, \tag{27.3}$$

$$\sum_{\alpha=1}^{\sigma} \delta V^{(\alpha)} = 0, \tag{27.4}$$

$$\sum_{\alpha=1}^{\sigma} \delta n_i^{(\alpha)} = 0 \quad (i = 1, \ldots, m). \tag{27.5}$$

Such an extremum formulation with secondary conditions may be treated by means of Lagrange's method of indeterminate multipliers. We multiply the secondary condition (27.3) by an indeterminate but constant factor λ_1, (27.4) by λ_2, and (27.5) by λ_{3i}. We now subtract the resulting equations from (27.2). Using (27.1) we then get

$$\sum_{\alpha} (T^{(\alpha)} - \lambda_1)\, \delta S^{(\alpha)} + \sum_{\alpha} (P^{(\alpha)} - \lambda_2)\, \delta V^{(\alpha)} + \sum_{i=1}^{m} \sum_{\alpha} (\mu_i^{(\alpha)} - \lambda_{3i})\, \delta n_i^{(\alpha)} = 0. \tag{27.5a}$$

We now assume the multiplicands fixed such that in each of the sums over α one bracket vanishes. The remaining variations are then freely adjustable and eq. (27.5a) can be satisfied only if all the remaining brackets also vanish. We have, therefore, for each α and each i

$$T^{(\alpha)} = \lambda_1, \quad P^{(\alpha)} = \lambda_2, \quad \mu_i^{(\alpha)} = \lambda_{3i}. \tag{27.5b}$$

It follows, therefore, that

$$T^{(\alpha)} = T^{(\beta)} = \ldots = T^{(\sigma)}, \tag{27.6}$$

$$P^{(\alpha)} = P^{(\beta)} = \ldots = P^{(\sigma)}, \tag{27.7}$$

$$\mu_i^{(\alpha)} = \mu_i^{(\beta)} = \ldots = \mu_i^{(\sigma)} \quad (i = 1, \ldots, m). \tag{27.8}$$

Under the given conditions, therefore, all the phases of a heterogeneous system at equilibrium must have the same temperature and pressure and the chemical potential of any component must be the same in all the phases. Equation (27.6) contains the condition for thermal equilibrium, eq. (27.7) that for mechanical equilibrium, and eq. (27.8) that for material equilibrium.

Equations (27.6)–(27.8) completely determine the state of the system since the fundamental equation together with (15.8) shows that the state is defined by the $(m+2)\,\sigma$ independent variables

$S^{(\alpha)}$, $V^{(\alpha)}$, $n_i^{(\alpha)}$. Considered as equations of state as in § 20 (e), eqs. (27.6)–(27.8) give $(m+2)(\sigma-1)$ equations. The secondary conditions (27.3)–(27.5) give a further $(m+2)$ equations so that we have as many equations as we have unknowns. We have, in fact, already introduced conditions (27.6)–(27.8) as facts of experience during the development of the laws of thermodynamics in Chapter I. The derivation of these conditions from Gibbs' general equilibrium condition shows the axiomatic aspect of the latter, as has already been mentioned in § 17.

The equilibrium conditions (27.8) may also be formulated by means of mean molar Gibbs free energies. If, for simplicity, we confine our discussion to two phases α and β, we get by introducing (26.20) and (26.22) into (27.8)

$$\left(\frac{\partial G^*}{\partial x_i}\right)^{(\alpha)} = \left(\frac{\partial G^*}{\partial x_i}\right)^{(\beta)} \quad (i = 1, \ldots, m-1), \tag{27.9}$$

$$\left(G^* - \sum_{i=1}^{m-1} x_i \frac{\partial G^*}{\partial x_i}\right)^{(\alpha)} = \left(G^* - \sum_{i=1}^{m-1} x_i \frac{\partial G^*}{\partial x_i}\right)^{(\beta)}. \tag{27.10}$$

§ 28 Membrane equilibria. Osmotic pressure

The usefulness of the extremum principles formulated in § 17 becomes apparent only in applications to more complicated problems. We shall give an example of great importance in the theory of liquid mixtures. Let us again consider a closed system with fixed work co-ordinates and consisting of m components and σ phases. We shall make the same assumptions we made in § 27 with the exception that all phases are here separated by rigid semipermeable membranes permeable only to s components ($s < m$). The volume of each phase is thus fixed. Furthermore, the variability of $(m-s)$ mole numbers vanishes for all the phases. The secondary conditions for the extremum formulation (27.2) are now, therefore,

$$\sum_{\alpha=1}^{\sigma} \delta S^{(\alpha)} = 0, \tag{28.1}$$

$$\delta V^{(\alpha)} = 0, \quad \delta V^{(\beta)} = 0, \quad \ldots, \quad \delta V^{(\sigma)} = 0, \tag{28.2}$$

$$\sum_{\alpha=1}^{\sigma} \delta n_i^{(\alpha)} = 0 \quad (i = 1, \ldots, s), \tag{28.3}$$

$$\delta n_j^{(\alpha)} = 0, \quad \delta n_j^{(\beta)} = 0, \quad \ldots, \quad \delta n_j^{(\sigma)} = 0 \quad (j = s+1, \ldots, m). \tag{28.4}$$

These, together with (27.1) and (27.2) give the equilibrium conditions

$$T^{(\alpha)} = T^{(\beta)} = \ldots = T^{(\sigma)}, \tag{28.5}$$

$$\mu_i^{(\alpha)} = \mu_i^{(\beta)} = \ldots = \mu_i^{(\sigma)} \quad (i = 1, \ldots, s). \tag{28.6}$$

The introduction of rigid semipermeable membranes thus means that co-existing phases can be under different pressures and that equality of chemical potential applies only to those components which can pass through the membranes.

The simplest and most important example of the application of these considerations concerns osmotic equilibrium (Fig. 15).

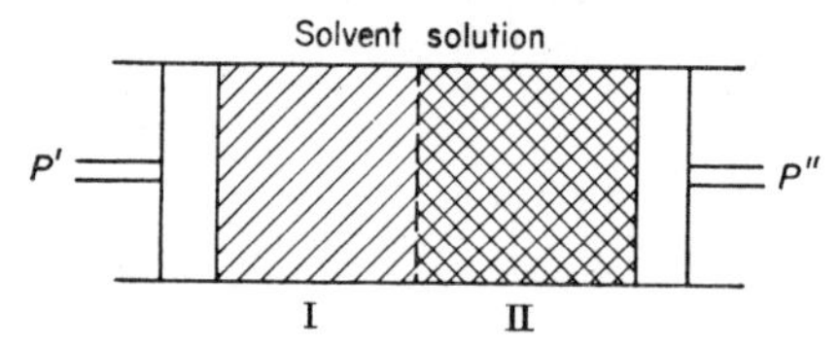

FIG. 15. Osmotic equilibrium

We here have $\sigma = 2$, one phase being pure solvent (component 1) and the second being a solution which we shall simply assume to be binary. The membrane is, therefore, permeable only to the solvent. The equilibrium conditions (28.5) and (28.6) now become

$$T' = T'' = T, \tag{28.7}$$

$$\mu_1'(T, P') = \mu_1''(T, P'', x_1). \tag{28.8}$$

Since the two phases are not in equilibrium when no membrane is present, i.e. $\mu_1(T, P, 1) \neq \mu_1(T, P, x_1)$, eq. (28.8) can only be satisfied if

$$P'' \neq P'. \tag{28.9}$$

The equilibrium conditions do not, however, give any information whether we have $P' < P''$ or $P' > P''$. This question can be answered by using the stability conditions and we have here a good example for their usefulness in thermodynamics. As will be shown later (Chapter VI) the stability conditions lead to the inequality

$$\left(\frac{\partial \mu_1}{\partial x_1}\right)_{T,P} > 0. \tag{28.10}$$

Further, according to (26.27) we have for the pure solvent

$$\left(\frac{\partial \mu_1}{\partial P}\right)_T' = v_1' > 0. \tag{28.11}$$

Since, according to (28.10), $\mu_1' > \mu_1''$ for $P' = P''$, the condition (28.8) can, because of (28.11), only be satisfied if $P'' > P'$. The excess pressure

$$\Pi = P'' - P' \tag{28.12}$$

under which the solution is in osmotic equilibrium is called the *osmotic pressure*.

There is a simple connexion between the osmotic pressure and the free energy of dilution defined by eq. (26.30). From eq. (28.11) we get (if P' is identified with normal pressure P_0)

$$\mu_1(P_0+\Pi, x_2) = \mu_1(P_0, x_2) + \int_{P_0}^{P_0+\Pi} v_1 \, \mathrm{d}P. \tag{28.13}$$

With (26.30) and (28.8) we get, therefore, that

$$\Delta\mu_1 = -\int_{P_0}^{P_0+\Pi} v_1 \, \mathrm{d}P, \tag{28.14}$$

or

$$\Pi = -\frac{\Delta\mu_1}{\bar{v}_1}, \tag{28.15}$$

where $\bar{v}_1$ is the mean partial molar volume between the pressures P_0 and $P_0+\Pi$. The osmotic pressure is, therefore, essentially a measure of the free energy of dilution.

Osmotic measurements are really meaningful only for very dilute solutions. For solutions of solutes of low molecular weight, Π at low molar concentration is of the order of 1–10 atm while for solutions of macromolecules it is 10^{-4}–10^{-3} atm. Measurements on solutes of low molecular weight are possible only in exceptional cases because of the difficulty of finding a semipermeable membrane for them. Since liquids have a very small compressibility the variation of the partial molar volume with pressure can, in practice, usually be neglected.

§ 29 The phase rule

Let us consider m components and σ phases forming a system at thermodynamic equilibrium. The conditions for equilibrium are given by eqs. (27.6)–(27.8). We now investigate what changes in the quantities of state are possible while the original heterogeneous equilibrium is maintained. In particular, the number of coexisting phases is to remain constant. The problem becomes more comprehensible if we first determine *how many* variables of state can be varied independently, as required above, for given values of m and σ.

We shall choose temperature, pressure, and chemical potentials as independent variables. The equilibrium conditions tell us that, for the changes we are considering, we have for each phase

$$\mathrm{d}T^{(\alpha)} = \mathrm{d}T, \quad \mathrm{d}P^{(\alpha)} = \mathrm{d}P, \quad \mathrm{d}\mu_i^{(\alpha)} = \mathrm{d}\mu_i \quad \text{(for all } \alpha, \text{ all } i). \tag{29.1}$$

As has been shown in § 20, the validity of the fundamental equation implies the validity of the Gibbs–Duhem equation for each phase. It is convenient here to write the Gibbs–Duhem equation in the form (26.12) and with (29.1) we get the set of equations

$$\left.\begin{array}{l} S^{*(\alpha)}\,\mathrm{d}T - V^{*(\alpha)}\,\mathrm{d}P + \sum_{i=1}^{m} x_i^{(\alpha)}\,\mathrm{d}\mu_i = 0, \\ \cdot \quad \cdot \quad \cdot \quad \cdot \quad \cdot \quad \cdot \\ S^{*(\sigma)}\,\mathrm{d}T - V^{*(\sigma)}\,\mathrm{d}P + \sum_{i=1}^{m} x_i^{(\sigma)}\,\mathrm{d}\mu_i = 0. \end{array}\right\} \tag{29.2}$$

We thus have $(m+2)$ independent variables whose variations are limited by the σ equations (29.2). The number f of freely adjustable variables is, therefore,

$$f = m+2-\sigma. \tag{29.3}$$

This equation constitutes *Gibbs' phase rule*; f is called the number of *thermodynamic degrees of freedom.*

According to the value of f we distinguish between invariant, univariant, bivariant, and multivariant equilibria.

From (29.3) we have the general conclusion that the number of thermodynamic degrees of freedom decreases by 1 when the number of co-existing phases increases by 1.

The derivation shows that the phase rule in the form of eq. (29.3) is bound by the assumptions made in § 27. We summarize these assumptions once more here since improper application of eq. (29.3) leads to apparent contradiction between theory and experiment. Thus, eq. (29.3) is based on the following assumptions:

(a) Boundary effects may be neglected.

(b) No chemical reactions occur.

(c) Volume is the only work co-ordinate.

(d) The boundaries between the phases are thermally conducting, deformable, and permeable to all components; in other words, entropy, volume, and mole numbers for each phase are freely variable within the limits set by the secondary conditions (27.3)–(27.5).

§ 30 Phase reactions

We have shown in § 27 that the state of a heterogeneous system is completely defined by the equilibrium conditions (27.6)–(27.8) together with the secondary conditions (27.3)–(27.5). The question now arises whether, and under what conditions, changes of state are possible in which only eqs. (27.5)–(27.8) are fixed while entropies and volumes of phases as well as those of the total system may change. Chemical potentials may be represented as functions of T, P, and x_i according to eqs. (26.2), (26.20), and (26.22); our question therefore concerns the possibility of and the conditions governing changes of state for which temperature, pressure, and mole fractions of the phases remain constant, i.e. changes in which the mass of certain phases increases at the expense of other phases while the total mass of each component remains constant. A change of state of this kind is called a *phase reaction*. It is always reversible since the heterogeneous equilibrium is maintained at all times.

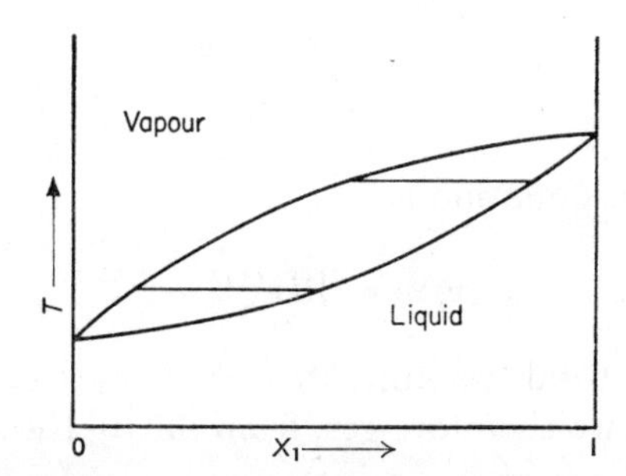

FIG. 16. Simple boiling point diagram of a binary liquid mixture

The possibility of phase reactions can be seen, for example, in the vaporization equilibrium of one-component systems. At constant T and P liquid may be freely changed to vapour and vice versa by the introduction or withdrawal of heat; the process is accompanied by a volume change. Phase reactions are, however, not always possible as can be seen from Fig. 16. The diagram represents a liquid–vapour equilibrium in a binary system at constant pressure, i.e. a boiling point diagram. Component 1 always has a higher concentration in the liquid than in the co-existing vapour. When vapour condenses, therefore, the concentration of 1 in the gas phase decreases and the equilibrium temperature changes as can be seen quite clearly in the diagram. A phase reaction is, therefore, impossible.

In order to derive the conditions for a phase reaction in a general form, let us consider a closed system at equilibrium consisting of m components and σ phases. Temperature, pressure, and the mole fractions in the phases are assumed to be fixed. Let the initial total number of moles of the individual phases be

$$n^{(\alpha)} = \sum_i n_i^{(\alpha)}. \tag{30.1}$$

Now let there be a phase reaction resulting in a state with unchanged $T, P, x_i^{(\alpha)}$ but with the mole numbers for the phases given by

$$n^{(\alpha)} + \nu^{(\alpha)} = \sum_i n_i^{(\alpha)} + \sum_i \nu_i^{(\alpha)}. \tag{30.2}$$

$\nu^{(\alpha)}$ is thus the change in mole number for the phase α for the phase reaction. Since the mole fractions for the phase must remain constant, we have

$$\frac{n_i^{(\alpha)}}{n} = \frac{n_i^{(\alpha)} + \nu_i^{(\alpha)}}{n^{(\alpha)} + \nu^{(\alpha)}}. \tag{30.3}$$

The solution of this equation is

$$\nu_i^{(\alpha)} = x_i^{(\alpha)} \nu^{(\alpha)}. \tag{30.3a}$$

By definition the total amount of each component of the system remains constant; we therefore get, from (30.3), the set of equations

$$\left.\begin{aligned} \sum_{\alpha=1}^{\sigma} x_1^{(\alpha)} \nu^{(\alpha)} &= 0, \\ \cdot \quad \cdot \quad \cdot \quad &\cdot \quad \cdot \quad \cdot \\ \sum_{\alpha=1}^{\sigma} x_m^{(\alpha)} \nu^{(\alpha)} &= 0. \end{aligned}\right\} \tag{30.4}$$

A phase reaction is, therefore, possible if and only if there is a solution of the homogeneous linear set of equations (30.4) for $\nu^{(\alpha)}$ which does not vanish.

We distinguish formally between the following cases:

(a) $m < \sigma$: There are fewer equations than unknowns and a solution other than zero thus always exists.

(b) $m = \sigma$: The number of equations and unknowns is the same. A solution different from zero exists if and only if we have

$$|x_i^{(\alpha)}| = 0 \tag{30.5}$$

for the coefficient determinant.

(c) $m > \sigma$: The number of equations is greater than the number of unknowns. The system is, therefore, overdefined and there is a solution different from zero only if all the equations are mutually consistent. This is so when all determinants of degree σ formed by the coefficient matrix vanish. Since the number of such determinants is $(1+m-\sigma)$ we have $(1+m-\sigma)$ conditional equations for the set (30.4) to be soluble.

These purely mathematical considerations lead to the following physical conclusions:

(a′) $m < \sigma$: According to (29.3) we have an invariant or univariant equilibrium. In either case, unlimited phase reactions are possible.

(b′) $m = \sigma$: According to (29.3) we have a bivariant equilibrium. A phase reaction can occur only in the case of those particular compositions of the phases for which eq. (30.5) is satisfied, i.e. the determinant formed from the mole fractions vanishes.

(c′) $m > \sigma$: According to (29.3) we have a multivariant equilibrium. Phase reactions are possible only for compositions for which the $(1+m-\sigma)$ determinants of degree σ formed from the mole fractions vanish.

The conditional equations mentioned in (b) and (c) are sometimes called *indifference conditions* in thermodynamics. We shall consider two simple examples:

(a″) Two-phase equilibrium of an m-component system: The $(m-1)$ indifference conditions are

$$\begin{vmatrix} x_i^{(\alpha)} & x_i^{(\beta)} \\ x_j^{(\alpha)} & x_j^{(\beta)} \end{vmatrix} = 0 \quad (i, j = 1, 2, \ldots, m;\ i \neq j). \tag{30.6}$$

These conditions can be satisfied only if

$$x_1^{(\alpha)} = x_1^{(\beta)}, \quad \ldots, \quad x_m^{(\alpha)} = x_m^{(\beta)}. \tag{30.7}$$

A phase reaction can, therefore, occur if and only if the two phases have the same composition. Such systems are called *azeotropic mixtures*.

(b″) Three-phase equilibrium in a ternary system. We have a bivariant equilibrium which, according to (b′) has the single indifference condition

$$\begin{vmatrix} x_1^{(\alpha)} & x_1^{(\beta)} & x_1^{(\gamma)} \\ x_2^{(\alpha)} & x_2^{(\beta)} & x_2^{(\gamma)} \\ x_3^{(\alpha)} & x_3^{(\beta)} & x_3^{(\gamma)} \end{vmatrix} = 0. \tag{30.8}$$

The geometric interpretation of this equation is that the compositions of the phases lie on a straight line when represented by Möbius' triangular co-ordinates (Fig. 17).

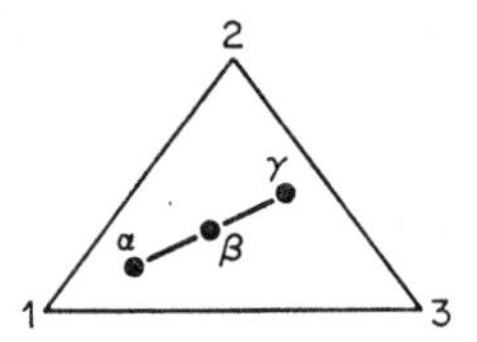

FIG. 17. Representation of the indifference condition for a three-phase equilibrium in a ternary system

A special case of this arises when two of the co-existing phases have the same composition since eq. (30.7) is then satisfied (coincidence of two points in the geometrical representation).

(c″) The examples lead directly to the following general theorem: For a phase reaction in bivariant and multivariant equilibria, the same composition of two co-existing phases is always a sufficient condition; for only two co-existing phases, it is also a necessary condition.

The definition of a non-variant equilibrium shows that a phase reaction can only occur at an appropriate point in the phase diagram. Since phase reactions in univariant equilibria are not subject to any limitations, they can occur anywhere along a $P(T)$-curve. Such a curve is called an *indifferent curve.*

For bivariant and multivariant equilibria the $(m+2-\sigma)$ degrees of freedom are limited by $(1+m-\sigma)$ indifference conditions. One degree of freedom therefore remains for phase reactions and the corresponding $P(T)$-curve is an indifferent curve as for univariant equilibria. In particular the curve on which lie the azeotropic mixtures of two-phase equilibria is an indifferent curve called the *azeotropic curve.*

§ 31 Invariant and univariant equilibria

Where $(m+2)$ phases co-exist at equilibrium, (29.3) requires that $f = 0$. Such an equilibrium is, therefore, possible only for a particular set of values of T, P, and μ_i and is represented by a point in the $(m+1)$-dimensional phase diagram. Such equilibria are called *invariant equilibria*.

Invariant equilibria in one-component systems are called *triple points*. Well-known examples are the triple points ice–water–water vapour, ice I–ice II–water, and the three triple points of sulphur: monoclinic–liquid–vapour, rhombic–monoclinic–liquid, rhombic–monoclinic–vapour.

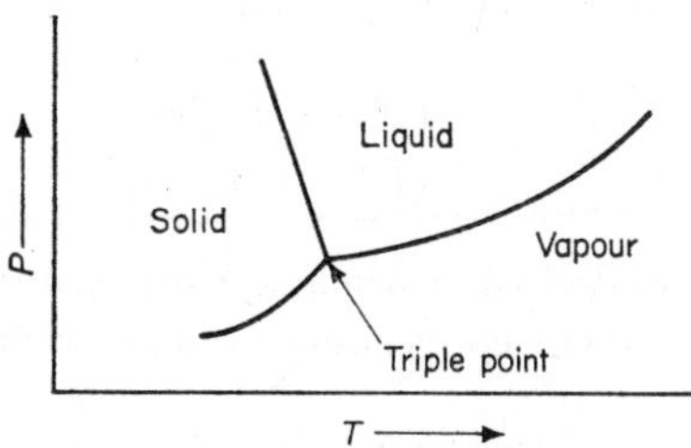

FIG. 18. Two-phase equilibria in a one-component system (schematic)

Invariant equilibria in binary systems involve the co-existence of four phases. Such equilibria are, therefore, called quadruple points. Well-known examples are systems of the type salt–salt hydrate–aqueous solution–water vapour in which gas phase, liquid, and two solid phases are in equilibrium.

The unlimited possibility of phase reactions in invariant equilibria (cf. § 30) is readily seen in the above examples.

A heterogeneous equilibrium involving $(m+1)$ co-existing phases has, according to (29.3), one degree of freedom and is thus called a univariant equilibrium. Let us examine the simplest example, i.e. two-phase equilibria in a one-component system, as represented in Fig. 18.

Since two phases co-exist along each curve, the curves are called *co-existence curves* whose differential equations we shall now derive.

According to (27.8) or (29.1) we have for movement along a co-existence curve

$$d\mu^{(\alpha)} = d\mu^{(\beta)}. \tag{31.1}$$

Since chemical potential and mean molar Gibbs free energy are equivalent for a one-component system we may write (31.1) as

$$dG^{*(\alpha)} = dG^{*(\beta)}. \tag{31.2}$$

This, with (26.8), (27.6), and (27.7), gives

$$-S^{*(\alpha)}\,dT + V^{*(\alpha)}\,dP = -S^{*(\beta)}\,dT + V^{*(\beta)}\,dP, \tag{31.3}$$

or

$$\frac{dP}{dT} = \frac{S^{*(\alpha)} - S^{*(\beta)}}{V^{*(\alpha)} - V^{*(\beta)}}. \tag{31.4}$$

However, (27.8) with (21.35) gives

$$H^{*(\alpha)} - TS^{*(\alpha)} = H^{*(\beta)} - TS^{*(\beta)}, \tag{31.5}$$

which, on introduction into (31.4), gives

$$\frac{dP}{dT} = \frac{H^{*(\alpha)} - H^{*(\beta)}}{T(V^{*(\alpha)} - V^{*(\beta)})}. \tag{31.6}$$

This relationship is called the *Clausius–Clapeyron equation* and is the differential equation of the co-existence curve for two-phase equilibria in one-component systems.

The enthalpy difference which appears in the numerator on the right-hand side is, according to eq. (21.23), the heat absorbed per mole during the phase reaction. It is convenient to denote the phase with the greater molar enthalpy by α and to write

$$L = H^{*(\alpha)} - H^{*(\beta)}. \tag{31.7}$$

The quantity L is called the *molar heat of phase change*, in particular the *molar heat of vaporization*, *molar heat of fusion*, etc. If we also write

$$\Delta V^* = V^{*(\alpha)} - V^{*(\beta)}, \tag{31.8}$$

the Clausius–Clapeyron equation appears in the simple form

$$\frac{dP}{dT} = \frac{L}{T\Delta V^*}. \tag{31.9}$$

Since $L > 0$ by definition, the gradient of the $P(T)$-curve has the same sign as the volume difference ΔV^*.

Let us now consider the general case of a univariant equilibrium. According to (29.3), we have $\sigma = m+1$ and the $P(T)$-curve is an indifferent curve according to § 30. In order to derive the differential equation of this curve, we start with the set of equations (29.2) which are now written

$$\left.\begin{array}{c} V^{*(\alpha)} \dfrac{\mathrm{d}P}{\mathrm{d}T} - \sum\limits_{i=1}^{m} x_i^{(\alpha)} \dfrac{\mathrm{d}\mu_i}{\mathrm{d}T} = S^{*(\alpha)}, \\ \cdot \quad \cdot \quad \cdot \quad \cdot \quad \cdot \quad \cdot \\ V^{*(m+1)} \dfrac{\mathrm{d}P}{\mathrm{d}T} - \sum\limits_{i=1}^{m} x_i^{(m+1)} \dfrac{\mathrm{d}\mu_i}{\mathrm{d}T} = S^{*(m+1)}, \end{array}\right\} \tag{31.10}$$

where all derivatives are taken along the equilibrium curve. Equations (31.10) are a set of inhomogeneous linear equations for the $(m+1)$ unknowns $\mathrm{d}P/\mathrm{d}T$, $\mathrm{d}\mu_1/\mathrm{d}T, \ldots, \mathrm{d}\mu_m/\mathrm{d}T$. The solution for $\mathrm{d}P/\mathrm{d}T$ is

$$\frac{\mathrm{d}P}{\mathrm{d}T} = \frac{D_S}{D_V}, \tag{31.11}$$

where

$$D_S = \begin{vmatrix} S^{*(\alpha)} & x_1^{(\alpha)} & \ldots & x_m^{(\alpha)} \\ \vdots & \vdots & \vdots & \vdots \\ S^{*(m+1)} & x_1^{(m+1)} & \ldots & x_m^{(m+1)} \end{vmatrix}, \tag{31.12}$$

$$D_V = \begin{vmatrix} V^{*(\alpha)} & x_1^{(\alpha)} & \ldots & x_m^{(\alpha)} \\ \vdots & \vdots & \vdots & \vdots \\ V^{*(m+1)} & x_1^{(m+1)} & \ldots & x_m^{(m+1)} \end{vmatrix}. \tag{31.13}$$

A Laplace expansion of the two determinants with respect to the elements of the first column gives

$$D_S = \sum_{\alpha} D^{(\alpha)} S^{*(\alpha)}, \tag{31.14}$$

$$D_V = \sum_{\alpha} D^{(\alpha)} V^{*(\alpha)} \tag{31.15}$$

if the co-factors (the same in both cases) are denoted by $D^{(\alpha)}$.†

† The sign conventions for $D^{(\alpha)}$ which we shall use from now on are:

$D^{(\alpha)}$ positive for odd α, negative for even α;

$D^{(m+1)}$ positive for even m, negative for odd m;

$D^{(\alpha)}/D^{(m+1)}$ negative when α and m are both even or both odd, otherwise positive.

We now write the general set of equations (30.4) in the form

$$\left.\begin{aligned}\sum_{\alpha=1}^{m} x_1^{(\alpha)}\,\nu^{(\alpha)} &= -x_1^{(m+1)}\,\nu^{(m+1)},\\ \cdot \quad \cdot \quad \cdot & \quad \cdot \quad \cdot \quad \cdot\\ \sum_{\alpha=1}^{m} x_m^{(\alpha)}\,\nu^{(\alpha)} &= -x_m^{(m+1)}\,\nu^{(m+1)}\end{aligned}\right\} \tag{31.16}$$

for our particular problem. If $\nu^{(m+1)}$ is known, eqs. (31.16) constitute a set of inhomogeneous linear equations for the m unknowns $\nu^{(\alpha)}$. If we use the above determinants $D^{(\alpha)}$ (i.e. the corresponding co-factors of the elements $S^{*(\alpha)}$ and $V^{*(\alpha)}$ of D_S and D_V) the solution of (31.16) is

$$\nu^{(\alpha)} = -\nu^{(m+1)}\,\frac{D^{(\alpha)}}{D^{(m+1)}}. \tag{31.17}†$$

If we now form the expression

$$\frac{D_S}{D_V} = \frac{\sum_\alpha D^{(\alpha)}\,S^{*(\alpha)}}{\sum_\alpha D^{(\alpha)}\,V^{*(\alpha)}} \tag{31.18}$$

and divide both numerator and denominator on the right-hand side by $-D^{(m+1)}$ we get, with (31.17),

$$\frac{D_S}{D_V} = \frac{\sum_\alpha \nu^{(\alpha)}\,S^{*(\alpha)}}{\sum_\alpha \nu^{(\alpha)}\,V^{*(\alpha)}}. \tag{31.19}$$

By the definition of $\nu^{(\alpha)}$ the numerator on the right-hand side represents the entropy change ΔS while the denominator represents ΔV for our phase reaction. Thus, eq. (31.11) becomes

$$\frac{\mathrm{d}P}{\mathrm{d}T} = \frac{\Delta S}{\Delta V}. \tag{31.20}$$

The enthalpy change for the phase reaction is

$$L = \Delta H = \sum \nu^{(\alpha)}\,H^{*(\alpha)}. \tag{31.21}$$

Since $\mathrm{d}T = 0$, $\mathrm{d}P = 0$, $\mathrm{d}n_i = 0$ for the phase reaction, we have

$$\Delta G = \Delta H - T\Delta S = L - T\Delta S = 0 \tag{31.22}$$

† The number of commutations which transforms the original determinant of the numerator into $D^{(\alpha)}$ is odd when α and m are both even or both odd, otherwise it is even.

so that we finally get

$$\frac{dP}{dT} = \frac{L}{T\Delta V}. \tag{31.23}$$

This relationship was first derived by Gibbs and constitutes the *generalized Clausius–Clapeyron equation* for any univariant equilibrium. For $m = 1$, $\sigma = 2$ eqs. (31.20) and (31.23) reduce to (31.4) and (31.9), the ordinary Clausius–Clapeyron equation.

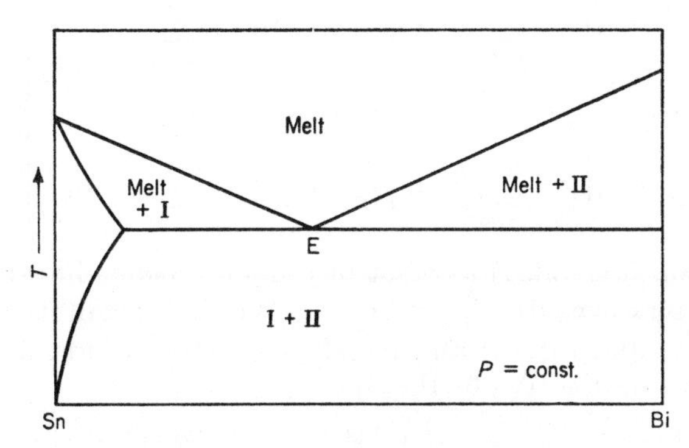

FIG. 19. Phase diagram of the system Sn-Bi (schematic) E = eutectic point

Equation (31.23) has a particularly clear meaning for vaporization equilibria of multi-component systems. If a system consists of m components, and m liquid phases are in equilibrium with the vapour, then eq. (31.23) represents the variation of the total vapour pressure with temperature or, alternatively, the variation of the boiling point with pressure.

So-called *eutectic mixtures* are an important special case of univariant equilibria in which two solid phases are in equilibrium with a liquid phase.

Figure 19 is an isobaric representation of such an equilibrium.† The variation of the eutectic temperature with pressure is given by eq. (31.23).

† Phase II does not really consist of pure Bi but contains about 0·1–0·2 atoms % of Sn near the eutectic point. The region of existence of this phase has been omitted from the diagram because of the scale on which the diagram is drawn.

The gas phase can generally be ignored in the discussion of isobaric phase diagrams of alloys since the pressure can be kept constant by means of an inert gas (e.g. argon) while the vapour pressure of the metals is usually negligible. The inclusion of the gas phase would, therefore, not change the number of thermodynamic degrees of freedom.

§ 32 Bivariant and multivariant equilibria

According to § 30 the main new factors for the case of bivariant and multivariant equilibria are the indifference conditions. Our particular problem is, therefore, the derivation of the differential equation for the indifferent $P(T)$-curve on which phase reactions can take place.

We start again with the set of equations (29.2) which we now conveniently write (for $m \geqslant \sigma$) as

$$\left.\begin{array}{l} V^{*(\alpha)}\,\mathrm{d}P - \sum\limits_{i=1}^{\sigma-1} x_i^{(\alpha)}\,\mathrm{d}\mu_i = S^{*(\alpha)}\,\mathrm{d}T + \sum\limits_{i=\sigma}^{m} x_i^{(\alpha)}\,\mathrm{d}\mu_i, \\ \quad\cdot \qquad \cdot \qquad \cdot \qquad \cdot \qquad \cdot \qquad \cdot \\ V^{*(\sigma)}\,\mathrm{d}P - \sum\limits_{i=1}^{\sigma-1} x_i^{(\sigma)}\,\mathrm{d}\mu_i = S^{*(\sigma)}\,\mathrm{d}T + \sum\limits_{i=\sigma}^{m} x_i^{(\sigma)}\,\mathrm{d}\mu_i. \end{array}\right\} \tag{32.1}$$

We shall consider (32.1) as a set of inhomogeneous linear equations for the σ unknowns $\mathrm{d}P, \mathrm{d}\mu_1, \ldots, \mathrm{d}\mu_{\sigma-1}$. We shall again use the definitions (31.12) and (31.13) and merely substitute σ and $\sigma-1$ for the indices $m+1$ and m. We further define

$$D_i = \begin{vmatrix} x_i^{(\alpha)} & x_1^{(\alpha)} & \ldots & x_{\sigma-1}^{(\alpha)} \\ \vdots & \vdots & \vdots & \vdots \\ x_i^{(\sigma)} & x_1^{(\sigma)} & \ldots & x_{\sigma-1}^{(\sigma)} \end{vmatrix}. \tag{32.2}$$

These determinants are identical with the determinants of degree σ mentioned in § 30(c) and formed from the coefficient matrix of (30.4). By using

$$\begin{vmatrix} A_{11}+B_{11} & A_{12} \\ A_{21}+B_{21} & A_{22} \end{vmatrix} = \begin{vmatrix} A_{11} & A_{12} \\ A_{21} & A_{22} \end{vmatrix} + \begin{vmatrix} B_{11} & A_{12} \\ B_{21} & A_{22} \end{vmatrix} \tag{32.3}$$

we solve (32.1) for $\mathrm{d}P$ and find

$$D_V\,\mathrm{d}P = D_S\,\mathrm{d}T + \sum_{i=\sigma}^{m} D_i\,\mathrm{d}\mu_i. \tag{32.4}$$

According to § 30(c) a phase reaction is possible only if the D_i vanish. We know from eq. (31.19) that D_V and D_S cannot be infinite for physical reasons nor are they generally zero. We therefore have the necessary and sufficient condition for a phase reaction for the case $m \geqslant \sigma$

$$\left(\frac{\partial P}{\partial \mu_i}\right)_{T,\mu_j} = 0 \quad (i,j = \sigma, \sigma+1, \ldots, m_i; i \neq j), \tag{32.5}$$

or

$$\left(\frac{\partial T}{\partial \mu_i}\right)_{P,\mu_j} = 0 \quad (i,j = \sigma, \sigma+1, \ldots, m_i; i \neq j). \tag{32.6}$$

The equilibrium pressure in the isothermal case, or the equilibrium temperature in the isobaric case, constitutes, according to (29.3), a function of $(m+1-\sigma)$ independent variables. We can choose the $(m+1-\sigma)$ chemical potentials as these variables and get, with (32.5) and (32.6), the theorem: If a phase reaction can occur in bivariant and multivariant equilibria, the equilibrium temperature at constant pressure has a stationary value as has the equilibrium pressure at constant temperature.

This theorem was found to hold for special cases by Gibbs and by Konowalow and is therefore known as the *generalized Gibbs–Konowalow* rule. It was first proved with complete generality by Saurel.

The differential equation of the indifferent curve is, according to (32.4),

$$\frac{\mathrm{d}P}{\mathrm{d}T} = \frac{D_S}{D_V}. \tag{32.7}$$

The right-hand side can, therefore, be calculated explicitly in the same way as in § 31. One simply introduces the first $(\sigma-1)$ equations of the set (30.4) instead of eq. (31.16). We thus find again that

$$\frac{\mathrm{d}P}{\mathrm{d}T} = \frac{\Delta S}{\Delta V} \tag{32.8}$$

and

$$\frac{\mathrm{d}P}{\mathrm{d}T} = \frac{L}{T\Delta V}. \tag{32.9}$$

This is the *generalized Clausius–Clapeyron equation* for the indifferent curve of bivariant and multivariant equilibria.

The simplest examples of the general rules developed here are found in the vaporization equilibria of binary mixtures of liquids. Figure 20 shows schematically the isobaric (boiling-point) diagram for the system benzene–ethanol. The mole fraction of ethanol has been chosen as the independent variable. The stationary point at which a phase reaction can occur (cf. the generalized Gibbs–Konowalow rule) and at which the two co-existing phases have the same composition [cf. § 30(c″)] is called an *azeotropic point* (cf. § 30). Equation (32.9) is the differential equation for the azeotropic curve.

The diagram constitutes a direct verification for the following rules (called *Konowalow's rules*) for boiling-point equilibria of binary liquid mixtures:

I. At constant pressure, the boiling point as a function of composition has a stationary value when and only when liquid and vapour have the same composition.

II. At constant pressure, the boiling point is raised by addition of that component whose concentration in the vapour is less than that in the liquid.

III. The compositions of liquid and co-existing vapour change in the same direction when the boiling point changes at constant pressure.

These rules can also be proved analytically. Rule I is merely a special case of the generalized Gibbs–Konowalow rule. Analogous rules apply for the isothermal representation.

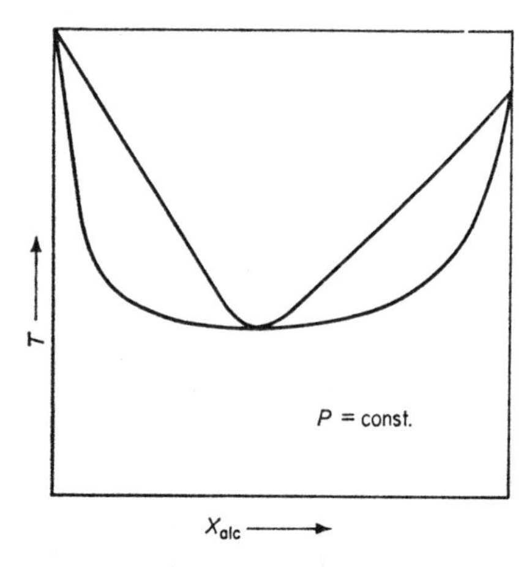

FIG. 20. Boiling point diagram of the system benzene–ethanol (schematic)

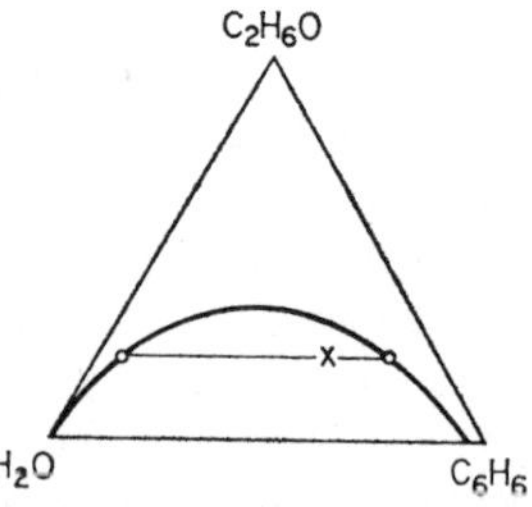

FIG. 21. Three-phase equilibrium in the ternary system ethanol–benzene–water (schematic) * = vapour phase

The ternary three-phase equilibrium ethanol–benzene–water is important for the manufacture of pure ethanol. The equilibrium is shown schematically in Fig. 21. The compositions of the three co-existing phases lie on a straight line as required according to § 30. The vapour pressure has a maximum at constant temperature. This maximum represents the highest vapour pressure of the whole vapour pressure surface of the system.

CHAPTER V

Chemical equilibria

§ 33 General equilibrium conditions

In Chapter IV we excluded chemical reactions from our considerations. Chemical equilibrium is a special case of internal equilibrium since it can be attained in a homogeneous system. In § 27 we assumed every phase to be in internal equilibrium; we may, therefore, allow chemical reactions within the phases as long as we ensure that chemical equilibrium is complete and that only mole numbers of independent components as defined in § 2 are introduced into the equilibrium conditions. Our earlier results will then remain unchanged but we can obviously obtain no information about the conditions for chemical equilibrium. Reactions between the phases (heterogeneous reactions) are permissible if equilibrium is maintained, only independent components are used, and either

(a) the equation for the reaction is irrelevant to the problem, or

(b) a phase reaction is identical with a chemical reaction.

So-called *dystectic points* are an example of (a). They are maxima on the isobaric melting-point curve. At these points the binary melt has the same composition as a solid chemical compound which is not miscible with the pure components. Figure 22 shows that two eutectic points E_1 and E_2 occur in the phase diagram. The question whether a chemical reaction occurs when the compound AB changes from the

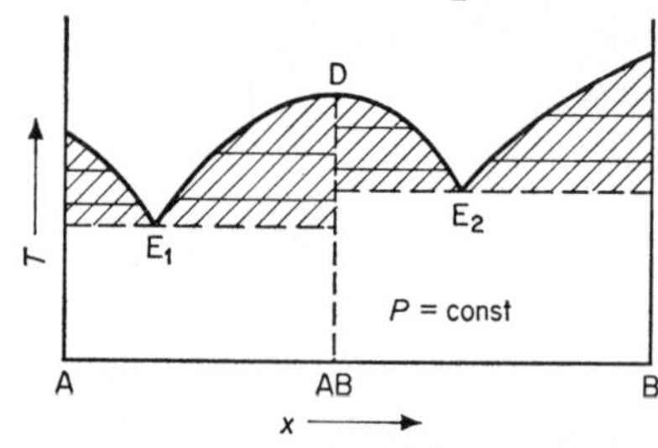

FIG. 22. Phase diagram of a binary system with one dystectic point D and two eutectic points E_1 and E_2

solid to the liquid phase is irrelevant as is a reaction between the three solid phases (which usually cannot occur, anyway, because of inhibition). A dystectic point is found, for example, in the system Sn–Mg where the solid intermetallic compound is $SnMg_2$.

An example of (b) is found in the decomposition of calcium carbonate according to the equation

$$CaCO_3 \rightleftharpoons CO_2 + CaO \tag{33.1}$$

which involves two independent components and three phases. It is, therefore, a univariant equilibrium according to (29.3). From a purely physical point of view, CO_2 and CaO are the independent components while $CaCO_3$ appears as a mixed phase co-existing with the other two phases. From § 30(a′) we have that unlimited phase reactions are possible here, and eq. (33.1) merely represents the phase reaction in a chemical way since the total amount of the components is conserved. The equilibrium pressure of CO_2 ('decomposition pressure') is, therefore, given by the generalized Clausius–Clapeyron equation (31.23).

The reaction

$$2NiO \rightleftharpoons 2Ni + O_2 \tag{33.2}$$

can be treated in a similar manner.

If, however, nickel is present in the form of a Ni–Pt alloy, we have three independent components and three phases and, therefore, a bivariant equilibrium. The chemical reaction can thus not be identified with a phase reaction (which is possible only for special compositions according to §§ 30 and 32). The methods of Chapter IV are, therefore, not applicable to this problem.

In order to derive completely general conditions for the chemical equilibrium we have to introduce the mole numbers of *all* substances occurring in the relevant chemical reaction equation into the fundamental equation as independent variables. Not all the m mole numbers of a phase are, therefore, independent of each other in the sense of § 2. For simplicity's sake we shall assume that only the one reaction

$$a\mathrm{A} + b\mathrm{B} \rightleftharpoons y\mathrm{Y} \tag{33.3}$$

occurs. It is irrelevant whether this reaction occurs within one phase, between several phases or in both ways. Together with these considerations, we apply the fundamental equations (27.1) and the equilibrium condition (27.2) which we summarize in the equation

$$\sum_{\alpha} T^{(\alpha)} \delta S^{(\alpha)} - \sum_{\alpha} P^{(\alpha)} \delta V^{(\alpha)} + \sum_{\alpha} \left[\sum_{i=1}^{m} \mu_i^{(\alpha)} \delta n_i^{(\alpha)} \right] = 0. \tag{33.4}$$

We again have the secondary conditions

$$\sum_{\alpha} \delta S^{(\alpha)} = 0, \tag{33.5}$$

$$\sum_{\alpha} \delta V^{(\alpha)} = 0. \tag{33.6}$$

Furthermore, we have without restriction

$$\sum_{\alpha} \delta n_i^{(\alpha)} = 0 \quad (i \neq \mathrm{A}, \mathrm{B}, \mathrm{Y}). \tag{33.7}$$

Since, however, A, B, and Y are related by the reaction equation (33.3) we *no longer necessarily* have

$$\sum_{\alpha} \delta n_{\mathrm{A}}^{(\alpha)} = 0, \quad \sum_{\alpha} \delta n_{\mathrm{B}}^{(\alpha)} = 0, \quad \sum_{\alpha} \delta n_{\mathrm{Y}}^{(\alpha)} = 0. \tag{33.8}$$

Instead of these secondary conditions we now have the equations

$$\left.\begin{aligned} \sum_{\alpha} \delta n_{\mathrm{A}}^{(\alpha)} + \frac{a}{y} \sum_{\alpha} \delta n_{\mathrm{Y}}^{(\alpha)} = 0, \\ \sum_{\alpha} \delta n_{\mathrm{B}}^{(\alpha)} + \frac{b}{y} \sum_{\alpha} \delta n_{\mathrm{Y}}^{(\alpha)} = 0. \end{aligned}\right\} \tag{33.9}$$

These conditions mean that the variations of the mole numbers of A, B, Y must either be related by the stoichiometric factors given by the reaction equation (33.3) or must each vanish according to (33.8) which merely means a change of substance from one phase to another. The conditions (33.8) therefore now represent a special case included in the more general conditions (33.9). The variations possible according to (33.8) therefore also satisfy (33.9). If we consider only the variations which satisfy (33.8) we find that we still need the equilibrium conditions (27.6)–(27.9). The possibility of a chemical reaction, however, adds a new condition. If we introduce (27.6)–(27.8) into (33.4) and take into account (33.5)–(33.7) we get

$$\mu_{\mathrm{A}} \sum_{\alpha} \delta n_{\mathrm{A}}^{(\alpha)} + \mu_{\mathrm{B}} \sum_{\alpha} \delta n_{\mathrm{B}}^{(\alpha)} + \mu_{\mathrm{Y}} \sum_{\alpha} \delta n_{\mathrm{Y}}^{(\alpha)} = 0. \tag{33.10}$$

Elimination of δn_{A} and δn_{B} by means of (33.9) gives

$$-a\mu_{\mathrm{A}} \sum_{\alpha} \delta n_{\mathrm{Y}}^{(\alpha)} - b\mu_{\mathrm{B}} \sum_{\alpha} \delta n_{\mathrm{Y}}^{(\alpha)} + y\mu_{\mathrm{Y}} \sum_{\alpha} \delta n_{\mathrm{Y}}^{(\alpha)} = 0. \tag{33.11}$$

Since the variations $\delta n_{\mathrm{Y}}^{(\alpha)}$ are completely arbitrary, i.e. $\delta n_{\mathrm{Y}}^{(\alpha)} \gtrless 0$ (cf. § 17), (33.11) is satisfied only if

$$a\mu_{\mathrm{A}} + b\mu_{\mathrm{B}} = y\mu_{\mathrm{Y}}. \tag{33.12}$$

In an analogous way it can be shown that for any chemical reaction

$$a\mathrm{A} + b\mathrm{B} + c\mathrm{C} + \ldots \rightleftharpoons y\mathrm{Y} + z\mathrm{Z} + \ldots \tag{33.13}$$

in thermodynamic equilibrium, the condition

$$a\mu_{\mathrm{A}} + b\mu_{\mathrm{B}} + c\mu_{\mathrm{C}} + \ldots = y\mu_{\mathrm{Y}} + z\mu_{\mathrm{Z}} + \ldots \tag{33.14}$$

must be satisfied. Equation (33.14) is the general condition for chemical equilibrium for both homogeneous and heterogeneous reactions. A convenient compact way of writing reaction (33.13) is

$$\sum \nu_i \mathrm{X}_i = 0, \tag{33.15}$$

where the products have the positive sign. The corresponding expression for the equilibrium condition is

$$\sum \nu_i \mu_i = 0. \tag{33.16}$$

§ 34 Homogeneous reactions. The law of mass action

This section deals with chemical reactions occurring within a phase or in a homogeneous system.

(a) *Homogeneous gas reactions*. The partial pressure of component i in a gaseous mixture is defined by the equation

$$p_i = x_i P. \tag{34.1}$$

For an ideal mixture of gases we have *Dalton's law of partial pressures*,

$$p_i = \frac{n_i}{V} RT. \tag{34.2}$$

The chemical potential of component i of an ideal gaseous mixture is

$$\mu_i = RT \ln p_i + \mu_i^0(T), \tag{34.3}$$

where $\mu_i^0(T)$ is a standard or reference value, i.e. the chemical potential for $p_i = 1$. Equation (34.3) gives, with (34.1),

$$\mu_i = RT \ln x_i + RT \ln P + \mu_i^0(T). \tag{34.4}$$

Let us now assume that the reaction (33.13) occurs in the ideal gaseous mixture. Introduction of (34.3) into the equilibrium condition (33.14) now gives

$$\frac{p_Y^y p_Z^z \cdots}{p_A^a p_B^b p_C^c \cdots} = K_p(T), \tag{34.5}$$

while introduction of (34.4) into (33.14) gives

$$\frac{x_Y^y x_Z^z \cdots}{x_A^a x_B^b x_C^c \cdots} = K_x(T, P). \tag{34.6}$$

Equations (34.5) and (34.6) are different forms of the *law of mass action* originally derived by Guldberg and Waage from kinetic considerations. The two 'equilibrium constants' K_p and K_x are connected by the relationship

$$K_x(T, P) = P^{a+b+c\ldots-y-z\ldots} K_p(T). \tag{34.7}$$

The law of mass action can also be formulated in a third way by introducing the quantities $c_i = n_i/V$ as concentration variables by means of eq. (34.2) which gives

$$\frac{c_Y^y c_Z^z \cdots}{c_A^a c_B^b c_C^c \cdots} = K_c(T) \tag{34.8}$$

and

$$K_c = (RT)^{a+b+c\ldots-y-z\ldots} K_p. \tag{34.9}$$

The 'equilibrium constant' K_c thus also depends only on T and not on P.

The ideal gas mixture has not been introduced merely for historical reasons: the above equations are usually a sufficient approximation for practical purposes when the total pressure does not exceed a few atmospheres.

If deviations from ideality have to be considered, an analogous formulation involving *fugacities* instead of partial pressures is available. The fugacities p_i^* are defined by the equation

$$\mu_i = RT \ln p_i^* + \mu_i^*(T), \tag{34.10}$$

where

$$\lim_{P\to 0} \frac{p_i^*}{x_i P} = \lim_{P\to 0} \frac{p_i^*}{p_i} = 1 \tag{34.11}$$

(*Gibbs–Dalton law*). We then have

$$\frac{p_Y^{*y}\, p_Z^{*z} \cdots}{p_A^{*a}\, p_B^{*b}\, p_C^{*c} \cdots} = K_p^*(T). \tag{34.12}$$

We have to remember that, on the right-hand side, we again have a pure temperature function but that fugacities are not simply concentration variables but are basically functions of all the variables of state.

Let us summarize the various forms of the mass action law in the compact notation of eqs. (33.15) and (33.16):

$$\left.\begin{aligned} \Pi p_i^{\nu_i} &= K_p(T), \\ \Pi x_i^{\nu_i} &= K_x(T, P), \\ \Pi c_i^{\nu_i} &= K_c(T), \end{aligned}\right\} \tag{34.13}$$

and, for real gases,

$$\Pi p_i^{*\nu_i} = K_p^*(T). \tag{34.14}$$

We shall use this notation exclusively from now on.

We still have to investigate the equilibrium constants. In order to obtain the most general results, we shall show first of all that the standard values of the chemical potentials occurring in eqs. (34.4) and (34.10) are identical. According to eqs. (26.27) and (34.10) we have

$$\left(\frac{\partial \mu_i}{\partial P}\right)_{T,x} = v_i = RT\left(\frac{\partial \ln p_i^*}{\partial P}\right)_{T,x}. \tag{34.15}$$

Since

$$RT\left(\frac{\partial \ln (Px_i)}{\partial P}\right)_{T,x} = \frac{RT}{P} \tag{34.16}$$

we can write (34.15) as

$$RT\left[\frac{\partial \ln (p_i^*/Px_i)}{\partial P}\right]_{T,x} = v_i - \frac{RT}{P}. \tag{34.17}$$

Integration of this equation at constant temperature and composition gives

$$\ln \frac{p_i^*}{Px_i} = \frac{1}{RT}\int_0^P \left(v_i - \frac{RT}{P}\right) \mathrm{d}P + \lim_{P\to 0} \ln \frac{p_i^*}{Px_i}. \tag{34.18}$$

If we now make use of the fact of experience that all real gas mixtures behave ideally at vanishingly small pressures, we find from (34.16) that

$$\lim_{P\to 0} \frac{Pv_i}{RT} = 1 \tag{34.19}$$

which proves the existence of the integral. Furthermore, the fact of experience together with (34.2) gives eq. (34.11), according to which the last term of (34.18) vanishes. Equation (34.10) can thus, by using (34.18), be written

$$\mu_i = RT\ln(Px_i) + \int_0^P \left(v_i - \frac{RT}{P}\right) \mathrm{d}P + \mu^*(T) \tag{34.20}$$

which gives the standard value as

$$\mu_i^*(T) = \lim_{P\to 0} [\mu_i - RT\ln(Px_i)]. \tag{34.21}$$

The limiting value of the first term is, according to the above fact of experience, given by eq. (34.4). We therefore have

$$\mu_i^0(T) = \mu_i^*(T). \tag{34.22}$$

From the derivation of eqs. (34.13) and (34.14) together with (34.22) we find

$$K_p(T) = K_p^*(T) \tag{34.23}$$

and

$$RT\ln K_p = -\sum \nu_i \mu_i^0 \equiv -\Delta G^0, \tag{34.24}$$

or

$$K_p = \exp\left(-\frac{\Delta G^0}{RT}\right) \tag{34.25}$$

if we use eq. (21.40). ΔG^0 is, by definition, the change in Gibbs free energy associated with one formular reaction (reaction of quantities of matter as written in the reaction equation) when all reactants and products are in their standard states.

We now have to consider the dependence of the equilibrium constants on temperature and pressure. Differentiation of (34.24) with respect to temperature gives

$$\frac{\mathrm{d}\ln K_p}{\mathrm{d}T} = -\sum_i \frac{\nu_i}{R} \frac{\mathrm{d}(\mu_i^0/T)}{\mathrm{d}T} \tag{34.26}$$

which, with eq. (26.29), becomes

$$\frac{\mathrm{d}\ln K_p}{\mathrm{d}T} = \sum \frac{\nu_i h_i^0}{RT^2}. \tag{34.27}$$

If the enthalpy of reaction for the standard state, ΔH^0, is defined by the equation

$$\Delta H^0 = \sum_i \nu_i h_i^0, \tag{34.28}$$

(34.27) becomes

$$\frac{\mathrm{d} \ln K_p}{\mathrm{d}T} = \frac{\Delta H^0}{RT^2} \tag{34.29}$$

which is known as the *van't Hoff equation*. It represents directly the temperature dependence of the chemical equilibrium for ideal gases. Since the partial molar enthalpies here are independent of composition, ΔH^0 is the true (measured) enthalpy of reaction. Conditions are more complicated for real gases and will, therefore, not be discussed.

Analogous relationships may be derived for the other equilibrium constants. We shall again confine our discussion to ideal gases.†

If we use the abbreviation $\sum \nu_i = \nu$, the isothermal–isobaric volume change for one formular reaction is

$$P\Delta V = \nu RT. \tag{34.30}$$

When we use the fact that the internal energy of an ideal gas mixture is independent of volume and pressure, we find that the enthalpy of reaction is given by

$$\Delta H = (\Delta U)_{T,P} + P(\Delta V)_{T,P} = (\Delta U)_{T,V} + P(\Delta V)_{T,P} = \Delta U + P\Delta V. \tag{34.31}$$

According to eq. (21.19), ΔU is the heat of reaction at constant volume. We thus have, with eq. (34.9),

$$\ln K_c = \ln K_p - \nu \ln RT. \tag{34.32}$$

From eqs. (34.29)–(34.32) we get

$$\frac{\mathrm{d} \ln K_c}{\mathrm{d}T} = \frac{\Delta U}{RT^2}. \tag{34.33}$$

This equation is also known as the van't Hoff equation or as the *reaction isochore*.

From (34.6) and (34.29) we get

$$\left(\frac{\partial \ln K_x}{\partial T}\right)_P = \frac{\Delta H}{RT^2}. \tag{34.34}$$

† We shall, therefore, omit the index ^{0} which might lead to misunderstanding.

Equations (34.6) and (34.30) give

$$\left(\frac{\partial \ln K_x}{\partial P}\right)_T = -\frac{\Delta V}{RT}. \tag{34.35}$$

This relationship was originally given by Planck and van Laar, and shows [just as eq. (34.6)] that the equilibrium constant K_x depends on pressure only if $\nu \neq 0$.

(b) *Homogeneous solution reactions in the liquid state.* Homogeneous reactions in the liquid phase may be treated by much the same methods as homogeneous gas reactions, i.e. the treatment leads to the law of mass action and the same equations for the dependence of the equilibrium constants on temperature and pressure. We can, therefore, keep our discussion much shorter. The main differences from gas reactions are found in the choice of concentration variables and standard states. From a theoretical point of view the most convenient concentration variable is the mole fraction.† The chemical potential of component i is, therefore, given by

$$\mu_i = RT \ln (f_i x_i) + \mu_i^0(T, P) \tag{34.36}$$

where the quantity $f_i(T, P, x_1, ..., x_{m-1})$ is called the *activity coefficient*.

There are two common definitions of the standard state:

$$(\alpha) \qquad \mu_i^0 = \mu_{i0}(T, P) \quad (i = 1, ..., m), \tag{34.37}$$

where μ_{i0} is the chemical potential of pure component i.

$$(\beta) \qquad \left.\begin{aligned} &\mu_1^0(T, P) = \mu_{10}(T, P), \\ &\mu_i^0(T, P) = \lim_{x_1 \to 1} (\mu_i - RT \ln x_i) \quad (i = 2, ..., m). \end{aligned}\right\} \tag{34.38}$$

Component 1 is called the solvent (cf. § 26). Definition (α) has the advantage of complete symmetry with respect to the components.

Definition (β) is closely connected with the gas theory and must be used when a formulation of the mass action law analogous to that in the gas theory is required. The choice of

† Experimental thermodynamics often uses the following other concentration variables: moles of solute per kg of solvent (molality); moles of solute per litre of solution (molarity). Molarity has the disadvantage that its numerical value changes with temperature.

standard state implies a corresponding standardization of the activity coefficients. We thus have

for (α)

$$\lim_{x_i \to 1} f_i = 1, \tag{34.39}$$

and for (β)

$$\lim_{x_1 \to 1} f_1 = 1, \quad \lim_{x_1 \to 1} f_i = 1 \quad (i = 2, \ldots, m). \tag{34.40}$$

If we introduce the *activities*

$$a_i = f_i x_i \tag{34.41}$$

together with eq. (34.37) into (33.16) we get

$$\Pi a_i^{\nu_i} = K(T, P) \tag{34.42}$$

where

$$RT \ln K = -\sum \nu_i \mu_i^0 = -\Delta G^0. \tag{34.43}$$

Standardization (β) gives for an ideal dilute solution†

$$\Pi x_i^{\nu_i} = K(T, P). \tag{34.44}$$

The dependence of the equilibrium constant on temperature and pressure is now given by

$$\left(\frac{\partial \ln K}{\partial T}\right)_P = \frac{\Delta H^0}{RT^2} \tag{34.45}$$

and, with (26.27), by

$$\left(\frac{\partial \ln K}{\partial P}\right)_T = -\frac{\Delta V^0}{RT} \tag{34.46}$$

where

$$\Delta V^0 = \sum \nu_i v_i^0. \tag{34.47}$$

The definition of standard states for chemical potentials discussed in this section is of general significance for the thermodynamics of liquid mixtures. It can easily be shown to be the basis for the definition of the free energy of dilution (§ 26).

The same formal treatment is usually applied to solid mixed phases for which, however, the definition (α) is always used. For pure solids we have, therefore, that $a_i = 1, f_i = 1$.

† This is a solution to which the equations derived from (34.36) together with (34.38) and (34.40) for $x_1 \to 1$ apply. Formally, these are limiting laws whose applicability to real systems depends on the nature of the system and the accuracy of measurement.

§ 35 Heterogeneous reactions

The general equilibrium condition (33.16) is, as already mentioned, valid for homogeneous as well as heterogeneous equilibria. The considerations of § 34 can, however, not generally be applied to heterogeneous equilibria. We shall confine our discussion to a few observations which clarify the nature of our present problems.

Let us again consider the reaction (33.15) but suppose that now the reactants and products appear not only in the gas phase but also in one or more condensed phases. A chemical reaction in the gas phase may be treated in the same way as in § 34(a). The assumption of ideality, therefore, leads again to eqs. (34.13). Under the given conditions, however, we have to consider for each component not only the chemical equilibrium but also the vaporization equilibrium. While the quotients on the left-hand side of the mass action law are functions of T or of T and P, we now have the situation where *each single partial pressure* p_i is a function of T and P fixed by the vaporization equilibrium. The advantage of the law of mass action for homogeneous reactions is twofold: on the one hand, the value of the equilibrium constants is a measure of the progress of the reaction under given initial conditions; on the other, changes in the initial concentration of individual components can be used to influence the course of the reaction. A well-known example of the latter is the prevention of the dissociation of HI according to

$$2HI \rightleftharpoons H_2 + I_2 \tag{35.1}$$

by the addition of I_2. Many further examples may be found in the methods used in chemical analysis. The two advantages mentioned disappear when the partial pressures appearing in the quotient of the mass action law are equilibrium vapour pressures. The formalism of the mass action law then becomes useless and we must go back to the general equilibrium condition (33.16).

Somewhat different considerations apply when states where certain participants in the reaction occur only in the gas phase are under discussion. Let us choose as our example the *Boudouard equilibrium*.

$$C_{(solid)} + CO_{2(gas)} \rightleftharpoons 2CO_{(gas)}. \tag{35.2}$$

Since μ_c in practice depends only on the temperature we can construct a mass action law expression (assuming ideal gas behaviour) in the form

$$\frac{p_{CO}^2}{p_{CO_2}} = K_p(T). \tag{35.3}$$

The equation for the equilibrium constant is, however, no longer (34.24) but [for eq. (35.2)]

$$RT \ln K_p = \mu_c + \mu^0_{CO_2} - 2\mu^0_{CO}. \tag{35.4}$$

Analogous considerations apply when the carbon is dissolved in solid or liquid iron. From the standardization (α) and eq. (34.36) we get

$$\frac{p^2_{CO}}{p_{CO_2} x_C f_C} = K_p(T), \tag{35.5}$$

where K_p is again defined by eq. (35.4). From eqs. (35.3) and (35.5) we find

$$x_C f_C = \frac{(p^2_{CO}/p_{CO_2})_{(\text{alloy})}}{(p^2_{CO}/p_{CO_2})_{(\text{graphite})}}. \tag{35.6}$$

The oxidation of nickel mentioned in § 33 can be treated in a similar way. For the oxidation of pure nickel we have

$$p_{O_2} = K_p(T) \tag{35.7}$$

with

$$RT \ln K_p = 2\mu_{NiO} - \mu^0_{O_2} - 2\mu_{Ni}. \tag{35.8}$$

If the nickel is in the form of a Ni–Pt alloy, eq. (34.36) with the standardization (α) gives

$$\frac{p_{O_2}}{x^2_{Ni} f^2_{Ni}} = K_p(T), \tag{35.9}$$

where K_p is again given by eq. (35.7). From (35.6) and (35.8) we get

$$x^2_{Ni} f^2_{Ni} = \frac{p_{O_2(\text{alloy})}}{p_{O_2(\text{nickel})}}. \tag{35.10}$$

Equations (35.6) and (35.10) show how activity coefficients of condensed mixed phases may be determined from measurements of heterogeneous equilibria.

When heterogeneous equilibria can be investigated over a considerable range of states, two cases may occur:

(a) A component initially present only in the gas phase reaches a partial pressure which leads to condensation under the appropriate conditions.

(b) A component initially present only in the condensed phase reaches a partial pressure less than the equilibrium vapour pressure so that it disappears from the condensed phase.

In neither case is a representation by means of the mass action law possible over the whole range of states and we have to go back to the general equilibrium condition (33.16).

Our final observation concerns the role of the phase rule in chemical reactions. The enumeration procedure is, in principle, the same as in § 29. We have thus again σ Gibbs–Duhem equations. The summation must now, however, be carried out over all substances (kinds of particle), for which we have r additional reaction equations of the type (33.15) and possibly b further conditional equations. The number of thermodynamic degrees of freedom is, therefore,

$$f = m + 2 - \sigma - r - b. \tag{35.11}$$

Comparison with the definition (2.1) of independent components shows that eq. (35.11) changes into the usual form (29.3) of the phase rule if m is taken to be the number of independent components.

If a reaction goes to completion so that one component disappears, m decreases by one as does r since the appropriate reaction equation need no longer be considered. The number of degrees of freedom therefore remains unchanged.

The only way in which m can decrease more rapidly than r is for the composition of the system to be exactly stoichiometric; f will thus become smaller.

§ 36 The progress variable. Affinity

We still have to discuss briefly a treatment of chemical reactions which is formally somewhat different from that of § 33. The treatment represents the second of the two methods discussed in §§ 16, 17, and 23. We shall exceptionally make use once more of the more general form

$$(\delta U)_{S,V,n} \geqslant 0 \tag{36.1}$$

of the equilibrium condition.

Let us consider a closed homogeneous system in which the reaction

$$\sum \nu_i \mathrm{X}_i = 0 \tag{36.2}$$

can occur. According to the sign convention of § 33 the products have the positive sign. The equilibrium is, therefore,

$$T\delta S - P\delta V + \sum_{i=1}^{m} \mu_i \delta n_i \geqslant 0 \tag{36.3}$$

with the secondary conditions

$$\delta S = 0, \quad \delta V = 0, \quad \delta n_i = 0 \tag{36.4}$$

for all i which do not appear in (36.2).

Only the variations of the mole numbers of participants in the reaction can thus differ from zero. As was shown in § 33, only one of these is independent. As was mentioned in § 14 a new independent variable can be introduced by means of the equation

$$\mathrm{d}n_i = \nu_i \,\mathrm{d}\xi. \tag{36.5}$$

The quantity ξ is called the *progress variable*. The equilibrium condition therefore becomes

$$\left(\sum_i \nu_i \mu_i\right) \delta\xi \geqslant 0. \tag{36.6}$$

When all the participants in the reaction are present in the system, we have $\delta\xi \gtrless 0$ and, according to § 17, the equality sign therefore applies. We therefore have

$$\sum_i \nu_i \mu_i = 0 \quad \text{(all reactants and products present)} \tag{36.7}$$

in agreement with eq. (33.16). If no reaction products are present we have $\delta\xi > 0$ and, therefore, from (36.6) that

$$\sum_i \nu_i \mu_i \geqslant 0 \quad \text{(no products present).} \tag{36.8}$$

If, however, only reaction products are present we have $\delta\xi < 0$ and, therefore,

$$\sum_i \nu_i \mu_i \leqslant 0 \quad \text{(only products present).} \tag{36.9}$$†

The quantity

$$A = -\sum_i \nu_i \mu_i = -\Delta G \tag{36.10}$$

was called the *affinity* of the reaction by de Donder. When all reactants and products are present the equilibrium condition is simply

$$A = 0. \tag{36.11}$$

The above considerations can easily be generalized to include heterogeneous systems. We have to remember that ξ is an internal parameter as defined in § 16 and not a thermodynamic variable of state.

§ 37 The thermodynamic calculation of chemical reactions

Concrete thermodynamic statements about chemical reactions obviously require an explicit knowledge of the relevant thermodynamic functions. These are available nowadays in the form of tables for the majority of homogeneous substances. The methods used

† A clear interpretation of the conditions (36.7)–(36.9) is obtained from (37.3).

for the determination of these functions cannot be discussed here. We shall, therefore, confine our discussion to a few general remarks.

Thermodynamics can make two kinds of fundamental statements about chemical reactions.

(a) For a given reaction, eq. (33.16) gives the values of the quantities of state corresponding to chemical equilibrium.

(b) Since, according to § 23, G for $T = \text{const}$, $P = \text{const}$ is a minimum at equilibrium, we can find, for given values of the quantities of state, whether a reaction is basically possible (i.e. if we neglect possible inhibitions).

The information in (b) is of particular importance for the many chemical reactions which do not result in equilibrium between all the participants in the reaction but go to completion so that at least one component disappears completely. The main examples of such behaviour are found in reactions between pure condensed phases, e.g.

$$2Ag + Hg_2Cl_2 \rightarrow 2AgCl + 2Hg \tag{37.1}$$

and in many reactions of organic compounds. However, cases are also found among heterogeneous reactions involving the gas phase, i.e.

$$TiC + O_2 \rightarrow TiO_2 + C. \tag{37.2}$$

We have quite generally†

$$\left.\begin{array}{lll} \Delta G = 0, & A = 0, & \text{chemical equilibrium,} \\ \Delta G < 0, & A > 0, & \text{reaction possible,} \\ \Delta G > 0, & A < 0, & \text{reaction impossible.} \end{array}\right\} \tag{37.3}$$

Thermodynamic statements about chemical reactions therefore always require a calculation of ΔG and thus a knowledge of the chemical potential of all the participants for the appropriate values of the quantities of state. Let us take a brief look at the most important special cases:

(α) *Pure condensed phases.* The chemical potentials are identical with the molar Gibbs free energies. We have, therefore,

$$\Delta G = \sum_i \nu_i H_i^* - T \sum_i \nu_i S_i^*. \tag{37.4}$$

H_i^* and S_i^* can be found from tables. Often, however, the tables only give the standard values H_i^{*0} and S_i^{*0} at 298·15 K and $P = 1$ atm. But, according to (21.16) we have

$$\left(\frac{\partial \Delta H}{\partial T}\right)_P = \Delta C_P \tag{37.5}$$

† The statements (37.3) constitute a clear explanation of the equilibrium conditions (36.7)–(36.9).

and, according to (25.4),

$$\left(\frac{\partial \Delta S}{\partial T}\right)_P = \frac{\Delta C_P}{T} \tag{37.6}$$

where

$$\Delta C_P = \sum_i \nu_i C_{P_i}. \tag{37.7}$$

We therefore find

$$\Delta G = \Delta H^0 - T\Delta S^0 + \int_{T^0}^{T} \Delta C_P \,\mathrm{d}T - T\int_{T^0}^{T} \frac{\Delta C_P}{T}\,\mathrm{d}T. \tag{37.8}$$

The integrals can be found by using either tables or interpolation formulae. Equation (37.8) is, of course, merely a solution of the Gibbs–Helmholtz equation (24.8). Equation (37.8) may also be written

$$\Delta G = \Delta H^0 - T\Delta S^0 - T\int_{T^0}^{T} \frac{\mathrm{d}T'}{T'^2}\int_{T'^0}^{T'} \Delta C_P \,\mathrm{d}T''. \tag{37.9}$$

This form is easily obtained by partial integration of (37.8) or by direct derivation from (24.10).

A direct experimental determination of ΔG is possible if the reaction can be used in the construction of a reversible galvanic cell (cf. § 72).

(β) *Homogeneous gas reactions.* The same principles as before apply to calculations concerned with homogeneous gas reactions but we must remember that the tables give values of H_i^{*0} and S_i^{*0} calculated for the ideal state at $P = 1$ atm. Corrections for departure from ideality may usually be neglected in chemical equilibrium calculations. We can thus find directly from the tables or by using (37.8) that

$$\Delta G^0 = -RT \ln K_p. \tag{37.10}$$

In principle it is always possible to force a homogeneous gas reaction to occur by a suitable choice of initial partial pressures although in practice limits are set by the available experimental conditions. If we wish to find ΔG for a given initial state we must add $RT \ln p_i$ to the values of μ_i obtained from the tables.

(γ) *Condensed mixed phases.* The calculation of chemical reactions involving condensed mixed phases are attended by much greater difficulties than the cases discussed previously; even for the simplest case, i.e. for reactions in ideal dilute solutions

(§ 34b), the standard state depends on the solvent when standardization (β) (which leads to the mass action law) is used, and systematic tables are not, therefore, available. If we avoid this difficulty by choosing standardization (α) we end up with the generalized mass action law (34.42). The equilibrium constant can now be calculated using the tables but we also need to know the activity coefficients if we want to find ΔG. A systematic tabulation of activity coefficients is, however, not practicable.

§ 38 Nernst's heat theorem. The unattainability of the absolute zero. Zero point entropies

Thermodynamic investigations of heterogeneous reactions led Nernst (1906) to the formulation of a general law now called the *Nernst heat theorem*. Further examination showed, however, that the original formulation of the theorem is not tenable and leads to serious problems whose investigation is possible only on the basis of quantum statistics and is not complete even today. On the other hand, Nernst's heat theorem is of outstanding importance for many aspects of applied thermodynamics. We shall, therefore, emphasize the applications while the problems will be only briefly indicated. A detailed discussion of the problems may be found in textbooks of statistical thermodynamics.

(a) *Nernst's heat theorem*. Let us consider an isothermal–isobaric process taking place between pure condensed phases.† Let the process be symbolized by the expression

$$\alpha \longrightarrow \beta. \tag{38.1}$$

This may be a heterogeneous reaction between pure crystalline solids, e.g.

$$\mathrm{Pb} + \mathrm{I_2} \longrightarrow \mathrm{PbI_2}, \tag{38.2}$$

$$\mathrm{Pb} + \mathrm{Hg_2Cl_2} \longrightarrow 2\mathrm{Hg} + \mathrm{PbCl_2}, \tag{38.3}$$

or a transition between allotropic modifications, such as

$$\mathrm{Sn\ (white)} \longrightarrow \mathrm{Sn\ (grey)}, \tag{38.4}$$

$$\mathrm{S\ (rhombic)} \longrightarrow \mathrm{S\ (monoclinic)}, \tag{38.5}$$

or a liquefaction process, e.g.

$$\mathrm{He\ (solid)} \longrightarrow \mathrm{He\ (liquid)}. \tag{38.6}$$

† i.e. we shall exclude a consideration of mixed phases.

We shall, for purposes of brevity, denote (38.1) quite generally as a reaction. We shall make the following

Assumption 1: No process which changes an internal parameter within a phase will occur with the reaction (38.1).

Experience shows that for certain processes in many substances internal equilibrium is no longer reached at low temperatures. An internal state, i.e. a certain value of the appropriate internal parameter, which corresponds to internal equilibrium at higher temperatures may remain the same at lower temperatures, i.e. it is 'frozen'. The most important examples of phases in frozen equilibrium are certain molecular crystals at very low temperatures (e.g. CO, NO, N_2O, H_2O) and inorganic and organic glasses.† Assumption 1 thus excludes processes associated with the 'thawing' of a frozen equilibrium.

We shall denote the difference between the values of a function of state Z in the states (α) and (β) by

$$\Delta Z \equiv Z(\beta) - Z(\alpha). \tag{38.7}$$

For chemical reactions this notation is more aptly expressed by eqs. (36.10), (37.4), and (37.7).

According to (24.8) we thus have

$$\Delta G = \Delta H + T\left(\frac{\partial \Delta G}{\partial T}\right)_P. \tag{38.8}$$

According to § 37 the solution of this equation may formally be written as

$$\Delta G = \Delta H_0 - T\Delta S_0 + \int_0^T \Delta C_P \, \mathrm{d}T - T\int_0^T \frac{\Delta C_P}{T} \mathrm{d}T, \tag{38.9}$$

or

$$\Delta G = \Delta H_0 - T\Delta S_0 - T\int_0^T \frac{\mathrm{d}T'}{T'^2} \int_0^{T'} \Delta C_P \, \mathrm{d}T'', \tag{38.10}$$

† A phase in frozen equilibrium is not necessarily a metastable phase in the sense of § 18. Ordinary polystyrene, for instance, is incapable of crystallizing. It can, therefore, never be a supercooled liquid. A kind of freezing does, however, occur at a reasonably reproducible temperature and turns the polystyrene into a 'glass'. Silicate glasses and glycerol, however, first form a supercooled liquid in internal equilibrium which freezes only at lower temperatures. In such cases the phase in frozen equilibrium is metastable with respect to the crystalline solid.

where ΔH_0 is the enthalpy and ΔS_0 the entropy of reaction at absolute zero. These quantities are defined by the equations

$$\Delta H = \Delta H_0 + \int_0^T \Delta C_P \, \mathrm{d}T, \tag{38.11}$$

$$\Delta S = \Delta S_0 + \int_0^T \frac{\Delta C_P}{T} \, \mathrm{d}T. \tag{38.12}$$

Since no measurements are possible at and in the immediate neighbourhood of the absolute zero,† the significance of the integrals occurring on the right-hand side of eqs. (38.9)–(38.12) still has to be found. We shall, for this purpose, make

Assumption 2: The integrals

$$\lim_{T^0 \to 0} \int_{T^0}^T C_P \, \mathrm{d}T \equiv \int_0^T C_P \, \mathrm{d}T, \tag{38.13}$$

$$\lim_{T^0 \to 0} \int_{T^0}^T \frac{C_P}{T} \, \mathrm{d}T \equiv \int_0^T \frac{C_P}{T} \, \mathrm{d}T \tag{38.14}$$

exist for all phases participating in the process (38.1). The assumption requires the specific heats to vanish for $T \to 0$‡ and this can be rigorously proved for simple models (e.g. the Debye crystal model) by means of quantum statistics. According to all our knowledge of condensed phases the assumption may be regarded without doubt as generally valid.§ The experimental evidence is, however, not quite as general. This is apparent from the typical example shown in Fig. 23.

Assumption 2, as given above, is merely a formal definition. To be useful it must be supplemented by mathematical rules. The main difficulty which we meet is that in many cases the specific heat in the region $T < 1$ K does not decrease monotonically to zero, a fact which is known both experimentally and theoretically. This knowledge, however, gives no pointer towards a method of extrapolation as can be seen, for example,

† The usual lower limit for which measurements are available is between about 1 K and 20 K.

‡ A sufficient condition for the integrals to converge is that $C_P \propto T^n$ for $T \to 0$, where $n > 0$. It is thus not necessary that $n \geqslant 1$.

§ This is justified even more by the experimental result that ΔH and ΔS are finite. The integrals must, therefore, converge.

from Fig. 23. Since such 'anomalies'† may still be found at temperatures far below those attainable by modern low-temperature techniques‡ an experimentally verifiable rule can be formulated only by completely neglecting these anomalies.§

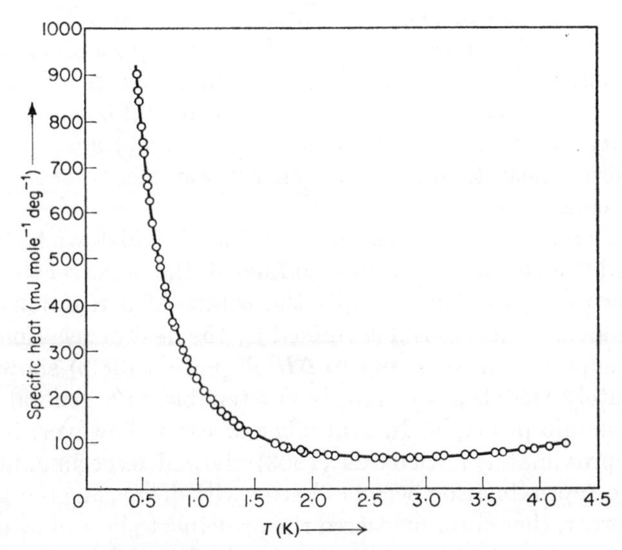

FIG. 23. The atomic heat of Tb at very low temperatures

We therefore make the further

Assumption 3: The decrease to zero of the specific heat, required by assumption 2, may be represented with sufficient accuracy by a simple potential law immediately following the last experimentally established decrease.

The usual practice is to use Debye's T^3 law for non-metallic crystals. For metals the electronic energy is taken into consideration and the expression used is

$$C_P = \alpha T^3 + \gamma T. \tag{38.15}$$

Experimental data and theoretical estimates at ordinary temperatures show that the error introduced by assumption 3

† Such anomalies are to be expected primarily in paramagnetic and ferromagnetic crystalline solids. They may, however, also occur in diamagnetic crystals.

‡ Certain anomalies deduced from theoretical considerations are expected to occur in the region of 10^{-6} K.

§ This is simply a way of 'legalizing' the method used before the development of modern low-temperature physics.

may, in view of the experimental errors involved, be completely neglected in the integral (38.13). The error in the integral (38.14) may be considerably greater and amount to more than 5% in unfavourable cases.

Let us now return to the integration of the Gibbs–Helmholtz equation. Since the quantity ΔH is directly measurable we can calculate ΔH_0 from calorimetric data by means of eq. (31.11). The quantity ΔS_0 then remains as an unknown constant of integration on the right-hand side of (38.9) and (38.10). The Nernst heat theorem is a general statement concerning this integration constant.

Nernst's ideas are related to a principle laid down by Thomsen and Berthelot about the middle of the nineteenth century. According to this principle the course of a reaction between condensed phases is determined by the heat of reaction, i.e. ΔG is supposed to be equal to ΔH. Equation (38.8) shows immediately that this theorem is not tenable as a general thermodynamic principle. In a number of cases, however, it is valid approximately; Richards (1903) showed experimentally that the approximation becomes better with decreasing temperature. Nernst, therefore, suspected the principle to be a limiting law at low temperatures such that not only does ΔG become equal to ΔH as $T \to 0$ [this follows from (38.8)]† but also that the two curves have contact of at least the first order. We therefore have

The Nernst heat theorem: If we accept assumptions 1–3, we have for all isothermal–isobaric processes involving pure condensed phases that

$$\lim_{T \to 0} \frac{\partial \Delta G}{\partial T} = \lim_{T \to 0} \frac{\partial \Delta H}{\partial T}. \qquad (38.16)$$

The behaviour required by (38.16) is shown diagrammatically in Fig. 24.

The quantity on the right-hand side of (38.16) is, according to (37.5), the limiting value of ΔC_P as $T \to 0$; according to assumption 2 the limiting value is zero. With (21.37) we therefore get

The Nernst heat theorem (*alternative formulation*): If assumptions 1–3 are correct, we have for all isothermal–isobaric

† It must be formally assumed that ΔS does not become infinite as $T \to 0$.

processes occurring amongst pure condensed phases that

$$\Delta S_0 = 0. \tag{38.17}$$

This formulation clearly shows the relation to the integration of the Gibbs–Helmholtz equation since ΔG can be calculated from purely calorimetric data if (38.17) is valid.

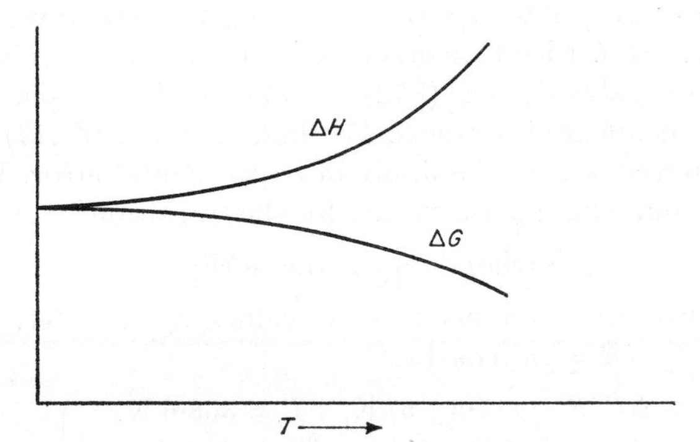

FIG. 24. The Nernst heat theorem

The following comments still need to be made:

(α) The quantities ΔH_0 and ΔS_0 are defined as limiting values obtained by extrapolation of experimental data into the experimentally inaccessible region. They thus have innate physical meaning only in conjunction with the Nernst heat theorem.†

(β) The physical assertion of the Nernst heat theorem may be re-stated in the following way: the quantity ΔS becomes so small at temperatures between 20 K and 1 K that an extrapolation to $T = 0$ done in conformity with assumption 3 yields eq. (38.17) within the limits of experimental error.

† This state of affairs may be made clearer by means of an analogy. The osmotic pressure of a binary solution (§ 28) at very high dilution is given by

$$\Pi = \frac{RT}{M_2} c_g \quad \text{(van't Hoff's law)},$$

where M_2 = molecular weight of solute, c_g = concentration of solute in g cm^{-3}. We have that $\Pi \to 0$ as $c_g \to 0$. The limiting value, $\lim_{c_g \to 0} \Pi/c_g$, is, however, finite and defines the molecular weight M_2 (cf. § 75). This limiting value is not directly measurable but is obtained by extrapolation into an experimentally inaccessible region.

(γ) We excluded mixed phases from the formulation developed above. We are, however, always dealing with mixtures of isotopes. We cannot discuss this problem here but will merely state that all the formulations remain valid as long as the isotopic composition of all the participating elements remains constant during the process (38.1).†

Appropriate methods for an experimental investigation of the Nernst heat theorem are measurements involving reversible galvanic cells (§ 72) and observations of phase changes in condensed systems. We find that eq. (38.17) is always obeyed within the limits of experimental error. We choose as our example the values for the transition

$$\text{S (rhombic)} \rightarrow \text{S (monoclinic)}$$

since we shall need these values again later. We find‡ (R = gas constant)

$$\left.\begin{aligned} \Delta H/R &= (47{\cdot}5 \pm 5)\ \text{K}, \quad T = 368{\cdot}6\ \text{K}, \\ \Delta S/R &= 0{\cdot}12 \pm 0{\cdot}01, \quad T = 368{\cdot}6\ \text{K}, \\ \Delta S_0/R &= 0{\cdot}01 \pm 0{\cdot}05. \end{aligned}\right\} \tag{38.18}$$

(b) *The unattainability of the absolute zero.* Another rule formulated by Nernst in 1912 is closely connected with the heat theorem. The rule is the *principle of the unattainability of the absolute zero*:

It is impossible to cool a system to the absolute zero by means of a finite number of operations of any process.

This theorem must be regarded as a fact of experience independent of the Laws of thermodynamics.§ In order to

† Hydrogen and other substances consisting of mixtures of *ortho-* and *para-*states involve further complications. For a discussion of these complications, the reader is referred to textbooks of statistical thermodynamics.

‡ E. D. Eastman and W. C. McGavock, *J. Am. Chem. Soc.*, **59**, 145 (1937). Later measurements by E. D. West [ibid. **81**, 29 (1959)] give somewhat different values for the transition temperature, ΔH, and ΔS. Since these measurements were not, however, made at low temperatures, we shall not use the results for reasons of consistency.

§ Nernst derived the unattainability principle from the Second Law by means of a cyclic process. This derivation was first criticized by Einstein. The controversies have continued to the present day; we cannot discuss them here. The most apt (though very abstract) representation of the nature of the problem is probably found in Landsberg's axiomatic formulation [*Rev. mod. Phys.*, **28**, 363 (1956)], which bases the unattainability principle on the status of the boundary points of an open set of points; in general, this status must be mathematically clarified. The validity of Nernst's derivation depends on this definition which, however, represents an additional postulate for each case.

make clear the empirical basis and the connexion with the Nernst heat theorem, we shall again consider a process

$$\alpha \longrightarrow \beta \tag{38.19}$$

taking place in condensed phases. We shall, however, use a somewhat different definition of the process from that in subsection (a). It should be obvious that the absolute zero could only be reached by means of an adiabatic process. According to the Second Law, however, the entropy increases during a non-static adiabatic process. Figure 25 shows immediately that a quasi-static adiabatic process would always be

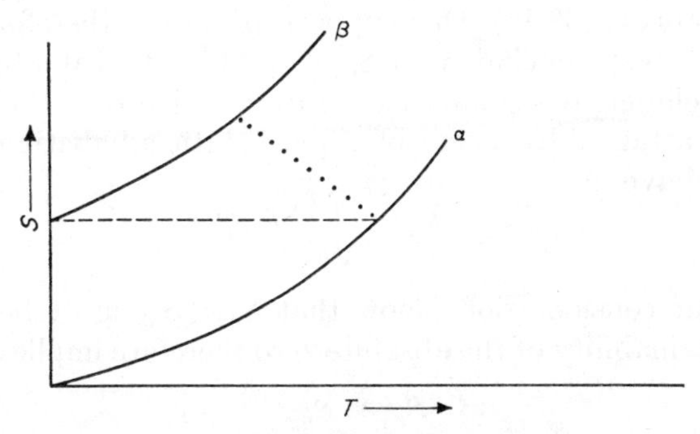

FIG. 25. Unattainability of the absolute zero

more favourable than a non-static adiabatic process for the purpose of reaching the absolute zero. We shall, therefore, assume from now on that the process (38.19) is quasi-static adiabatic, i.e. reversible. In particular, we may be dealing with the change of a work co-ordinate of a homogeneous condensed system, a chemical reaction in a reversible galvanic cell, or a phase transition in the condensed state. We shall, however, ignore the last of these for the present. If we denote the initial temperature by T' and the final temperature by T'', the entropies of the system in the two states are

$$S_\alpha = S_{0\alpha} + \int_0^{T} \frac{C_\alpha}{T} \mathrm{d}T, \tag{38.20}$$

$$S_\beta = S_{0\beta} + \int_0^{T''} \frac{C_\beta}{T} \mathrm{d}T, \tag{38.21}$$

where C_α and C_β are the heat capacities. The equation

$$S_{0\alpha}+\int_0^{T'}\frac{C_\alpha}{T}\,\mathrm{d}T = S_{0\beta}+\int_0^{T''}\frac{C_\beta}{T}\,\mathrm{d}T \tag{38.22}$$

must be true for a quasi-static adiabatic process. For $T'' = 0$, we have, therefore,

$$S_{0\beta}-S_{0\alpha} = \int_0^{T'}\frac{C_\alpha}{T}\,\mathrm{d}T. \tag{38.23}$$

If $S_{0\beta}-S_{0\alpha} > 0$ there must always be a temperature T' which satisfies (38.23), since C_α must be positive,† i.e. starting from a temperature T' the absolute zero can be reached by means of the process (38.19) Our supposition must, therefore, be false and we may conclude that $S_{0\beta} \leqslant S_{0\alpha}$. If state β at a temperature T' is chosen as the initial state from which the absolute zero is to be attained by means of a quasi-static adiabatic process, we must have

$$S_{0\alpha}-S_{0\beta} = \int_0^{T'}\frac{C_\beta}{T}\,\mathrm{d}T. \tag{38.24}$$

Similar considerations show that $S_{0\alpha} \leqslant S_{0\beta}$ must be true. The unattainability of the absolute zero therefore implies that

$$S_{0\alpha} = S_{0\beta}. \tag{38.25}$$

Figure 26, which should require no further explanation, gives a diagrammatic representation of the above considerations.

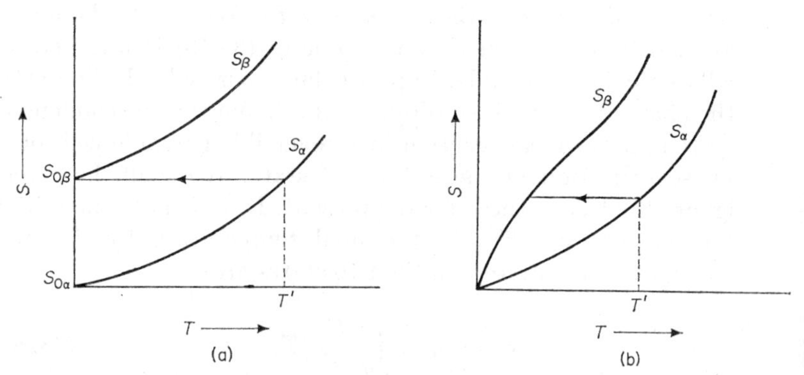

FIG. 26. Entropy as a function of absolute temperature at very low temperatures (schematic). a. The absolute zero is attainable by a quasi-static adiabatic process. b. The absolute zero is unattainable.

† The proof will be given in § 41.

Let us now take a brief look at the three most important special cases.

(α) The process (38.19) consists of a change in a work co-ordinate y.

Equation (38.25) can then be written

$$\lim_{T\to 0}\left(\frac{\partial S}{\partial y}\right)_T = 0, \tag{38.26}$$

or, because of the Maxwell relation analogous to (24.20), as

$$\left(\frac{\partial S}{\partial y}\right)_T = \left(\frac{\partial Y}{\partial T}\right)_y, \tag{38.27}$$

$$\lim_{T\to 0}\left(\frac{\partial Y}{\partial T}\right)_y = 0. \tag{38.28}$$

If we now put $y = V$, $Y = -P$, we get from (25.5), (25.6), and (25.9)

$$\left(\frac{\partial P}{\partial T}\right)_V = \frac{\alpha}{\kappa}. \tag{38.29}$$

If we assume the compressibility to remain finite for $T \to 0$, (38.28) becomes

$$\lim_{T\to 0} \alpha = 0. \tag{38.30}$$

This condition is fulfilled for solids and for liquid helium according to the available experimental data.

If we put $y = \boldsymbol{M}$, $Y = \boldsymbol{H}$, where $\boldsymbol{M}$ is the total magnetization and $\boldsymbol{H}$ the magnetic field strength (cf. § 65), we have

$$\left(\frac{\partial \boldsymbol{H}}{\partial T}\right)_{\boldsymbol{M}} = \frac{(\partial \boldsymbol{M}/\partial T)_{\boldsymbol{H}}}{(\partial \boldsymbol{M}/\partial \boldsymbol{H})_T}. \tag{38.31}$$

If we assume that the numerator on the right-hand side remains finite for $T \to 0$, we get from (38.28)

$$\lim_{T\to 0}\left(\frac{\partial \boldsymbol{M}}{\partial T}\right)_{\boldsymbol{H}} = 0. \tag{38.32}$$

This condition is also fulfilled according to the experimental data.

The proof of eqs. (38.30) and (38.32) forms the real empirical basis of the unattainability principle. Equation (38.32) is, however, by far the most important since it is related to the process of adiabatic demagnetization mentioned in § 12, the only known way of generating extremely low temperatures (cf. § 66). A theoretical analysis of the molecular mechanism (which we cannot discuss here) leads to the same result, i.e. that the absolute zero cannot be reached in this way.

(β) The process (38.19) consists of a chemical reaction between pure condensed phases.

Equation (38.25) is now identical with the Nernst heat theorem (38.17). The requirement that the process (38.19) shall take place in a quasi-static adiabatic manner implies assumption 1 since the 'thawing' of a frozen equilibrium is necessarily an adiabatic irreversible process.

(γ) The process (38.19) is a phase change in a condensed system.

A reversible transition is possible only along the co-existence curve and the above equations are no longer directly applicable. A recent careful examination of the problem by Haase, however, leads to an analogous result with the assumption that the co-existence curve at finite pressures extends as far as $T = 0$. This condition is, for example, fulfilled for the solid–liquid equilibrium of helium but has not been established for the previously mentioned transition S (rhombic) $\rightarrow$ S (monoclinic). In the latter case, the Nernst heat theorem can thus not be derived from the unattainability principle although the experimental data show it to be valid.

The unattainability principle thus goes further than the Nernst heat theorem in one respect, since it requires the existence of a zero point entropy independent of the work co-ordinates for all homogeneous substances in the condensed state; on the other hand, the principle does not cover the whole region of applicability of the Nernst heat theorem since the latter can be deduced for phase changes only with certain restrictive conditions. The Nernst heat theorem leads to the conclusion that the zero point entropy is independent of the crystalline modification; this conclusion cannot, however, be obtained with general validity from the unattainability principle which cannot, therefore, be simply regarded as equivalent to the Nernst heat theorem but must be treated as an independent theorem of experience.

(c) *Zero point entropies.* One of the most important applications of the rules discussed in (a) and (b) is the construction of the tables of thermodynamic functions mentioned in § 37(α). A brief discussion of this will also serve to amplify our previous considerations in certain respects.

Let us first of all consider enthalpy. The function H contains, according to the First Law, an arbitrary additive constant. Only the difference

$$\Delta H = H_\beta - H_\alpha \tag{38.33}$$

is measurable, where α and β denote two equilibrium states of the system which are connected by means of an arbitrary process.† If, therefore, we take α to be a standard state to be defined later for which the additive constant is fixed, it is also fixed by (38.33) for all states which can be reached from α by any process. It is therefore sufficient to define the additive constant for each element in a standard state. The following conventions are now generally accepted and form the basis of the tables:

Standard state for enthalpy:

$$T = 298{\cdot}15\ \mathrm{K}, \quad P = 1\ \mathrm{atm}. \tag{38.34}$$

Standardization of the enthalpy: For each element in its stable form in the standard state (38.34) we have

$$H^{*0} = 0. \tag{38.35}$$

The standard enthalpies of all chemical compounds thus become numerically equal to the enthalpies of formation defined by (38.33). These enthalpies are experimentally accessible. Comprehensive tabulations of standard enthalpies are available. Other untabulated enthalpies must be calculated by the methods outlined in § 37(α). The zero point enthalpy H_0 is obtained from standard values in an analogous way. For molar quantities we have

$$H^{*0} = H_0^* + \int_0^{298\cdot15} \Delta C_P \, \mathrm{d}T \quad (P = 1\ \mathrm{atm}). \tag{38.36}$$

We have the same basic situation for the case of entropy. According to the Second Law, S contains an arbitrary constant, and only the difference

$$\Delta S = S_\beta - S_\alpha \tag{38.37}$$

† We exclude any kind of nuclear reaction both here and later.

is experimentally accessible. The standardization (38.35) is, however, not convenient for entropy since the entropy of formation is not directly measurable† and difficulties arise when the Nernst heat theorem has to be taken into consideration. The following conventions have, therefore, been adopted:

Standard state for entropy:

$$T = 298{\cdot}15\ \mathrm{K}, \quad P = 1\ \mathrm{atm}. \tag{38.38}$$

Standardization of the entropy: For each element in its stable condensed form as $T \to 0$ we have

$$S_0^* = 0, \quad (T \to 0). \tag{38.39}$$

This standardization is *independent of Nernst's heat theorem.* The standardization means that the zero point entropy for all chemical compounds is numerically equal to the entropy of formation at the absolute zero. But we have, for any homogeneous substance, that

$$S^{*0} = S_0^* + \int_0^{298{\cdot}15} \frac{C_P}{T}\,\mathrm{d}T. \tag{38.40}$$

The only experimentally accessible quantity in this equation is the integral on the right-hand side. The standardization (38.39) is therefore not sufficient to determine the standard entropy S^{*0} numerically.

This difficulty is overcome by using the Nernst heat theorem. This use explains the extraordinary importance of the theorem in thermodynamic applications. Equation (38.17) and the unattainability principle together with the standardization (38.39) lead the formulation first put forward by Planck:

The Nernst heat theorem (*Planck's formulation*): For all homogeneous substances forming stable condensed phases we have

$$S_0^* = 0 \tag{38.41}$$

irrespective of the values of the work co-ordinates, the crystalline modification, and the state of aggregation. The standard entropy S^{*0} can be calculated from purely calorimetric data by means of (38.40) and (38.41). The calculation of the entropy for other conditions has been outlined in § 37(α). The quantity S_0^* is called the *conventional zero point entropy*.

† $\Delta G = 0$ and thus $\Delta H/T = \Delta S$ only at equilibrium.

We still have to make some remarks concerning the formulation (38.41). Giauque has shown that it makes possible a much more exact experimental verification than earlier formulations. We make the assumption that the substance under consideration may be transformed from the crystalline state into the dilute gas state.† We then have for the entropy of the gas

$$S_g^*(T,P) - S_0^* = \lim_{T^0 \to 0} \int_{T^0}^{T} \frac{dQ}{T}, \tag{38.42}$$

where dQ is the amount of heat introduced including any heats of transition. The integration path is a reversible path from crystalline solid at T^0 and $P \to 0$ to gas at temperature T and pressure P. The passage to the limit $T^0 \to 0$ must be carried out in the manner explained under subsection (a), assumption 3. The right-hand side of eq. (38.42) can then be determined by calorimetric measurements. The quantity $S_g^*(T,P) - S_0^*$ is therefore called the *calorimetric entropy*.‡ The quantity $S_g^*(T,P)$ for a dilute gas can be shown by the methods of statistical thermodynamics to be calculable from spectroscopic data if certain requirements are met.§ It is therefore called the *spectroscopic entropy*. Equation (38.42) shows that a comparison of calorimetric and spectroscopic entropies constitutes a direct test of eq. (38.41).

Equation (38.42) has been used to investigate about thirty substances (diatomic and simple polyatomic molecules). In most cases eq. (38.41) was found to be obeyed within the limits of experimental error.|| Certain substances, however, show definite deviations from eq. (38.41). This proves that Planck's formulation of the Nernst heat theorem must not be regarded as a universally exact statement. However, the deviations are usually associated with frozen equilibria of molecular crystals, a fact which has already been mentioned in subsection (a) in connexion with assumption 1. We have, in fact, ignored

† This assumption is not valid for high melting crystals (e.g. C, Al_2O_3, TiC) which have no measurable vapour pressure nor does it apply to organic compounds which decompose on heating (e.g. sucrose).

‡ This concept is obviously not confined to the gaseous state.

§ Details can be found in textbooks of statistical thermodynamics and of molecular spectroscopy. We may note, however, that the requirements are not met, for instance, in polyatomic molecules with internal rotations (e.g. ethane).

|| The difference between calorimetric and spectroscopic entropies was usually less than 0·5%.

assumption 1 in the formulation of (38.41). Two ways of changing (38.41) into an exact rule have been proposed. One is to additionally specify internal equilibrium for the substance as $T \to 0$, the second to replace the zero on the right-hand side by a finite positive quantity $R \ln o$. The defect in the first way is that the concept of internal equilibrium makes sense only with respect to defined processes.† Comparison of calorimetric and spectroscopic entropies shows that, for the second method, either $o = 1$ or o is of the order of magnitude 2. This comparison can, however, be done only with a relatively small number of substances. Apart from this, we have to rely on supposition. In practice, therefore, the only general method is to rely on eq. (38.41) and to accept that the normal entropies calculated in this way are uncertain to an extent $R \ln 2$. This method finds further justification in the fact that the uncertainties inherent in the extrapolation $T \to 0$ [subsection(a), assumption 3] are of the same order of magnitude. This order of magnitude is insignificant for most applications. The formulation as a 'rule'‡ obeyed in practice with sufficient accuracy not only corresponds best to the nature of the Nernst heat theorem as obtained from the consideration of subsection (a), but also accords with modern theories of statistical thermodynamics. The investigations of Klein and Casimir have shown that previous attempts to formulate a 'Third Law' on the basis of quantum statistics are untenable. We are still far from understanding the Nernst heat theorem on a statistical basis and only this could form the foundation of an exact formulation.

† Even with the limitation to defined processes, the requirement of internal equilibrium makes the practical use of eq. (38.41) illusory. This is best illustrated by means of the following example: at room temperature TiO crystals have a concentration of unoccupied sites of about 15% and this concentration is reasonably reproducible. We do not, however, know of any method for deciding whether the concentration and spatial distribution of empty sites for $T \to 0$ corresponds to internal equilibrium.

‡ The most important example of such a 'rule', which is not exact but of great practical significance, is the 'principle of corresponding states'. This postulates a universal thermal equation of state for real gases when the equation is expressed in terms of dimensionless ('reduced') quantities of state (e.g. T/T_c, P/P_c, V/V_c, where the subscript c refers to the critical point). Statistical thermodynamics shows that this principle cannot be exact. It is, however, obeyed with sufficient accuracy for many substances.

CHAPTER VI

Stability conditions

§ 39 Statement of the problem. Gibbs' criterion

We have so far, with certain exceptions (e.g. in § 37), used the Second Law only in the form of the statement that the thermodynamic potentials have a stationary value at equilibrium. As has been mentioned briefly in §§ 18 and 23, the stability conditions are contained in the statement that this stationary value is a minimum. The object of the present chapter is to perform for the stability conditions a task analogous to that performed for the equilibrium conditions in Chapters IV and V, i.e. to draw explicit conclusions from the general formulation of the stability conditions in §§ 18 and 23 together with the fundamental equation. We have thus defined the boundaries of our present subject. Formally, we have to investigate variations of higher order for the thermodynamic potentials. Within the framework of thermodynamics an explicit treatment needs to consider, exactly as we did for the equilibrium conditions, only virtual displacements which can be represented by means of quantities of state. This limitation allows, for the equilibrium conditions of homogeneous systems, only the treatment of equilibria which may be represented by internal parameters. Even though we exclude internal equilibria, stability conditions for homogeneous equilibria immediately present us with a non-trivial problem. We shall show that all stability problems other than those concerning chemical equilibria may be referred to one basic problem, i.e. the stability of a homogeneous phase. The question of the stability of a chemical equilibrium is relatively simple and a few later remarks about it will be sufficient. For the present we shall exclude chemical reactions.†

† For simplicity we shall use the expression 'chemical reaction' here, and later, to include all processes which can be represented by means of internal parameters.

Let us, therefore, consider a homogeneous system of m components in the absence of chemical reactions; let this system be at thermodynamic equilibrium. The limitation discussed above still allows us to ask the following question:

What conditions must be obeyed by the system to prevent it from becoming heterogeneous owing to arbitrarily small local disturbances, i.e. prevent it from separating into at least two (macroscopic) phases which differ from the original phase?

This question forms the basis of thermodynamic stability theory as long as chemical reactions are excluded. We may well ask whether this limitation prejudices the general validity of our deductions. This is, however, not so since the concept of stability like that of equilibrium can have a physical meaning only with respect to defined processes. The general validity would become questionable only if there were virtual displacements which could be represented by quantities of state and were not contained in the above formulation or, at least, could not be reduced to it.

The above formulation still leaves two possibilities open:

(a) The new phases differ from the original phase only by infinitesimal changes in the quantities of state (neighbouring phases).

(b) The new phases differ from the original phase by finite (discontinuous) changes in the quantities of state.

Possibility (b) leads to the concept of metastable equilibrium briefly mentioned in § 18. If the phase diagram is known we can, in this way, find out which phases are metastable with respect to others. General rules, however, cannot be found in this way. We shall, therefore, confine our attention to case (a) which refers directly to the principle of virtual displacements and forms the real content of thermodynamic stability theory.

As in Chapter II we shall assume that the system under consideration is closed and that the values of the work co-ordinates are fixed. We shall denote the original phase by $'$ and the two phases formed by virtual displacements by * and **. According to our assumptions and eq. (20.43) we have for the original phase that

$$U' - TS' + PV' - \sum_{i=1}^{m} \mu_i n_i' = 0. \tag{39.1}$$

We shall assume that we can write for the two new phases

$$\left.\begin{aligned} U^* - TS^* + PV^* - \sum_{i=1}^{m} \mu_i n_i^* \geqslant 0, \\ U^{**} - TS^{**} + PV^{**} - \sum_{i=1}^{m} \mu_i n_i^{**} \geqslant 0. \end{aligned}\right\} \quad (39.2)\dagger$$

We further have the secondary conditions

$$\left.\begin{aligned} S^* + S^{**} &= S', \\ V^* + V^{**} &= V', \\ n_i^* + n_i^{**} &= n_i' \quad (i = 1, ..., m). \end{aligned}\right\} \quad (39.3)$$

If the equality sign is operative in (39.2) we get, with (39.1) and (39.3),

$$U^* + U^{**} = U'. \quad (39.4)$$

The formation of the new phases thus does not change the internal energy as long as the secondary conditions (39.3) are obeyed. According to eq. (18.6) this means that the equilibrium is neutral with respect to the formation of the new phases, i.e. that they can co-exist with the original phase.

If the inequality sign is operative in (39.2), we have

$$U^* + U^{**} > U'. \quad (39.5)$$

Since the secondary conditions (39.3) apply the internal energy now increases if the new phases are formed. According to the Second Law such a process cannot be spontaneous. According to eq. (18.2) the original phase is thus stable with respect to the new phases. If, finally, $>$ is replaced by $<$ in (39.2) we get

$$U^* + U^{**} < U'. \quad (39.6)$$

With the secondary conditions (39.3) the formation of the new phase decreases the internal energy. According to (18.4) the original phase is, therefore, unstable with respect to the new phases. The original homogeneous system can, therefore, not exist as such but would immediately become heterogeneous.

We thus have the *theorem:* If, in the expression

$$U - TS + PV - \sum_{i=1}^{m} \mu_i n_i \quad (39.7)$$

† The quantities with asterisks are here not mean molar quantities.

the intensive parameters T, P, μ_i can be given values such that the expression is zero for the phase under consideration and positive for all other phases (consisting of the same components) then the phase under consideration is absolutely stable with respect to separation into other phases. If the expression (39.7) is zero for certain other phases as well, the phase under consideration is in neutral equilibrium with respect to these other phases; they can co-exist with the original phase and T, P, μ_i are the common values of temperature, pressure, and chemical potentials required by the equilibrium conditions.

This theorem constitutes *Gibbs' stability criterion.*

Two points arising from it merit some further discussion. The fact that the intensive parameters have the same values as those of the original phase may seem arbitrary for statement (39.2). Actually the statement follows directly from the supposition that the original phase is in thermodynamic equilibrium, since, according to § 17, the first-order variations for all virtual displacements must then vanish. Since it is in principle of no consequence whether a phase is already present or is formed by the displacement, the method of § 27 is directly applicable. We thus find that the necessary and sufficient condition for the first-order variations to vanish is the equality of the intensive parameters for all phases (real or virtual) (cf. § 40).

Furthermore, we should note that Gibbs' stability criterion constitutes a very general and, therefore, necessarily a very abstract formulation. [It also includes metastable equilibria since we have not explicitly used the limitation to the possibility (a).] Its significance is twofold: it is obtained directly from the Second Law by means of simple considerations and the derivation of all other general and special stability conditions is possible by purely mathematical methods.

§ 40 The stability conditions in the energy representation

From now on we shall confine our attention to the possibility (a) mentioned in § 39. Let us look at an example which should clarify the relations between the general formulations. We can ignore the case of neutral equilibrium (co-existence of neighbouring phases) since we are dealing only with neighbouring phases. Let us, therefore, take a homogeneous one-component system at thermodynamic equilibrium and investigate the virtual separation into two new phases with volumes $\frac{1}{2}V$, entropies $\frac{1}{2}(S+\delta S)$, $\frac{1}{2}(S-\delta S)$, and the mole numbers

$\frac{1}{2}(n+\delta n)$, $\frac{1}{2}(n-\delta n)$. The statements (39.2) now appear in the form

$$\left.\begin{aligned} U^* - \tfrac{1}{2}T(S+\delta S) + \tfrac{1}{2}PV - \tfrac{1}{2}\mu(n+\delta n) > 0, \\ U^{**} - \tfrac{1}{2}T(S-\delta S) + \tfrac{1}{2}PV - \tfrac{1}{2}\mu(n-\delta n) > 0. \end{aligned}\right\} \tag{40.1}$$

The secondary conditions (39.3) have already been included in (40.1). We therefore find that

$$U^* = \frac{1}{2}\left[U + \frac{\partial U}{\partial S}\delta S + \frac{\partial U}{\partial n}\delta n + \frac{1}{2}\frac{\partial^2 U}{\partial S^2}(\delta S)^2 + \frac{\partial^2 U}{\partial S\,\partial n}\delta S\,\delta n + \frac{1}{2}\frac{\partial^2 U}{\partial n^2}(\delta n)^2 + \ldots\right] \tag{40.2}$$

and, correspondingly,

$$U^{**} = \frac{1}{2}\left[U - \frac{\partial U}{\partial S}\delta S - \frac{\partial U}{\partial n}\delta n + \frac{1}{2}\frac{\partial^2 U}{\partial S^2}(\delta S)^2 + \frac{\partial^2 U}{\partial S\,\partial n}\delta S\,\delta n + \frac{1}{2}\frac{\partial^2 U}{\partial n^2}(\delta n)^2 - \ldots\right]. \tag{40.3}$$

If eqs. (40.2) and (40.3) are added together, the first-order terms cancel. If the result of the addition is introduced into (40.1) and the result subtracted from (39.1), we find that

$$U^* + U^{**} > U' \tag{40.4}$$

if, and only if,

$$\frac{\partial^2 U}{\partial S^2}(\delta S)^2 + 2\frac{\partial^2 U}{\partial S\,\partial n}\delta S\,\delta n + \frac{\partial^2 U}{\partial n^2}(\delta n)^2 > 0. \tag{40.5}$$

The stability condition thus states that the quadratic form on the left-hand side of (40.5) must be positive-definite. We know from the theory of quadratic forms that this is possible if, and only if,

$$\begin{vmatrix} \dfrac{\partial^2 U}{\partial S^2} & \dfrac{\partial^2 U}{\partial n\,\partial S} \\[2ex] \dfrac{\partial^2 U}{\partial S\,\partial n} & \dfrac{\partial^2 U}{\partial n^2} \end{vmatrix} \quad \text{with all principal minors} > 0. \tag{40.6}$$

This is the form of the stability condition most convenient for our further discussion.

Before we go into further detail, we should derive the condition (40.6) in a general form and for systems of m components.

According to Gibbs' criterion the inequality

$$U^* - TS^* + PV^* - \sum_{i=1}^{m} \mu_i n_i^* > 0 \tag{40.7}$$

must be true for each phase neighbouring to that under consideration. If eq. (39.1) is subtracted from (40.7) we find that

$$\Delta U > T\Delta S - P\Delta V + \sum_{i=1}^{m} \mu_i \Delta n_i \tag{40.8}$$

for the virtual displacements; the symbol Δ here represents a variation of any order, and we have $\Delta x \to \delta x$ for the independent variables. We choose S, V, n_i as independent variables. According to the fundamental equation we then have

$$\delta U = T\delta S - P\delta V + \sum_{i=1}^{m} \mu_i \delta n_i. \tag{40.9}$$

First-order variations thus cancel in (40.8). We shall assume, for the time being, that second-order terms are not zero. Compared with the second-order terms, all later terms are then small of higher order and can be neglected. We therefore have

$$\delta^2 U(S, V, n_i) > 0. \tag{40.10}$$

The discussion of the fundamental equation has, however, already shown (§ 20) that the thermodynamics of a homogeneous system needs only $m+1$ independent variables. The above example shows explicitly that one of the independent variables (namely V) is used merely to describe the ratio of the amounts of the phases formed in the virtual displacement. This independent variable does not, however, play any part in the stability condition since the ratio of quantities has obviously nothing whatsoever to do with the stability problem. This problem is met quite simply by keeping V, n_m or $\sum n_i$ constant in (40.8) and (40.9). We shall choose the last of these variables which means (cf. § 20) that we write the fundamental equation for (mean) molar quantities and that we use $m-1$ mole fractions instead of m mole numbers as independent variables. Thus (40.10) becomes (in the notation of § 26)†

$$\delta^2 U^*(S^*, V^*, x_i) > 0. \tag{40.11}$$

The stability condition thus tells us that the second-order variation (which is, of course, a quadratic form) must be positive-definite. This

† From here to the end of the section, the asterisk will indicate mean molar quantities.

is possible if, and only if,

$$\begin{vmatrix} \dfrac{\partial^2 U^*}{\partial S^{*2}} & \cdots & \dfrac{\partial^2 U^*}{\partial x_{m-1}\,\partial S^*} \\ \dfrac{\partial^2 U^*}{\partial S^*\,\partial V^*} & \cdots & \cdots \\ \vdots & \vdots & \vdots \\ \dfrac{\partial^2 U^*}{\partial S^*\,\partial x_{m-1}} & \cdots & \dfrac{\partial^2 U^*}{\partial x_{m-1}^2} \end{vmatrix} \quad \text{with all principal minors} > 0. \tag{40.12}$$

This is the general formulation of the stability conditions in the energy representation. We can see that the stability conditions constitute statements about the signs of the second derivatives of the characteristic functions, and that these statements have to be fulfilled for every stable phase. Furthermore, (40.12) shows that there are $m+1$ independent stability conditions for a homogeneous system of m components.

Let us look explicitly at two special conditions arising from (40.12). We have immediately that

$$\left(\frac{\partial^2 U^*}{\partial S^{*2}}\right)_{V^*,x} > 0; \tag{40.13}$$

but, according to the fundamental equation,

$$\left(\frac{\partial^2 U}{\partial S^{*2}}\right)_{V^*,x} = \left(\frac{\partial T}{\partial S^*}\right)_{V^*,x}. \tag{40.14}$$

The expression (40.13) can, by using (25.7), therefore be written

$$\left(\frac{\partial S^*}{\partial T}\right)_{V^*,x} > 0. \tag{40.15}$$

However, from the fundamental equation and the definition (20.21a) we get

$$T\left(\frac{\partial S^*}{\partial T}\right)_{V^*,x} = \left(\frac{\partial U^*}{\partial T}\right)_{V^*,x} = C_V. \tag{40.16}$$

Since T is positive by definition we finally get

$$C_V > 0. \tag{40.17}$$

For a stable phase, therefore, the molar heat at constant volume is always positive. The expression (40.17) is called the *thermal stability condition.*

Since the order of the variables can be chosen arbitrarily in the construction of the determinant (40.12), it must also be true that

$$\left(\frac{\partial^2 U^*}{\partial V^{*2}}\right)_{S^*,x} > 0. \tag{40.18}$$

With the fundamental equation and the relationship (25.7), this becomes

$$-\left(\frac{\partial V^*}{\partial P}\right)_{S^*,x} > 0. \tag{40.19}$$

The quantity

$$\kappa_S = -\frac{1}{V}\left(\frac{\partial V}{\partial P}\right)_{S,x} \tag{40.20}$$

is called the *adiabatic* or, more correctly, the *isentropic compressibility*. We have, therefore, that

$$\kappa_S > 0, \tag{40.21}$$

i.e. the isentropic compressibility is always positive for a stable phase. The expression (40.21) is called the *mechanical stability condition.*

The derivation of further explicit statements (of more than merely formal interest) is possible, but complicated because of the structure of the fundamental equation discussed in § 21. We are thus faced with a problem, analogous to that of § 23, of expressing stability conditions by means of Legendre transforms of the fundamental equation.

§ 41 Transformation of the stability conditions

We now wish to express the stability conditions with the aid of an arbitrary characteristic function. For simplicity we shall confine our calculations to the energy representation and only give the result for the entropy representation (arrived at in an analogous manner).

According to the general concepts developed in § 39, the virtual phase is also in a thermodynamic state, i.e. in internal equilibrium. The fundamental equation must, therefore, apply but the intensive parameters are no longer those of the original phase [as in (40.7)] but are determined by the fundamental equation of the virtual phase. We therefore have the integrated form.†

$$U^* = T^* S^* - P^* V^* + \sum_{i=1}^{m} \mu_i^* n_i^*. \tag{41.1}$$

† From here on, the asterisk again denotes the virtual phase.

The condition (40.7) can therefore be written

$$-TS^* + PV^* - \sum_{i=1}^{m} \mu_i n_i^* + T^* S^* - P^* V^* + \sum_{i=1}^{m} \mu_i^* n_i^* > 0. \quad (41.2)$$

If the 'asterisked' phase is in internal equilibrium according to (41.1), we may regard it as the original phase and consider its stability with respect to a spontaneous formation of the 'unasterisked' phase. This condition is obtained directly from (41.2) by interchanging the asterisked and unasterisked quantities. We have, therefore,

$$-T^* S + P^* V - \sum_{i=1}^{m} \mu_i^* n_i + TS - PV + \sum_{i=1}^{m} \mu_i n_i > 0. \quad (41.3)$$

Addition of (41.2) and (41.3) gives

$$\Delta T \Delta S - \Delta P \Delta V + \sum_{i=1}^{m} \Delta \mu_i \Delta n_i > 0. \quad (41.4)$$

The expression on the left is, however, simply the second-order variation $\delta^2 U$. The inequality (41.4) therefore merely constitutes a re-arrangement of the condition (40.10) in a form convenient for the transformation.

Let us introduce, as we did in § 21, the generalized quantities of state X_i and P_i. We can now write for the stability condition (41.4)

$$\sum_{i=1}^{r} \Delta P_i \Delta X_i > 0. \quad (41.5)$$

We now want to express this abstract formulation with the aid of the second derivatives of the thermodynamic potential

$$\Psi_k = \Psi_k(P_i, X_j) \quad (i = 1, \ldots, k;\ j = k+1, \ldots, r). \quad (41.6)$$

For this purpose we have to represent the ΔX_s $(s = 1, \ldots, k)$ and the ΔP_t $(t = k+1, \ldots, r)$ as functions of the P_i and X_j. Since, in (41.5) we only go as far as the second order, we obtain the set of linear equations

$$\left.\begin{aligned} \Delta X_s &= \sum_{i=1}^{k} \frac{\partial X_s}{\partial P_i} \Delta P_i + \sum_{j=k+1}^{r} \frac{\partial X_s}{\partial X_j} \Delta X_j, \\ \Delta P_t &= \sum_{i=1}^{k} \frac{\partial P_t}{\partial P_i} \Delta P_i + \sum_{j=k+1}^{r} \frac{\partial P_t}{\partial X_j} \Delta X_j. \end{aligned}\right\} \quad (41.7)$$

According to eq. (21.9) we have

$$\frac{\partial X_s}{\partial P_i} = -\frac{\partial^2 \Psi_k}{\partial P_s \partial P_i} \quad (41.8)$$

and according to (21.10)

$$\frac{\partial P_l}{\partial X_j} = \frac{\partial^2 \Psi_k}{\partial X_l \partial X_j}. \tag{41.9}$$

Furthermore, according to Maxwell's relation (24.11) we have

$$\frac{\partial X_s}{\partial X_j} = -\frac{\partial P_l}{\partial P_i}. \tag{41.10}$$

If we now introduce (41.7) into (41.5) we get a quadratic form in ΔP_i and ΔX_j whose coefficients are the second derivatives of Ψ_k. From the eqs. (41.8)–(41.10) we can extract directly the following rules:

(a) All terms of the form $\Delta P_i \Delta X_j$, i.e. terms made up of mixed intensive and extensive parameters, cancel out. The original expression therefore separates into two quadratic forms, one dependent only on intensive, the other only on extensive parameters.

(b) The quadratic form for intensive parameters has the negative sign, that for extensive parameters the positive sign.

We therefore get

$$[\delta^2 \Psi_k(X_j)]_{P_i} - [\delta^2 \Psi_k(P_i)]_{X_j} > 0. \tag{41.11}$$

This inequality can be fulfilled if, and only if, the first quadratic form is positive-definite, the second negative-definite. If we further eliminate variations which merely change the mass of the virtual phase by putting $X_r = \text{const}$, we get the stability conditions in the form

$$[\delta^2 \Psi_k(P_1, \ldots, P_k)]_{X_j} < 0, \tag{41.12}$$

$$[\delta^2 \Psi_k(X_{k+1}, \ldots, X_{r-1})]_{P_i, X_r} > 0 \tag{41.13}$$

valid for all thermodynamic potentials. The equivalent and, in practice, more important formulation is

$$\left| \frac{\partial^2 \Psi_k}{\partial P_i \partial P_s} \right| \begin{cases} < 0 & \text{for the principal minors of odd order,} \\ > 0 & \text{for the principal minors of even order,} \end{cases} \tag{41.14}$$

$$\left| \frac{\partial^2 \Psi_k}{\partial X_j \partial X_l} \right| \quad \text{with all principal minors} > 0. \tag{41.15}$$

The above method of derivation is due to Gibbs while the general formulation (41.12), (41.13) as well as the rules (a) and (b) were first given by Schottky, Ulich, and Wagner.

Let us examine the most important special case of the conditions (41.14) and (41.15), i.e. the Gibbs free energy, a little more closely.

The conditions (41.14) are here explicitly (for mean molar quantities)†

$$\begin{vmatrix} \dfrac{\partial^2 G^*}{\partial T^2} & \dfrac{\partial^2 G^*}{\partial P\,\partial T} \\ \dfrac{\partial^2 G^*}{\partial T\,\partial P} & \dfrac{\partial^2 G^*}{\partial P^2} \end{vmatrix} > 0, \quad \frac{\partial^2 G^*}{\partial T^2} < 0, \quad \frac{\partial^2 G^*}{\partial P^2} < 0. \tag{41.16}$$

It can easily be seen that only two of these conditions are independent, e.g. the third inequality follows from the first two.

The second inequality with (26.7) and (25.4) gives

$$C_P > 0, \tag{41.17}$$

i.e. the molar heat at constant pressure is always positive for a stable phase (thermal stability).

The third inequality with (25.6) gives

$$\kappa > 0, \tag{41.18}$$

i.e. the isothermal compressibility is always positive for a stable phase (mechanical stability).

The first inequality with (26.7) and (25.4)–(25.6) and (25.30) gives

$$\frac{\partial^2 G^*}{\partial T^2}\frac{\partial^2 G^*}{\partial P^2} - \left(\frac{\partial^2 G^*}{\partial T\,\partial P}\right)^2 = \frac{\kappa v C_V}{T} > 0 \tag{41.19}$$

which, with (41.18), gives

$$C_V > 0. \tag{41.20}$$

This has already been derived in § 20 by the use of internal energy.

The above inequalities yield some further noteworthy conclusions. From (41.17), (41.18), and (41.20) with eq. (25.30) we get

$$C_P > C_V. \tag{41.21}$$

The molar heat capacity at constant pressure is always greater than the molar heat capacity at constant volume for a stable phase if $\alpha \neq 0$. This condition is not fulfilled for, e.g., water at 4 °C in which case $\alpha = 0$.

By means of the identities [cf. eq. (25.9)]

$$\left(\frac{\partial V^*}{\partial P}\right)_{S^*} = -\frac{(\partial S^*/\partial P)_{V^*}}{(\partial S^*/\partial V^*)_P}, \tag{41.22}$$

$$\left(\frac{\partial V^*}{\partial P}\right)_T = -\frac{(\partial T/\partial P)_{V^*}}{(\partial T/\partial V^*)_P}, \tag{41.23}$$

† From here on the asterisk again indicates mean molar quantities.

we obtain from the definitions (25.6) and (40.20) together with (25.8)

$$\frac{\kappa_S}{\kappa} = \frac{(\partial S^*/\partial T)_V}{(\partial S^*/\partial T)_P}. \tag{41.24}$$

This, with (24.4) and (40.16) becomes

$$\frac{\kappa_S}{\kappa} = \frac{C_V}{C_P}. \tag{41.25}$$

Finally, because of (41.21), we get

$$\kappa > \kappa_S, \tag{41.26}$$

i.e. the isothermal compressibility is always greater than the isentropic compressibility for a stable phase if $\alpha \neq 0$.

Let us now turn to a consideration of the stability conditions (41.15). We shall alter the general formulation somewhat by keeping $\sum n_i$ constant instead of X_r. We shall carry this out explicitly for the sake of clarity. Let us first write (41.13) without the limitation $X_r = \text{const}$ in the form

$$[\Delta G]_{T,P} > \sum_{i=1}^{m} \mu_i \Delta n_i. \tag{41.27}$$

For $\sum n_i = \text{const}$ this becomes

$$[\Delta G^*]_{T,P} > \sum_{i=1}^{m} \mu_i \Delta x_i \tag{41.28}$$

or, because of (26.4),

$$[\Delta G^*]_{T,P} > \sum_{i=1}^{m-1} (\mu_i - \mu_m) \Delta x_i. \tag{41.29}$$

Since, according to (26.20), we have

$$\left(\frac{\partial G^*}{\partial x_i}\right)_{T,P,x_j} = \mu_i - \mu_m, \tag{41.30}$$

the right-hand side of (41.29) represents the first-order variation of G^* at constant T and P. We thus get

$$[\delta^2 G^*(x_1, \ldots, x_{m-1})]_{T,P} > 0 \tag{41.31}$$

as the relationship equivalent to (41.13), or

$$\left| \frac{\partial^2 G^*}{\partial x_i \partial x_j} \right| \text{ with all principal minors } > 0 \; (i,j = 1, \ldots, m-1). \tag{41.32}$$

For *binary systems* (41.32) reduces to the condition

$$\left(\frac{\partial^2 G^*}{\partial x_1^2}\right)_{T,P} > 0 \tag{41.33}$$

but from eqs. (26.16) and (41.30) we get

$$\left(\frac{\partial^2 G^*}{\partial x_1^2}\right)_{T,P} = \frac{1}{1-x_1}\left(\frac{\partial \mu_1}{\partial x_1}\right)_{T,P}. \tag{41.34}$$

The stability condition (41.33) can thus also be written

$$\left(\frac{\partial \mu_1}{\partial x_1}\right)_{T,P} > 0, \tag{41.35}$$

an expression which we have used previously in § 28.

For ternary systems the stability conditions (41.32) become

$$\begin{vmatrix} \dfrac{\partial^2 G^*}{\partial x_1^2} & \dfrac{\partial^2 G^*}{\partial x_1\,\partial x_2} \\ \dfrac{\partial^2 G^*}{\partial x_2\,\partial x_1} & \dfrac{\partial^2 G^*}{\partial x_2^2} \end{vmatrix} > 0, \quad \frac{\partial^2 G^*}{\partial x_1^2} > 0, \quad \frac{\partial^2 G^*}{\partial x_2^2} > 0. \tag{41.36}$$

Here, too, only two of the conditions are independent since, for example, the third inequality is a consequence of the first two. These inequalities state that for T = const, P = const, the G^*-surface is double convex towards the x_1–x_2-plane. In other words: any vertical section through the G^*-surface gives a sectional curve which is convex when viewed from below.

Completely analogous results are obtained for the entropy representation of the fundamental equation and for the Massieu–Planck functions. These results differ from those for thermodynamic potentials mainly by a change of sign, i.e. we have for the entropy

$$[\delta^2 S(X_1, \ldots, X_{r-1})]_{X_r} < 0 \tag{41.37}$$

and for the Massieu–Planck functions

$$[\delta^2 \Phi_k(P_1, \ldots, P_k)]_{X_j} > 0, \tag{41.38}$$

$$[\delta^2 \Phi_k(X_{k+1}, \ldots, X_{r-1})]_{P_i, X_r} < 0. \tag{41.39}$$

§ 42 Stability conditions for heterogeneous systems

Our discussions of stability conditions have so far been limited to homogeneous systems. We now want to show that heterogeneous systems are necessarily stable if the stability conditions are fulfilled for the individual phases. For simplicity we shall discuss only a one-component system with two phases, α and β. Virtual displacements which merely change the amounts of the phases are eliminated by keeping the mole numbers $n^{(\alpha)}$ and $n^{(\beta)}$ of the phases constant. We are thus assuming the phases to be separated by a frictionless moveable diathermic wall. We shall choose the energy representation and investigate the virtual displacement ΔU with the secondary conditions

$$\delta n^{(\alpha)} = 0, \quad \delta n^{(\beta)} = 0, \tag{42.1}$$

$$n^{(\alpha)}\,\delta s^{(\alpha)} + n^{(\beta)}\,\delta s^{(\beta)} = 0, \tag{42.2}$$

$$n^{(\alpha)}\,\delta v^{(\alpha)} + n^{(\beta)}\,\delta v^{(\beta)} = 0, \tag{42.3}$$

in which we have used the definitions (20.36). Since we have assumed equilibrium the first-order variations vanish and we obtain the stability condition

$$\begin{aligned}\delta^2 U = \tfrac{1}{2} n^{(\alpha)2}\Bigg[&\left(\frac{1}{n^{(\alpha)}}\frac{\partial^2 u^{(\alpha)}}{\partial s^{(\alpha)2}} + \frac{1}{n^{(\beta)}}\frac{\partial^2 u^{(\beta)}}{\partial s^{(\beta)2}}\right)(\delta s^{(\alpha)})^2 \\ &+ 2\left(\frac{1}{n^{(\alpha)}}\frac{\partial^2 u^{(\alpha)}}{\partial s^{(\alpha)}\,\partial v^{(\alpha)}} + \frac{1}{n^{(\beta)}}\frac{\partial^2 u^{(\beta)}}{\partial s^{(\beta)}\,\partial v^{(\beta)}}\right)(\delta s^{(\alpha)}\,\delta v^{(\alpha)}) \\ &+ \left(\frac{1}{n^{(\alpha)}}\frac{\partial^2 u^{(\alpha)}}{\partial v^{(\alpha)2}} + \frac{1}{n^{(\beta)}}\frac{\partial^2 u^{(\beta)}}{\partial v^{(\beta)2}}\right)(\delta v^{(\alpha)})^2\Bigg] > 0.\end{aligned} \tag{42.4}$$

The quadratic form must thus be positive-definite. This must be true for any values of $n^{(\alpha)}$ and $n^{(\beta)}$ and so we can put $n^{(\alpha)} = n^{(\beta)} = 1$ and obtain the explicit equilibrium conditions

$$\frac{\partial^2 u^{(\alpha)}}{\partial s^{(\alpha)2}} + \frac{\partial^2 u^{(\beta)}}{\partial s^{(\beta)2}} > 0, \tag{42.5}$$

$$\frac{\partial^2 u^{(\alpha)}}{\partial v^{(\alpha)2}} + \frac{\partial^2 u^{(\beta)}}{\partial v^{(\beta)2}} > 0, \tag{42.6}$$

$$\left(\frac{\partial^2 u^{(\alpha)}}{\partial s^{(\alpha)2}} + \frac{\partial^2 u^{(\beta)}}{\partial s^{(\beta)2}}\right)\left(\frac{\partial^2 u^{(\alpha)}}{\partial v^{(\alpha)2}} + \frac{\partial^2 u^{(\beta)}}{\partial v^{(\beta)2}}\right) - \left(\frac{\partial^2 u^{(\alpha)}}{\partial s^{(\alpha)}\,\partial v^{(\alpha)}} + \frac{\partial^2 u^{(\beta)}}{\partial s^{(\beta)}\,\partial v^{(\beta)}}\right) > 0. \tag{42.7}$$

The stability of the individual phases is obviously a necessary condition for the stability of the complete system. According to

(40.12), therefore,

$$\frac{\partial^2 u^{(\alpha)}}{\partial s^{(\alpha)2}} > 0, \quad \frac{\partial^2 u^{(\beta)}}{\partial s^{(\beta)2}} > 0, \tag{42.8}$$

$$\frac{\partial^2 u^{(\alpha)}}{\partial v^{(\alpha)2}} > 0, \quad \frac{\partial^2 u^{(\beta)}}{\partial v^{(\beta)2}} > 0, \tag{42.9}$$

$$\left.\begin{aligned} &\frac{\partial^2 u^{(\alpha)}}{\partial s^{(\alpha)2}} \frac{\partial^2 u^{(\alpha)}}{\partial v^{(\alpha)2}} - \left(\frac{\partial^2 u^{(\alpha)}}{\partial s^{(\alpha)} \partial v^{(\alpha)}}\right)^2 > 0, \\ &\frac{\partial^2 u^{(\beta)}}{\partial s^{(\beta)2}} \frac{\partial^2 u^{(\beta)}}{\partial v^{(\beta)2}} - \left(\frac{\partial^2 u^{(\beta)}}{\partial s^{(\beta)} \partial v^{(\beta)}}\right)^2 > 0 \end{aligned}\right\} \tag{42.10}$$

must be true. It is easy to see that the inequalities (42.5) and (42.6) follow from the inequalities (42.8) and (42.9).

But we know that

$$\begin{aligned}(a_1+a_2)(c_1+c_2)-(b_1+b_2)^2 &= [1+(a_1/a_2)](a_2c_2-b_2^2) \\ &+ [1+(a_2/a_1)](a_1c_1-b_1^2)+(a_1b_2-a_2b_1)^2/a_1a_2 \end{aligned} \tag{42.11}$$

and, therefore, that the inequality (42.7) follows directly from (42.8)–(42.10). We therefore have the

Theorem: The fulfilment of the stability conditions for the individual phases is the necessary and sufficient condition for the stability of a heterogeneous equilibrium.

§ 43 The Le Châtelier–Braun principle

A statement often called the principle of least constraint is closely connected with the stability conditions. The origins of this principle go back to the 18th century (d'Alembert, Gauss). The modern formulation is due to Le Châtelier (1884) and Braun (1887). It is therefore called the *Le Châtelier–Braun principle.*

The simplest formulation (usually called Le Châtelier's principle) is: If an external constraint is applied to a system in equilibrium, the equilibrium reacts in such a way as to minimize the external constraint.

It is easy to see that the simpler stability conditions are summarized in this principle. If, for instance, heat is introduced into a system, the temperature difference between system and heat reservoir is the driving force. The driving force is, therefore, minimized if the introduction of heat increases the temperature which requires that $C_V > 0$ or $C_P > 0$, in agreement with (40.17) and (41.17). The principle includes the condition (41.18) in a similar way. We shall mention further examples when we come to deal with the stability of chemical equilibria.

Le Châtelier's principle is thus equivalent to the stability conditions for simple cases. Its use as a purely qualitative statement is found mainly in chemistry. It can, however, lead to wrong conclusions just because of this imprecise formulation. For example, the statement

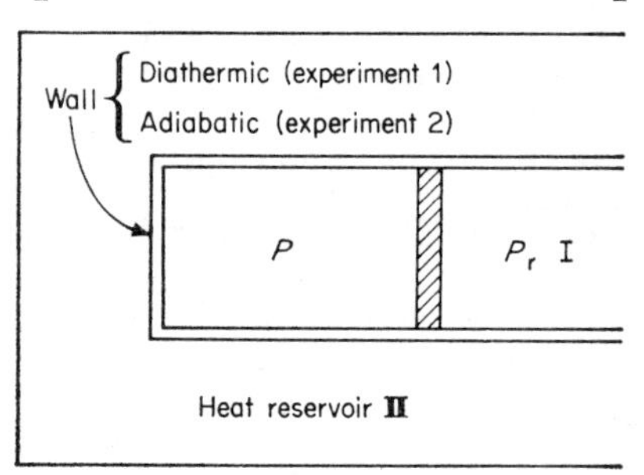

Fig. 27. The Le Châtelier–Braun principle

'exothermic solution causes a reduction in solubility with increasing temperature' may be wrong. The temperature dependence of the solubility is determined not only by the stability condition (41.35) but also by the differential heat of solution at the saturation point† and *not* by the integral heat of solution; the two heats of solution may have opposite signs. More complicated stability conditions such as the first inequality of (41.36) are quite unobtainable from Le Châtelier's principle which, therefore, has no significance for thermodynamics and has been mentioned only because of its frequent use.

Much more important is the actual Le Châtelier–Braun principle. We shall explain its content by means of a simple example before formulating the actual principle. Let a cylinder filled with gas be in pressure equilibrium with a reservoir I *via* a moveable piston. In a first experiment, let the temperature of the cylinder be controlled by means of a diathermic wall and a heat reservoir II. A pressure change dP then simply causes a volume change $d_1 V$. In a second experiment, let the cylinder be adiabatically isolated. The same pressure change now causes a volume change $d_2 V \neq d_1 V$ since a temperature change $d_2 T$ is also induced. The Le Châtelier–Braun principle states that

$$d_2 V < d_1 V. \tag{43.1}$$

† No name appears to be given to this heat of solution in the English language. In German, the term '*letzte Lösungswärme*' (= final heat of solution) is used. (*Translator.*)

The correctness of the statement for the above example is easily verified by reference to the isothermals and adiabatics for a gas.

The usual qualitative formulation of the Le Châtelier–Braun principle states that the above result may be generalized for any pairs of variables of state. A quantitative analysis by Ehrenfest has shown that the principle does not hold to such a general extent. It must be supplemented by the requirement that one of the two parameters must be extensive and the other intensive. We shall now give Ehrenfest's proof which yields the correct version of the principle.

Let P_j be the intensive parameter regulated by reservoir I, X_j the directly regulated conjugate extensive parameter, and P_i the intensive parameter kept constant by reservoir II in experiment 1; in experiment 2 $\mathrm{d}X_i = 0$ and, therefore, a change in P_i is induced by the change in P_j.

The Le Châtelier–Braun principle now states that

$$|\,\mathrm{d}_2 X_j\,| < |\,\mathrm{d}_1 X_j\,|. \tag{43.2}$$

Proof: We have

$$\mathrm{d}P_j = \left(\frac{\partial P_j}{\partial X_j}\right)_{P_i} \mathrm{d}_1 X_j, \tag{43.3}$$

$$\mathrm{d}P_j = \left(\frac{\partial P_j}{\partial X_j}\right)_{P_i} \mathrm{d}_2 X_j + \left(\frac{\partial P_j}{\partial P_i}\right)_{X_j} \mathrm{d}_2 P_i \tag{43.4}$$

and, furthermore, for the second experiment

$$\mathrm{d}_2 X_i = \left(\frac{\partial X_i}{\partial X_j}\right)_{P_i} \mathrm{d}_2 X_j + \left(\frac{\partial X_i}{\partial P_i}\right)_{X_j} \mathrm{d}_2 P_i = 0. \tag{43.5}$$

Elimination of $\mathrm{d}P_j$ and $\mathrm{d}_2 P_i$ gives

$$\left[\left(\frac{\partial P_j}{\partial X_j}\right)_{P_i}\left(\frac{\partial X_i}{\partial P_i}\right)_{X_j} - \left(\frac{\partial P_j}{\partial P_i}\right)_{X_j}\left(\frac{\partial X_i}{\partial X_j}\right)_{P_i}\right] \mathrm{d}_2 X_j = \left[\left(\frac{\partial P_j}{\partial X_j}\right)_{P_i}\left(\frac{\partial X_i}{\partial P_i}\right)_{X_j}\right] \mathrm{d}_1 X_j, \tag{43.6}$$

or, using Maxwell's relation (24.11),

$$\left[\left(\frac{\partial P_j}{\partial X_j}\right)_{P_i}\left(\frac{\partial X_i}{\partial P_i}\right)_{X_j} + \left(\frac{\partial X_i}{\partial X_j}\right)^2_{P_i}\right] \mathrm{d}_2 X_i = \left[\left(\frac{\partial P_j}{\partial X_j}\right)_{P_i}\left(\frac{\partial X_i}{\partial P_i}\right)_{X_j}\right] \mathrm{d}_1 X_j. \tag{43.7}$$

According to the stability conditions (41.14) and (41.15), and eq. (21.9) we have

$$\left(\frac{\partial P_j}{\partial X_j}\right)_{P_i} > 0, \quad \left(\frac{\partial X_i}{\partial P_i}\right)_{X_j} > 0 \tag{43.8}$$

and the inequality (43.2) follows directly.

The Le Châtelier–Braun principle clearly states that the induced parameter change $d_2 P_i$ 'relieves' the directly affected parameter X_j. Although the principle in the above form is strictly valid it has not gained great significance in thermodynamics. It has, however, proved fruitful for the theory of irreversible processes.

§ 44 The stability of chemical equilibria

We still have to discuss briefly the stability of chemical equilibria and we shall simplify the discussion by assuming throughout that $T = \text{const}$, $P = \text{const}$. The stability conditions may now be formulated by means of the Gibbs free energy. If we confine our discussion to a reaction (33.15) we have, according to § 36, the progress variable ξ as the only independent variable.

The inequality (41.27) thus becomes

$$[\Delta G]_{T,P} > \left(\sum_i \nu_i \mu_i\right) \Delta\xi. \tag{44.1}$$

According to (36.6), the right-hand side represents the first-order variation. The stability condition therefore becomes

$$\left[\frac{\mathrm{d}}{\mathrm{d}\xi} \sum_i \nu_i \mu_i\right]_{T,P} > 0. \tag{44.2}$$

We shall make some deductions from this and, at the same time, illustrate Le Châtelier's principle. Let us consider the shifting of a chemical equilibrium (whose position is specified by the value ξ_e) with temperature at constant pressure and with pressure at constant temperature. We shall start with the equation

$$\mathrm{d}G = -S\,\mathrm{d}T + V\,\mathrm{d}P + \left(\sum_i \nu_i \mu_i\right) \mathrm{d}\xi \tag{44.3}$$

where we have, at equilibrium, that

$$\left(\frac{\partial G}{\partial \xi}\right)_{T,P} = \sum_i \nu_i \mu_i = 0. \tag{44.4}$$

From (44.3) we get for the total differential of $(\partial G/\partial \xi)_{T,P}$ that

$$\mathrm{d}\left(\frac{\partial G}{\partial \xi}\right)_{T,P} = \mathrm{d}\left(\sum \nu_i \mu_i\right) = -\left(\frac{\partial S}{\partial \xi}\right)_{T,P} \mathrm{d}T + \left(\frac{\partial V}{\partial \xi}\right)_{T,P} \mathrm{d}P + \left(\frac{\partial^2 G}{\partial \xi^2}\right)_{T,P} \mathrm{d}\xi. \tag{44.5}$$

The left-hand side vanishes at equilibrium and we obtain, for $\mathrm{d}P = 0$,

$$\left(\frac{\partial \xi_e}{\partial T}\right)_P = \frac{(\partial S/\partial \xi)_{T,P}}{(\partial^2 G/\partial \xi^2)_{T,P}}. \tag{44.6}$$

According to (24.8) we have, however that

$$\left(\frac{\partial G}{\partial \xi}\right)_{T,P} = \left(\frac{\partial H}{\partial \xi}\right)_{T,P} + T\left(\frac{\partial S}{\partial \xi}\right)_{T,P}. \tag{44.7}$$

The left-hand side is zero at equilibrium and the quantity $(\partial H/\partial \xi)_{T,P}$ represents, according to (21.23), the heat of reaction at constant pressure or enthalpy of reaction. Thus (44.6) becomes

$$\left(\frac{\partial \xi_e}{\partial T}\right)_P = \frac{(\partial H/\partial \xi)_{T,P}}{T(\partial^2 G/\partial \xi^2)_{T,P}}. \tag{44.8}$$

Since the stability condition requires the numerator to be positive, eq. (44.8) states:

An increase in temperature at constant pressure shifts the chemical equilibrium in the direction in which the heat of reaction is positive (i.e. the reaction is endothermic).

This statement agrees with Le Châtelier's principle.

In an analogous way we get

$$\left(\frac{\partial \xi_e}{\partial P}\right)_T = -\frac{(\partial V/\partial \xi)_{T,P}}{(\partial^2 G/\partial \xi^2)_{T,P}}. \tag{44.9}$$

An increase in pressure at constant temperature shifts the chemical equilibrium in the direction in which the volume change for the reaction is negative (at T = const, P = const). This statement also agrees with Le Châtelier's principle.

We want to emphasize that the above statements are completely general; they can be obtained from the van't Hoff equation or the Planck–van Laar equation respectively only for the special case of ideal gases. This is based on the fact that the stability condition (44.2) is automatically fulfilled for chemical equilibrium in an ideal gaseous mixture.

CHAPTER VII

Critical phases
Transitions of higher order

§ 45 Definition and properties of critical phases

Experimental experience shows that variation of the quantities of state often causes two co-existing phases to vary in such a way that they ultimately become identical. The phase for which this occurs is called a *critical phase*. We shall see that critical phases of one-component systems have no thermodynamic degrees of freedom. They are therefore represented by points in phase space called *critical points*. If, as is often the case, the dependence of the critical phase on the quantities of state is irrelevant, we also speak of critical points for multi-component systems. Let us start by giving a list of the most important cases.

(a) *One-component systems.* A vapour–liquid equilibrium always has a critical point. It is impossible to liquefy a gas above a certain temperature T_c, the critical temperature. As Andrews showed in 1869 for CO_2, there is a continuous change from the gaseous to the liquid state. This can easily be seen from the isotherms of the P–v-diagram for a real gas shown in Fig. 28.

The curve on which the co-existing liquid and gaseous phases lie is called a *co-existence curve* or *binodal curve*. The broken lines joining two co-existing phases are called *connodals*. These names are also used for other heterogeneous equilibria.

The equilibrium vapour–crystalline solid ends at the triple point (Fig. 18). This equilibrium does not exist for ^{3}He and ^{4}He for which the crystalline solid is in equilibrium only with the liquid. No triple point therefore exists for ^{3}He and ^{4}He.

In spite of much effort, no critical point has ever been found experimentally for solid–liquid equilibria. Certain theoretical considerations indicate that no such critical point can exist. A satisfactory explanation has, however, not yet been found.

(b) *Binary systems.* The equilibrium vapour–liquid shows a critical point which is of interest because of the strange phenomena of retrograde condensation and vaporization. We shall return briefly to these at the end of § 47.

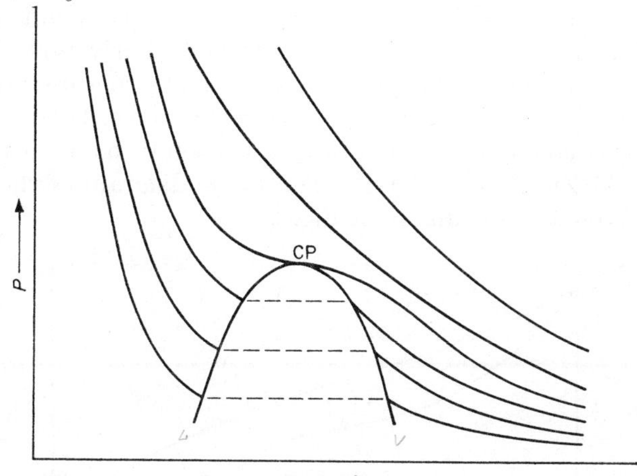

FIG. 28. Isotherms of a real gas (schematic)

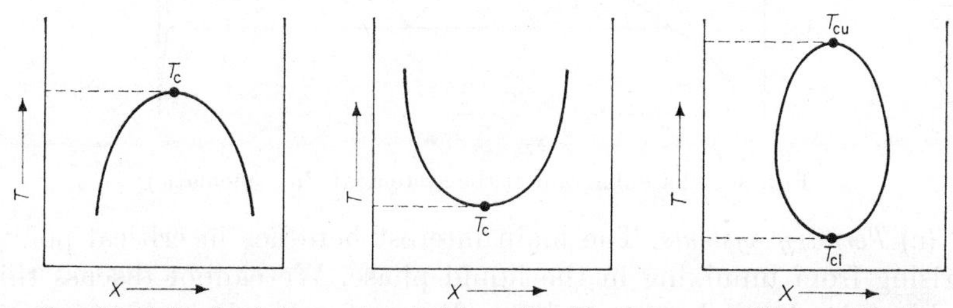

FIG. 29. Critical solution points in binary liquid mixtures. a. Upper critical point. 9. Lower critical point. c. Closed miscibility gap with two critical points

More important are the critical points found when a homogeneous solution separates into two liquid phases. The representation for the case corresponding to Fig. 28 would be given by the isotherms of the diagram of the chemical potential of one component (say μ_A) against the mole fraction (x_1 or x_2), for P = const. It is often sufficient, however, to employ the T–x-diagrams already used several times in Chapter IV. The latter representation is used in Fig. 29 to show diagrams of the observed types. Case I called the *upper critical solution point* or *upper consolute point* is the one found most often; it is observed for many mixtures of organic liquids. Examples of the lower consolute point shown in Case II are the systems water–diethylamine and water–triethylamine. The closed miscibility gap shown in Case III has been found in the system water–nicotine. Both an upper (T_{cu}) and a lower (T_{cl}) consolute point are present.

In contrast with one-component systems, binary systems show critical points in the solid state as well. These critical points are consolute points as for liquids. They are actually fairly rare since the co-existing phases usually have different crystal structures and therefore cannot become identical for geometrical reasons. There are, however, some binary alloys which have a consolute point in the solid state, e.g. Al–Zn, Au–Ni, Au–Pt. The phase diagram of the system Al–Zn is given as an example in Fig. 30.

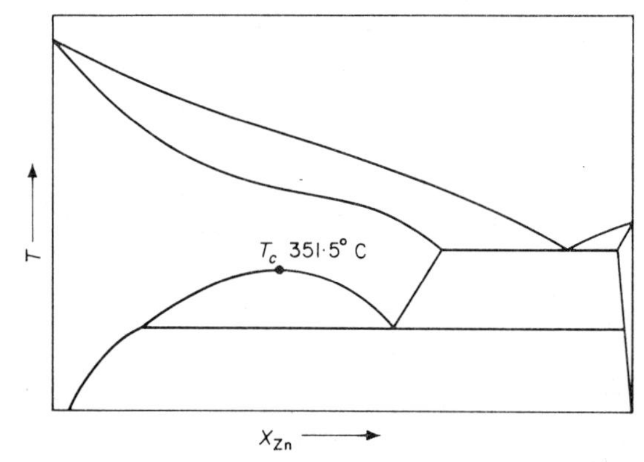

FIG. 30. Phase diagram of the system Al–Zn (schematic)

(c) *Ternary systems.* The main interest here lies in critical points arising from unmixing in the liquid phase. We cannot discuss this subject in detail but two observations of general significance are worth mentioning.

We have seen that, for one-component systems, the critical temperature is the highest temperature at which the system can become heterogeneous. For binary systems the critical phase has one degree of freedom. If, however, we fix the pressure as is usual in experimental arrangements, the critical temperature is again the highest temperature (or the lowest in the case of a lower consolute point) at which the homogeneous solution (or the homogeneous mixed crystals) can become heterogeneous. The situation is quite different in systems of three or more components. We shall show that the critical phase of a ternary system has two degrees of freedom. If we again fix the pressure, a representation is obtained by plotting isothermal G^*-surfaces above the x_1–x_2-plane in a three-dimensional diagram.

Such a diagram is shown schematically in Fig. 31. We can see that each G^*-surface is associated with a binodal curve with a critical point; the concept of critical temperature therefore no longer has any meaning. We are, however, left with one degree of freedom after fixing the pressure and if we use this degree of freedom for fixing the ratio

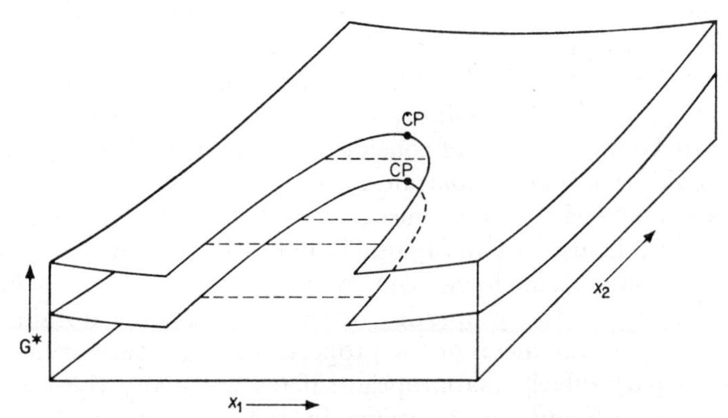

FIG. 31. Isothermal–isobaric G^*-surfaces for ternary systems. CP = critical phase

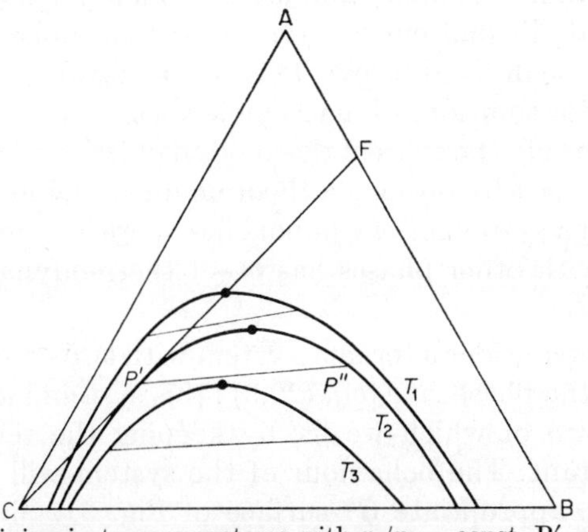

FIG. 32. Unmixing in ternary systems with x_1/x_2 = const. P′ = Phase with the highest unmixing temperature. P″ = phase co-existing with phase P′. Critical points indicated by dots

x_1/x_2 we can again define a critical point. The relevant 'critical temperature' is, however, no longer the highest temperature at which the system can become heterogeneous. This can be seen most clearly by using Möbius' triangular co-ordinates (Fig. 32) which we also met previously in Chapter IV. The diagram shows binodal curves at various temperatures. The critical points are indicated by dots. All solutions containing the components in the fixed ratio x_1/x_2 lie on the straight line FC. This line can pass through only one critical point which is thus completely defined. The highest temperature at which the system can become heterogeneous does not, however, correspond to this binodal curve but corresponds to the binodal curve to which the straight line FC is the tangent. The corresponding phase is represented by the point of contact P′. We can see that this is not generally a critical point, but is in equilibrium with a phase P″. The above ideas are due to Tompa and are of considerable importance for the theory of the solubility of macromolecules.

Our second consideration concerns a case often met in practice, i.e. the use of a mixture of two organic liquids as a solvent. In the older literature such systems have been treated in various ways as binary systems in which the properties of the solvent were obtained by somehow taking the mean of the properties of the components. This method is particularly inappropriate if (as is usually the case) the solvent has different compositions in the co-existing phases. In particular, it is not permissible to determine the consolute point by the simple methods of calculation used for binary systems. The much more complicated equations for ternary systems must be used. These equations will be derived below. This consideration, too, is of significance mainly for solutions of macromolecules.

Let us now apply the rules of thermodynamics to what we have just discussed and start by proving a theorem due to Gibbs.

Theorem. In a system of m components, a critical phase which does not co-exist with other phases has $m-1$ thermodynamic degrees of freedom.

Proof. Let us consider a ternary system with two co-existing phases. According to the phase rule [eq. (29.3)] this system has three degrees of freedom, two of which we fix by keeping the temperature and pressure constant. The behaviour of the system will then be represented by the appropriate G^*-surface of Fig. 31. Since we are left with one degree of freedom we can vary, for example, the chemical potential μ_1 without making either of the two phases disappear. Figure 31 shows that corresponding to these variations there is a continuous series of co-existing phases ending in a critical point. If we now give to the quantities T and P other values which differ only infinitesimally from the original ones, we obtain in a similar manner a series of pairs of phases which again ends in a critical point. The position of this critical point will also differ only infinitesimally from that of the original one.† This argument can be continued to show that the critical phase has two degrees of freedom in this case.

Let us again consider two co-existing phases but now for the general case of a system of m components. The phase rule tells us that this heterogeneous system has m degrees of freedom. We shall choose as independent variables m quantities from the set of intensive parameters $T, P, \mu_1, ..., \mu_m$ and fix $m-1$ of these by giving them values corresponding to one critical point among all the critical phases which we may consider. We can vary the remaining intensive parameter (pressure, temperature, or one of the chemical potentials) without causing either of the two phases to disappear. In this way we again obtain a continuous series of co-existing phases ending in a critical point. If we now change the values of the initially fixed $m-1$ intensive parameters by infinitesimal amounts we again obtain a series of co-existing phases by varying the free parameter. Now each pair of co-existing phases in the first series must correspond to an infinitesimally differing pair in the second series and vice versa. The second series must, therefore, end in a critical point whose position differs only infinitesimally from that of the first series. The argument can again be continued in the same manner. If, therefore, the values of the $m-1$ intensive parameters given by the position of a critical phase are freely varied, each set of varied values again corresponds to one and only one critical phase. A critical phase in a system of m components thus has $m-1$ thermodynamic degrees of freedom and the theorem is proved.

The previously stated results follow from the theorem: for one-component systems the critical phase is a singular point in phase space; critical phases of binary systems have one degree of freedom which is generally used to fix the pressure in binary systems.‡

In ternary systems we already have two degrees of freedom and, therefore, a much more complicated state of affairs. As stated previously, the above statements are valid in the form given only if no other phase co-exists with the critical phase. Since, according to the phase rule, every new phase reduces the number of degrees of freedom by one, we conclude that no other phase can co-exist with the critical phase of a one-component system, while only one other phase can co-exist with the critical phase of a binary system. Generally, a critical

† Since Fig. 31 is an isobaric diagram, only the variation of T can be shown.

‡ The dependence of the consolute temperature on pressure is, in any case, very small for condensed systems. It can be completely ignored under the usual experimental conditions.

phase can co-exist with at most $m-1$ other phases. Comparison with eq. (29.3) shows that a critical phase must be counted threefold in the sense of the phase rule.

§ 46 Gibbs' equations for critical phases

We shall not discuss here any cases of the co-existence of the critical phase with other phases. According to the phase rule an ordinary phase has $m+1$ degrees of freedom while, according to the above theorem a critical phase has $m-1$ degrees of freedom. A direct consequence is, therefore, that a critical phase is defined by two equations.

These equations for one-component systems can be easily obtained by a consideration of Fig. 28. According to the equilibrium conditions (27.6) and (27.7), co-existing phases must always lie on the same isotherm and the connodals must run parallel to the abscissa. The critical point is defined by the coincidence of the co-existing phases and must, therefore, be a point of inflexion of the isotherms with a horizontal tangent. We thus have immediately that

$$\left(\frac{\partial P}{\partial v}\right)_T = 0, \quad \left(\frac{\partial^2 P}{\partial v^2}\right)_T = 0. \tag{46.1}$$

An analogous representation to Fig. 28 for the critical solution point of a binary system is, as already mentioned, given by the isotherms of a μ–x-diagram. An analogous consideration of this diagram gives for the critical solution point of a binary system that

$$\left(\frac{\partial \mu_1}{\partial x_1}\right)_{T,P} = 0, \quad \left(\frac{\partial^2 \mu_1}{\partial x_1^2}\right)_{T,P} = 0. \tag{46.2}$$

The conditions for systems with more than two components are much less obvious and it is necessary, as shown by Gibbs, to formulate the theory in a much stricter and more general way. The basis of this formulation is the stability theory developed in Chapter IV. The equations for the critical phase can then be deduced in various ways leading to different forms of the equations. We shall, in this section, use the method developed by Gibbs. A more recent method will be given in §§ 49–51.

It is, however, convenient to give here a few preliminary remarks which should contribute to the understanding of the more recent method. Let us first consider a one-component system (Fig. 28) and note that on each connodal joining two co-existing phases there are necessarily states for which the homogeneous system is absolutely unstable. This statement is made clearer by assuming that the isotherms are analytic functions over the whole region of state. The isotherms must then necessarily have the form known from van der Waals' equation of state (Fig. 33). The states lying on each isotherm between the minimum and the maximum are necessarily unstable. The curve connecting these minima and maxima therefore bounds the unstable region and is called a *spinodal curve*. The region between the spinodal and binodal curves corresponds to metastable states (superheated liquid, supersaturated vapour). This representation is frequently used in thermodynamics as a useful device. It is, however,

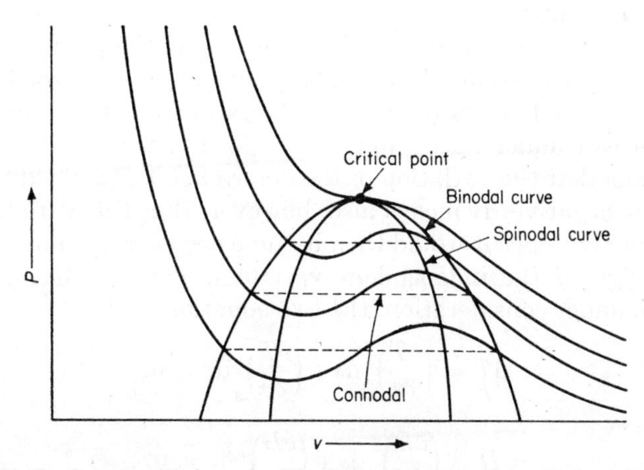

FIG. 33. Isotherms with 'van der Waals loop' for a real gas (schematic)

introduced here only as an illustration and not used formally in the derivation since the basic hypothesis can be shown to be untenable and thus purely fictitious.†

We now picture the two co-existing phases as near as we like to the critical point. Since there must still be absolutely unstable states between the two co-existing phases, it follows that the critical point must lie on the boundary between the stable and the absolutely unstable region. In the representation introduced above this means that the binodal and spinodal curves touch at the critical point.

The boundary between the stable and the unstable region is now given by the requirement that at least one of the second derivatives

† It can mathematically be strictly shown that an exact statistical calculation of thermodynamic functions can never lead to unstable states. Isotherms calculated in this way can, therefore, never show the 'van der Waals loop' but will always yield the diagram shown in Fig. 28.

of the internal energy becomes zero. This means, however, that the quadratic form $\delta^2 u(s, v)$ is positive semidefinite at this place. We shall show in §§ 50 and 51 that for this to be true it is necessary and sufficient that the condition

$$D \equiv \begin{vmatrix} \dfrac{\partial^2 u}{\partial s^2} & \dfrac{\partial^2 u}{\partial s\, \partial v} \\ \dfrac{\partial^2 u}{\partial v\, \partial s} & \dfrac{\partial^2 u}{\partial v^2} \end{vmatrix} = 0 \tag{46.3}$$

is fulfilled. Equation (46.3) is, therefore, already the first equation for the critical point.

In order to derive the second equation, we fix the temperature and consider the variation of the critical phase on the isotherm. An examination of Fig. 28 or Fig. 33 shows immediately that such a variation can under no circumstances go into the unstable region. This means that the variation of the determinant D defined by (46.3) cannot be negative. It means also, however, that this variation cannot be positive since it would then become negative by simply reversing the sign of the independent variables. Thus we have, for the variation under consideration, the two equations

$$\mathrm{d}T = \left(\frac{\partial T}{\partial s}\right)_v \mathrm{d}s + \left(\frac{\partial T}{\partial v}\right)_s \mathrm{d}v = 0, \tag{46.4}$$

$$\mathrm{d}D = \left(\frac{\partial D}{\partial s}\right)_v \mathrm{d}s + \left(\frac{\partial D}{\partial v}\right)_s \mathrm{d}v = 0. \tag{46.5}$$

This is a set of homogeneous linear equations for the unknowns $\mathrm{d}s$ and $\mathrm{d}v$. The condition for the equations to be soluble is

$$D' \equiv \begin{vmatrix} \left(\dfrac{\partial T}{\partial s}\right)_v & \left(\dfrac{\partial T}{\partial v}\right)_s \\ \left(\dfrac{\partial D}{\partial s}\right)_v & \left(\dfrac{\partial D}{\partial v}\right)_s \end{vmatrix} = 0 \tag{46.6}$$

and this is the second equation for the critical point.

Equations (46.3) and (46.6) are ultimately based on the fundamental equation and suffer, therefore, from the disadvantages already mentioned several times. In particular, the proof of the identity of these equations with eq. (46.1) is complicated. We might, therefore, think that it would be better to choose as our starting point a formulation of the stability conditions with the aid of a suitable thermodynamic potential. The potential most appropriate to the present problem is the Helmholtz free energy. The stability conditions then are, according to (41.14) and (41.15),

$$\left(\frac{\partial s}{\partial T}\right)_v > 0, \quad \left(\frac{\partial P}{\partial v}\right)_T < 0. \tag{46.7}$$

We can see immediately that the first of these conditions is irrelevant for the determination of the critical point. If considerations analogous to those above are applied to the second condition, the direct result is eq. (46.1).

The generalization of the above derivations for systems of m components is not difficult. We shall, first of all, write the abbreviation

$$D^* \equiv \begin{vmatrix} \frac{\partial^2 U^*}{\partial S^{*2}} & \frac{\partial^2 U^*}{\partial S^* \partial V^*} & \cdots & \frac{\partial^2 U^*}{\partial S^* \partial x_{m-1}} \\ \frac{\partial^2 U^*}{\partial V^* \partial S^*} & \cdot & & \vdots \\ \vdots & & \cdot & \vdots \\ \frac{\partial^2 U^*}{\partial x_{m-1} \partial S^*} & \cdots & \cdots & \frac{\partial^2 U^*}{\partial x_{m-1}^2} \end{vmatrix}. \tag{46.8}$$

We shall further denote by $D^{*\prime}$ a determinant which differs from D^* in that the elements of an arbitrarily chosen row in it are replaced by the partial derivatives occurring in the expression

$$\mathrm{d}D^* = \frac{\partial D^*}{\partial S^*}\mathrm{d}S^* + \frac{\partial D^*}{\partial V^*}\mathrm{d}V^* + \frac{\partial D^*}{\partial x_1}\mathrm{d}x_1 + \ldots + \frac{\partial D^*}{\partial x_{m-1}}\mathrm{d}x_{m-1}. \tag{46.9}$$

The boundary of the stability region is again given by the requirement that the quadratic form (40.11) is positive semidefinite. As we shall show in §§ 50 and 51, the necessary and sufficient condition for this is that the determinant D^* vanishes and we have thus found the first equation for the critical phase.

In order to get the second equation the critical phase must again be varied. The secondary conditions for this are, however, that of the $m+1$ derivatives

$$\frac{\partial U^*}{\partial S^*}, \quad \frac{\partial U^*}{\partial V^*}, \quad \frac{\partial U^*}{\partial x_1}, \quad \ldots, \quad \frac{\partial U^*}{\partial x_{m-1}} \tag{46.10}$$

m quantities must be kept constant. The choice of these quantities is arbitrary and defines, as can be seen from eqs. (27.6)–(27.8), in each case one path on an m-dimensional hypersurface. The path connects two co-existing phases. A variation of the critical phase can, therefore, never go into the unstable region. If we suppose, for instance, that the first m derivatives of (46.10) are kept constant, we have the following secondary conditions for the variation of the critical phase:

$$\left.\begin{aligned}
\mathrm{d}\left(\frac{\partial U^*}{\partial S^*}\right) &= \frac{\partial^2 U^*}{\partial S^{*2}}\mathrm{d}S^* + \frac{\partial^2 U^*}{\partial S^*\,\partial V^*}\mathrm{d}V^* + \frac{\partial^2 U^*}{\partial S^*\,\partial x_1}\mathrm{d}x_1 \\
&\qquad + \ldots + \frac{\partial^2 U^*}{\partial S^*\,\partial x_{m-1}}\mathrm{d}x_{m-1} = 0, \\
\mathrm{d}\left(\frac{\partial U^*}{\partial V^*}\right) &= \frac{\partial^2 U^*}{\partial V^*\,\partial S^*}\mathrm{d}S^* + \frac{\partial^2 U^*}{\partial V^{*2}}\mathrm{d}V^* + \frac{\partial^2 U^*}{\partial V^*\,\partial x_1}\mathrm{d}x_1 \\
&\qquad + \ldots + \frac{\partial^2 U^*}{\partial V^*\,\partial x_{m-1}}\mathrm{d}x_{m-1} = 0, \\
\mathrm{d}\left(\frac{\partial U^*}{\partial x_1}\right) &= \frac{\partial^2 U^*}{\partial x_1\,\partial S^*}\mathrm{d}S^* + \frac{\partial^2 U^*}{\partial x_1\,\partial V^*}\mathrm{d}V^* + \frac{\partial^2 U^*}{\partial x_1^2}\mathrm{d}x_1 \\
&\qquad + \ldots + \frac{\partial^2 U^*}{\partial x_1\,\partial x_{m-1}}\mathrm{d}x_{m-1} = 0 \\
&\cdot\quad\cdot\quad\cdot\quad\cdot\quad\cdot\quad\cdot\quad\cdot\quad\cdot \\
\mathrm{d}\left(\frac{\partial U^*}{\partial x_{m-2}}\right) &= \frac{\partial^2 U^*}{\partial x_{m-2}\,\partial S^*}\mathrm{d}S^* + \frac{\partial^2 U^*}{\partial x_{m-2}\,\partial V^*}\mathrm{d}V^* + \frac{\partial^2 U^*}{\partial x_{m-2}\,\partial x_1}\mathrm{d}x_1 \\
&\qquad + \ldots + \frac{\partial^2 U^*}{\partial x_{m-2}\,\partial x_{m-1}}\mathrm{d}x_{m-1} = 0.
\end{aligned}\right\} \tag{46.11}$$

Since the variation cannot go into the unstable region, we can show by reasoning in a way completely analogous to our previous discussion that we must have

$$\mathrm{d}D^* = \frac{\partial D^*}{\partial S^*}\mathrm{d}S^* + \frac{\partial D^*}{\partial V^*}\mathrm{d}V^* + \frac{\partial D^*}{\partial x_1}\mathrm{d}x_1 + \ldots + \frac{\partial D^*}{\partial x_{m-1}}\mathrm{d}x_{m-1} = 0. \tag{46.12}$$

The system of homogeneous linear equations (46.11), (46.12) thus has a non-trivial solution only if the determinant of the coefficients vanishes. This gives us the general equations

$$D^* = 0, \quad D^{*\prime} = 0 \tag{46.13}$$

for a critical phase of a system of m components.

A different choice of the secondary conditions (46.11) naturally leads to a different form of the determinants $D^{*\prime}$. This possibility was taken into account in the definition so that eqs. (46.13) are not affected by it. It is easily seen that eqs. (46.13) reduce to eqs. (46.3) and (46.6) for one-component systems. Further equivalent formulations are obtained by choosing other ways of writing the basic conditions (40.11) and (40.12), for example by keeping the volume constant (as was done by Gibbs).

§ 47 Transformation of the equations for critical phases. Further properties of critical phases

For purposes of application even in the general case, the equations for critical phases are derived more conveniently from suitable thermodynamic potentials than from the fundamental equation. This possibility depends on the fact that certain stability conditions remain valid even on the boundary of the stable region. For the transformation of the stability conditions (§ 41) a separation into conditions for intensive parameters and conditions for extensive parameters was carried out. This makes it possible to isolate the required conditions right from the start and thus to simplify the equations for the critical phase. We shall give two examples. We have stated above that, for the liquid–vapour equilibrium of one-component systems, the thermal stability condition $C_V > 0$ remains valid on the boundary of the region of stability. The same applies for liquid–vapour equilibria of multicomponent systems. The use of the Helmholtz free energy is, therefore, particularly appropriate here, since it allows the separation of just this condition. We then apply completely analogous reasoning to that used above to the remaining extensive parameters $V^*, x_1, \ldots, x_{m-1}$. We can, therefore, confine our discussion to a simple statement of the result. If we define D_F^* by

$$D_F^* \equiv \begin{vmatrix} \dfrac{\partial^2 F^*}{\partial V^{*2}} & \dfrac{\partial^2 F^*}{\partial V^* \partial x_1} & \cdots & \dfrac{\partial^2 F^*}{\partial V^* \partial x_{m-1}} \\ \dfrac{\partial^2 F^*}{\partial x_1 \partial V^*} & \cdot & & \vdots \\ \vdots & & \cdot & \vdots \\ \dfrac{\partial^2 F^*}{\partial x_{m-1} \partial V^*} & \cdots & \cdots & \dfrac{\partial^2 F^*}{\partial x_{m-1}^2} \end{vmatrix} \tag{47.1}$$

and denote by $D_F^{*\prime}$ the determinant formed from (47.1) by replacing the elements of an arbitrarily chosen row by

$$\frac{\partial D_F^*}{\partial V^*}, \quad \frac{\partial D_F^*}{\partial x_1}, \quad \ldots, \quad \frac{\partial D_F^*}{\partial x_{m-1}}, \tag{47.2}$$

the equations for the critical phase are

$$D_F^* = 0, \quad D_F^{*\prime} = 0. \tag{47.3}$$

It should be noted that all partial derivatives which occur here must be formed at constant temperature. For one-component systems the eqs. (47.3) reduce to eqs. (46.1).

In the unmixing of condensed phases a further simplification occurs since here (unlike liquid–vapour equilibria) the mechanical stability condition $(\partial^2 F/\partial V^2)_T > 0$ is also generally obeyed at the boundary of the stability region.† In these cases, therefore, we usually use the formulation by means of the Gibbs free energy which starts with the conditions (41.31) and (41.32). The derivation itself is again carried out in the same way. If we define D_G^* by

$$D_G^* \equiv \begin{vmatrix} \dfrac{\partial^2 G^*}{\partial x_1^2} & \dfrac{\partial^2 G^*}{\partial x_1 \partial x_2} & \cdots & \dfrac{\partial^2 G^*}{\partial x_1 \partial x_{m-1}} \\ \dfrac{\partial^2 G^*}{\partial x_2 \partial x_1} & \cdot & & \vdots \\ \vdots & & \cdot & \vdots \\ \dfrac{\partial^2 G^*}{\partial x_{m-1} \partial x_1} & \cdots & \cdots & \dfrac{\partial^2 G^*}{\partial x_{m-1}^2} \end{vmatrix} \tag{47.4}$$

and denote by $D_G^{*\prime}$ the determinant formed from (47.4) by replacing the elements of an arbitrarily chosen row by the partial derivatives

$$\frac{\partial D_G^*}{\partial x_1}, \quad \frac{\partial D_G^*}{\partial x_2}, \quad \ldots, \quad \frac{\partial D_G^*}{\partial x_{m-1}}, \tag{47.5}$$

the equations for the critical phase are

$$D_G^* = 0, \quad D_G^{*\prime} = 0. \tag{47.6}$$

All derivatives must here be taken with respect to the mole fractions at constant T and P. The equations (47.6) differ from the previous

† This means that exceptions are of an accidental nature and have no thermodynamic connexion with the occurrence of the critical phase. There is, however, a certain class of liquid–vapour equilibria for which the mechanical stability condition cannot be fulfilled for the critical phase.

ones in that they cannot be reduced to the equations for a one-component system. They have found interesting applications in the theory of solutions of macromolecules.

Equations (46.13) and (47.6) were originally derived by Gibbs in a somewhat different form. The formulation by means of the Helmholtz free energy [eq. (47.3)] was used largely by van der Waals in his study of vaporization equilibria.

Critical phases are of special interest since they show a number of remarkable manifestations collectively known as 'critical phenomena'. We shall give only a few examples. A remarkable phenomenon observed even in one-component systems is that the molar heat at constant pressure, the coefficient of thermal expansion, and the compressibility all approach infinity as the critical point is approached. These facts depend, as we shall see more clearly in § 51, on the position of the critical point on the boundary of the region of stability. The behaviour of the compressibility means that exact measurements are extraordinarily difficult near the critical point because of the influence of gravity.

In binary systems the phenomena of 'retrograde vaporization' and 'retrograde condensation' may appear near the vaporization critical point. A detailed discussion of these phenomena is outside the scope of this book. We shall, therefore, only explain these concepts briefly and show by means of a simple example how they can be caused. Let us consider a one-component system whose temperature $T < T_c$ is fixed (cf. Fig. 28). Compression of the gas is represented by movement along the isotherm. Condensation occurs at a certain density; further compression causes an increase in the amount of the liquid phase until the gas phase finally vanishes. In binary gaseous systems, on the other hand, the following can occur near the critical point. If the gaseous mixture is compressed isothermally, a pressure is reached at which the liquid phase appears. Under certain conditions of temperature and composition, however, further compression causes a reduction in the amount of liquid and, finally, the disappearance of the liquid phase. This phenomenon was first observed by Cailletet on a mixture of CO_2 and air and is called *retrograde condensation of the first kind*. Figure 34 represents schematically an isothermal P–x-diagram and shows how this phenomenon can occur. The dot indicates the critical point. The compression occurs along the broken line. Condensation starts at point A; further compression causes the amount of liquid to increase, then to decrease again until the liquid phase finally vanishes again at point B. Analogous behaviour is observed if the temperature is varied at constant pressure. This behaviour is described as *retrograde condensation of the second kind*.

The most remarkable and strange critical phenomenon is called *critical opalescence*. This was found by Avenarius (1874) only a few years after the discovery of the critical point. If a gas of critical density is cooled it starts to emit a bluish opalescent light at about 1 °C above the critical temperature. The intensity of this light

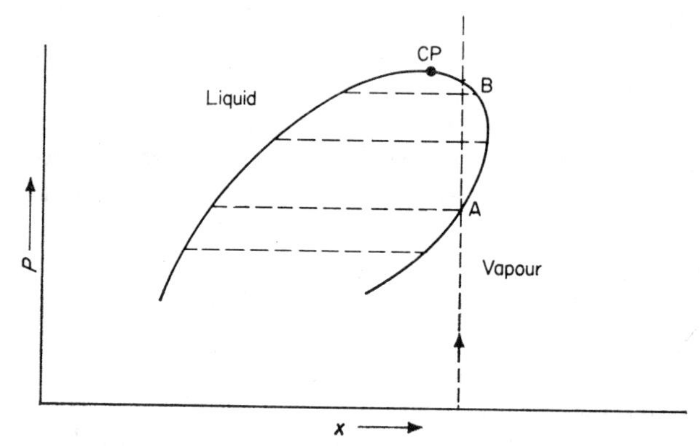

FIG. 34. Explanation of retrograde condensation. CP = critical point

increases rapidly on approach to the critical point although the system is still homogeneous. This phenomenon is caused by a large increase in light-scattering intensity, particularly in a forward direction, on approaching the critical point. The same manifestation is observed near the critical point of unmixing in liquid and solid systems. In the latter case, of course, X-rays are required for the demonstration of the phenomenon. Theoretical analysis shows that the critical opalescence is likewise a direct consequence of the fact that the critical point lies on the boundary of the stability region.

§ 48 Transitions of higher order. The Ehrenfest equations

The characteristic functions introduced in Chapter III are continuous functions of the quantities of state and can, in general, be differentiated as many times as we like. When this is not so for certain well-defined values of the quantities of state, we speak of a (thermodynamic) transition. We shall confine our discussion to the Gibbs free energy of one-component systems since the appropriate conditions are particularly easily comprehensible. The simplest example of transitions is found in phase changes† in which a new phase appears while the original phase finally disappears (e.g. melting, evaporation). The region of state in which the two phases co-exist constitutes the heterogeneous equilibrium which was discussed in detail in Chapter IV. We there used the fact of experience that the molar volumes, the molar entropies, and therefore also the molar enthalpies of the two co-existing phases always differ by finite amounts. In connexion with the present section we are, however, not

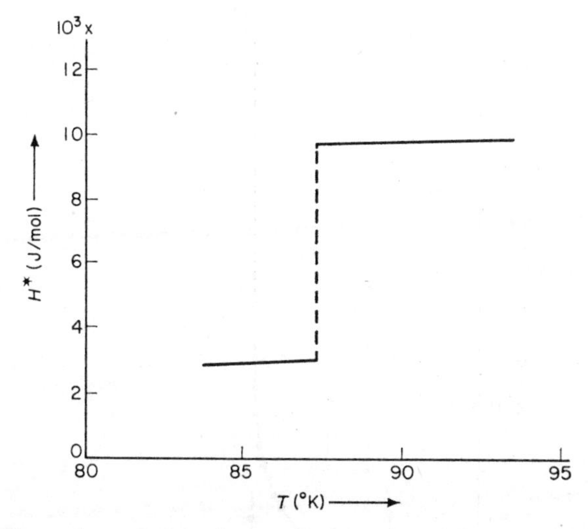

FIG. 35. The molar enthalpy of argon in the region of the normal boiling point

interested in the co-existence of the phases but in the behaviour of the characteristic function over a larger region of state. The state of affairs under discussion then means, according to (21.37), that at the place where a phase change takes place the first derivatives of the molar Gibbs free energy show a discontinuity and thus cannot be differentiated.‡ The discontinuity of the molar enthalpy can best be seen from eq. (24.9). We may regard these statements as the thermodynamic definition of a phase change. Since they are usually easily verified experimentally and, furthermore, other physical properties of the phases (e.g. crystal structure) can be used in addition, it nearly always is possible to identify phase changes with certainty. Figure 35

† In the literature the expression 'phase change' is often also used in the more general context of thermodynamic transitions.

‡ We can say alternatively that at the place where a phase change takes place the second derivatives of the Gibbs free energy become infinite.

shows the behaviour of the molar enthalpy of argon in the vicinity of the boiling point at normal pressure (1 atm) as an example.

Many transitions have, however, been found for which entropy and volume are continuous at the transition point whereas the second or possibly higher derivatives of the Gibbs free energy are discontinuous. We shall call these collectively *transitions of higher order* and shall, for purposes of clarification, show the dependence of the molar heat on temperature at constant pressure for a few systems.

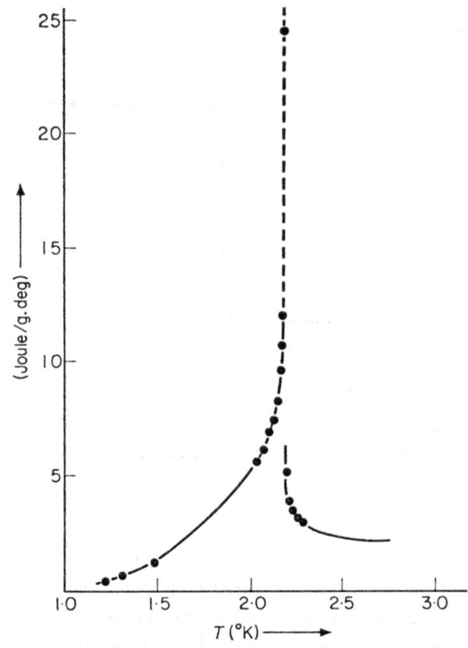

FIG. 36. The molar heat capacity of ^{4}He in the region of the λ-point

One of the best known examples is the λ-transition which occurs at 2·19 K in liquid ^{4}He under normal pressure. The name is due to the shape of the $C(T)$-curve which is shown in Fig. 36 (taken from the older literature).†

Transitions of higher order have also been found in alloys (substitution mixed crystals). They are usually called *hyperstructure transitions* since they are related to the appearance or disappearance of the

† For experimental reasons the molar heat of liquid helium is measured at the equilibrium vapour pressure.

so-called hyperstructure lines in the X-ray diffraction diagram. Figure 37 shows the $C_P(T)$-curve for β-brass.†

Many crystals consisting of molecules or polyatomic ions also show higher-order transitions. Such transitions are usually called *rotation transitions*.‡ As examples we show the behaviour of the molar heats of ethylene dibromide (Fig. 38) and hydrogen iodide (Fig. 39).

We finally come to electric and magnetic transition points (Curie point, Néel point, etc.) which also belong to the type under discussion. We shall, however, only give one example (Fig. 40) since we shall return to these transitions in Chapters IX and X.

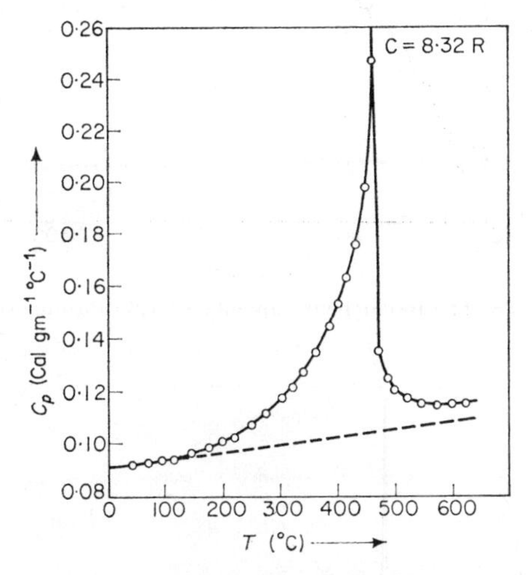

FIG. 37. The specific heat of β-brass

The first attempt to classify these manifestations and to develop a thermodynamic theory for them was made by Ehrenfest. His scheme implies a generalization of the definition of a phase change formulated above and defines a phase change of the nth order generally by the statement that the Gibbs free energy and its derivatives up to the order $n-1$ are continuous while the nth derivatives show finite discontinuities. The case $n = 1$ includes phase changes which are, therefore, also called first-order transitions. It is easy to see that relationships analogous to the Clausius–Clapeyron equation (31.6)

† The region of existence of the β-phase lies between 45·8 and 48·9 atom % Zn in the temperature range under discussion.

‡ The name is taken from an older hypothesis due to Pauling, according to which free rotation of molecules or ions sets in at the transition point. Actually we are dealing with a difference in degree of orientation.

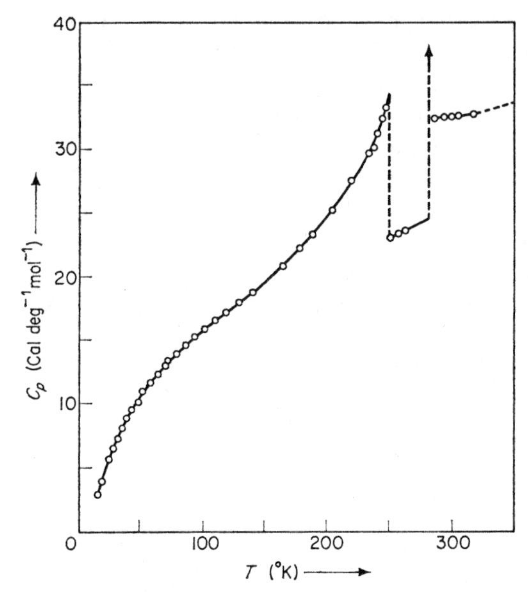

FIG. 38. The molar heat capacity of 1,2-dibromoethane

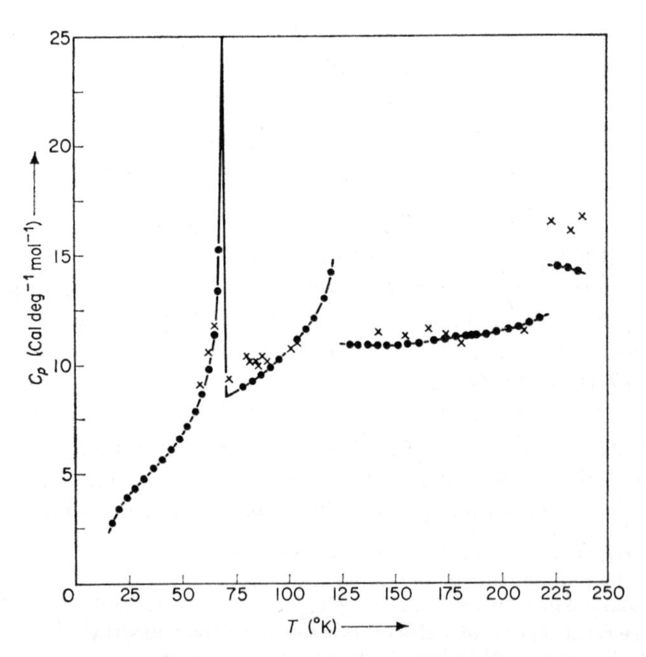

FIG. 39. The molar heat capacity of hydrogen iodide

must exist for the higher-order transitions defined above. To keep the discussion simple, we shall confine our attention to the most important case which is $n = 2$. We must first make the requirements more

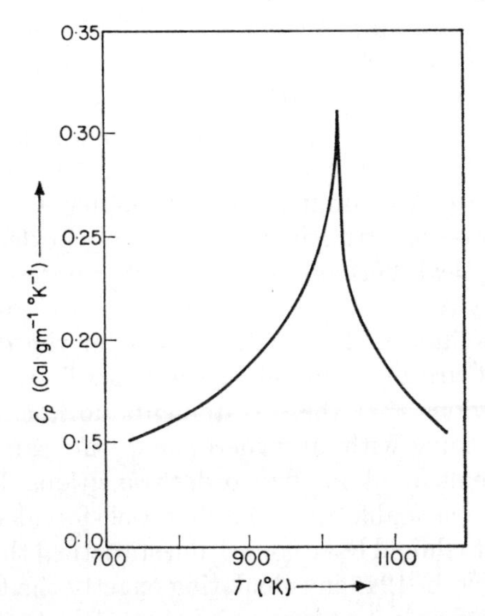

FIG. 40. The specific heat of iron in the region of the Curie point

precise by assuming that the first derivatives of G^* are continuous over a finite stretch of the equilibrium curve (not merely at one isolated point). We then have

$$\mathrm{d}\left(\frac{\partial G^*}{\partial T}\right) = -\mathrm{d}S^* = \frac{\partial^2 G^*}{\partial T^2}\mathrm{d}T + \frac{\partial^2 G^*}{\partial P\,\partial T}\mathrm{d}P \tag{48.1}$$

and

$$\mathrm{d}\left(\frac{\partial G^*}{\partial P}\right) = \mathrm{d}V^* = \frac{\partial^2 G^*}{\partial T\,\partial P}\mathrm{d}T + \frac{\partial^2 G^*}{\partial P^2}\mathrm{d}P. \tag{48.2}$$

From the definitions (25.4)–(25.6) we get

$$\frac{\partial^2 G^*}{\partial T^2} = -\frac{C_P}{T}, \quad \frac{\partial^2 G^*}{\partial T\,\partial P} = V^*\alpha, \quad \frac{\partial^2 G^*}{\partial P^2} = -V^*\kappa. \tag{48.3}$$

If we denote the temperature and pressure of the second-order transition by T_{II} and P_{II} respectively, then

$$\frac{\mathrm{d}T_{\mathrm{II}}}{\mathrm{d}P_{\mathrm{II}}} = \frac{V^*T(\alpha''-\alpha')}{C_P''-C_P'} \tag{48.4}$$

and

$$\frac{dT_{II}}{dP_{II}} = \frac{\kappa'' - \kappa'}{\alpha'' - \alpha'} \tag{48.5}$$

must be true because of the assumed continuity of S^* and V^* and because of (48.1)–(48.3). Equations (48.4) and (48.5) are the *Ehrenfest equations for second-order transitions.*

The Ehrenfest equations (just like the theory of critical phases discussed in § 45) constitute an application of thermodynamics to a set of previously defined circumstances. Whether these circumstances correspond to those occurring in nature can only be decided by experiment or by statistical thermodynamics. Higher-order transitions were originally usually regarded as second-order transitions in the sense of the Ehrenfest definition. Numerical values for the discontinuity in the specific heat are often found in the older literature. It should, however, be obvious that these statements do not follow from the experimental results with anywhere near the same certainty as analogous problems involving first-order transitions. In view of these difficulties it is reasonable that the first well-founded objections to the older concepts should have come from statistical thermodynamics. Onsager succeeded in 1945 in calculating exactly the thermodynamic functions for a simple two-dimensional model (the so-called Ising model); he also showed that the specific heat tends logarithmically to infinity at the Curie point and at the transition point of a hyperstructure transition. Figures 41 and 42 respectively show the course of the enthalpy and of the specific heat in the vicinity of a transition of the Onsager type.†

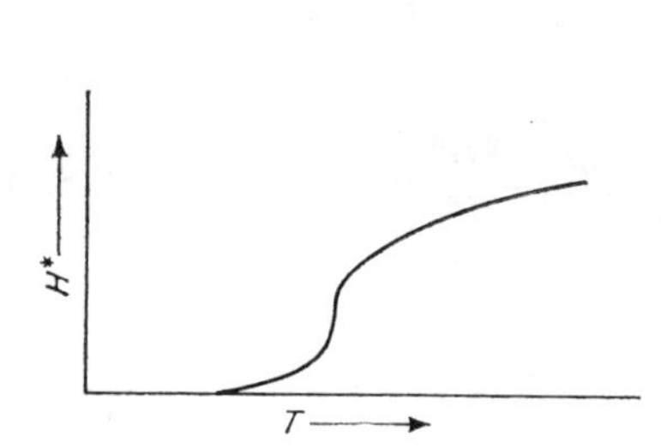

FIG. 41. Enthalpy in the region of an Onsager transition (schematic)

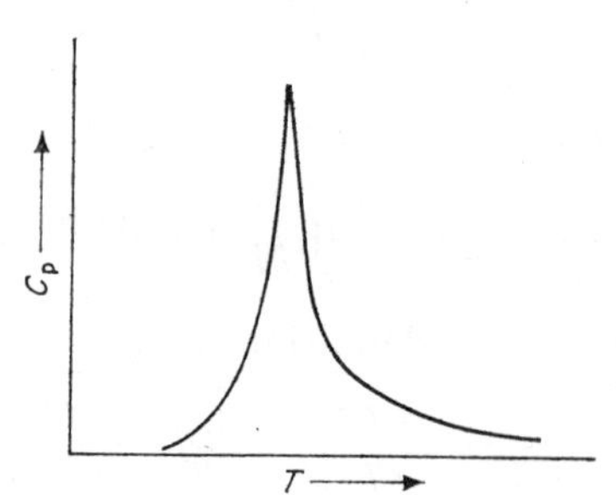

FIG. 42. The specific heat in the region of an Onsager transition

† The Ising model is based on a rigid lattice. The compressibility and the coefficient of thermal expansion are therefore zero and the theory yields the internal energy and the molar heat at constant volume. Figures 41 and 42 are scaled to the other figures in this section.

Nobody has so far succeeded in solving the resulting three-dimensional problem. Methods of successive approximation which have been much improved in recent years make it practically certain that the specific heat at the transition point here also tends to infinity.

The first experimental indication of the correctness of the new concepts was obtained by Fairbank *et al.* (1957) by means of precision measurements on liquid helium where the temperature was measured and controlled to better than 10^{-6} degrees. The results shown in

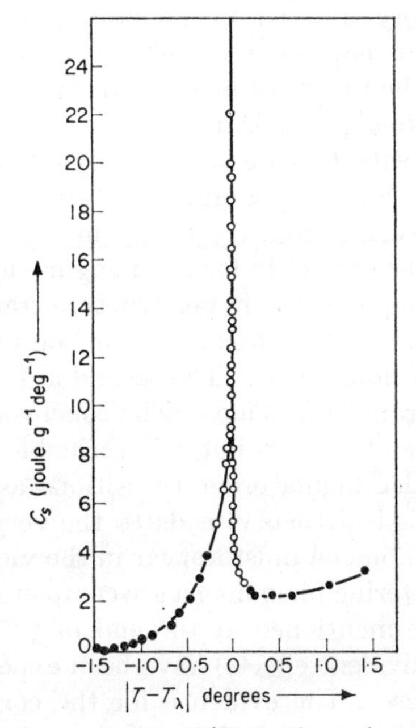

Fig. 43. The molar heat capacity of ^{4}He in the region of the λ-point from recent precision measurements

Fig. 43 can be represented very accurately by a formula which becomes logarithmically infinite at the λ-point. It is, therefore, almost beyond doubt that the λ-transition of helium has essentially the characteristics of a transition of the Onsager type.

A survey of the experimental results for other higher-order transitions shows that in the great majority of cases the experimental data are compatible with the picture outlined above. According to opinions generally held at present, the only case established with reasonable certainty of a second-order transition of the Ehrenfest type is the beginning of superconduction at vanishing magnetic field strength. This transition will be discussed in § 67 and we shall, therefore, give no details here. In most other cases we are probably dealing with transitions of the Onsager type. The Ehrenfest theory is, therefore, not applicable and we are faced once more with the problem of finding a purely thermodynamic treatment of higher-order transitions.

§ 49 Tisza's theory. I: The general basis

A new purely thermodynamic theory of higher-order transitions, which takes Onsager's results into account, was developed by Tisza (1951). Tisza's theory is based on the assumption that most higher-order transitions do not differ essentially from the critical points discussed in § 45. The theory of higher-order transitions is, therefore, obtained by a suitable generalization of the theory of critical phases. Before proceeding with this we shall briefly describe the arguments which support the above assumption. Two results are of special significance in this connexion, the first being of a purely theoretical nature. We can show explicitly for the Ising model (cf. § 48) that the ferromagnetic Curie point, the hyperstructure transition, the critical condensation point, and the critical solution point of a binary solution are mathematically equivalent.† The second argument is based on a conclusion drawn from Tisza's ideas. The conclusion goes beyond the framework of thermodynamics but it is verifiable directly by experiment. If we consider higher-order transitions as a special kind of critical point, statistical theory leads to the very reasonable result that scattering phenomena must appear in the vicinity of the transition point. The scattering phenomena are completely analogous to the critical opalescence mentioned at the end of § 47. These scattering phenomena (or equivalent effects) have been experimentally demonstrated in many cases. The evidence for the correctness of Tisza's theory can, therefore, be accepted as sufficiently certain.

We now want to generalize the theory of critical phases so that it will include most higher-order transitions. For this purpose it is necessary first of all to remind ourselves of the method used in § 46. We started from a purely phenomenological definition of a critical phase and then reduced this definition to an abstract statement about the stability conditions. This reduction was possible only because we had in § 39 particularized the concept of the stability of a homogeneous phase as a stability with respect to separation into at least two phases different from the original phase. A generalization of the theory must, therefore, necessarily start with the formulation of the concept of

† The equivalence derives from the abstractly defined parameters of the Ising model being physically interpreted in different ways.

stability. According to the discussion in § 39, the question arises whether this can be done within the formalism of thermodynamics. We can actually do this since we used the formalism only partially in § 39. First of all, we used only one work co-ordinate, the volume. This limitation is not required by thermodynamics as can be seen from §§ 14 and 15. The limitation will not be admissible for many of the systems in question here as can be seen from the examples given in § 46. We shall here give only one illustrative example since the introduction of further work co-ordinates will be discussed in Chapters VIII–X. Let us consider a ferromagnetic crystalline solid above the Curie point. The total magnetization is here an additional work co-ordinate (extensive parameter) and the magnetic field strength is the conjugate generalized force (intensive parameter). If we choose the latter as the independent variable, the magnetization tends to zero with vanishing field strength corresponding to a disordered distribution (over considerable regions) of the spin orientations. The comparison state needed, according to § 40, for the derivation of the stability conditions (the asterisked state) is now a state in which the magnetization has a finite value even at a vanishingly small field strength, i.e. in which there is spontaneous magnetization as there is in the stable state below the Curie point. This state corresponds to a preferred parallel orientation of the spins. The state does not generally constitute a phase in the thermodynamic sense but may be treated as such for the purpose of deriving the stability conditions.† ‡ We shall not carry out a detailed derivation but merely note that the most important result for our purpose is the statement that the reciprocal magnetic susceptibility is always positive for a stable phase.

A considerable number of higher-order transitions is not covered by the introduction of additional work co-ordinates; among them are, in particular, the λ-transition of helium, hyperstructure transitions, and rotational transitions. For these cases we can use a method used so far almost exclusively in the treatment of chemical reactions (§§ 36 and 44), i.e. the introduction of internal parameters. These are, as explained in § 14, not variables of state; on the contrary, we showed in § 17 explicitly by means of an example that their equilibrium value is determined by the values of the variables of state.

† The magnetization must obviously here be chosen as the independent variable.

‡ The theorem proved in § 45 concerning the number of degrees of freedom does not apply to the critical point (i.e. the Curie point in this case) since we are not dealing with a phase in the thermodynamic sense.

Since, however, internal parameters have independent significance apart from the equilibrium, they can, as shown in §§ 14 and 39, be introduced as independent variables into the fundamental equation for the treatment of equilibrium and stability problems.

We choose as our illustrative example the case of β-brass which is particularly easy to understand. For simplicity we shall assume an atom ratio of 1 : 1.† β-Brass has a body-centred cubic crystal lattice. The minimum potential energy corresponds to a distribution of atoms such that each Cu atom has only Zn atoms as nearest neighbours and vice versa. At the absolute zero, therefore, each centre of a cube is

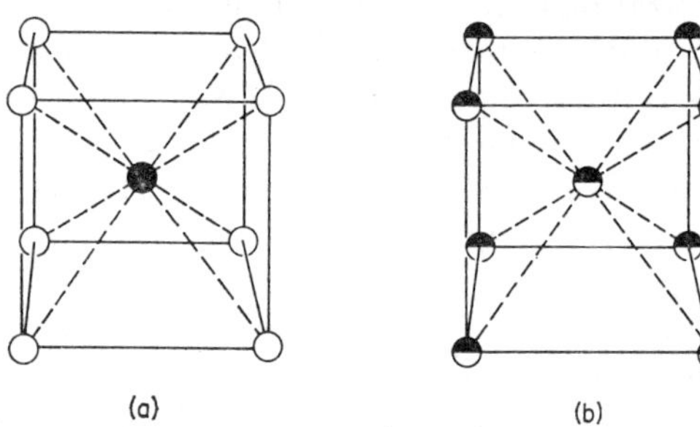

FIG. 44. Distribution of atoms in β-brass. a, ordered. b, disordered

occupied by a Cu atom while the corners are occupied by Zn atoms (or vice versa) (Fig. 44a). If we imagine the body-centred cubic lattice to be formed from two interpenetrating simple cubic lattices, one simple lattice will, at the absolute zero, be occupied only by Cu atoms, the other only by Zn atoms. With increasing temperature, increasing numbers of atoms migrate to 'wrong' lattice positions. The simple Cu lattice thus contains increasing numbers of Zn atoms and the Zn lattice corresponding numbers of Cu atoms. Above the transition point the ratio of atoms is 1 : 1 in each simple lattice also, i.e. half the centres and half the corners of cubes are occupied by Cu atoms, the other halves by Zn atoms (Fig. 44b). This behaviour can be described, according to Bragg and Williams, by an internal parameter called the *degree of long-range order* η. If we denote the number of Cu atoms in the 'right' lattice positions by $N_1(1)$, and the number of Cu atoms in the 'wrong' lattice positions by $N_1(2)$ we have

$$\eta = \frac{N_1(1)-N_1(2)}{N_1(1)+N_1(2)}. \tag{49.1}$$

† Deviations from this 'stoichiometric' composition are usually not significant for the purpose of theoretical considerations.

This definition shows immediately that the degree of long-range order at stable equilibrium at the absolute zero is unity whereas it is zero above the transition point. If we again consider the region above the transition point, we see that the comparison state necessary for the derivation of the stability conditions (the asterisked state) is characterized by a finite value of the degree of long-range order. We thus find that the second derivative of the internal energy with respect to the degree of long-range order must be positive for a stable phase.

We need not here consider the question how the internal parameter is to be defined in each individual case. The question must be answered by statistical thermodynamics in connexion with the discussion of the molecular mechanism. We may without hesitation assume the possibility of such a definition since it is known explicitly for a number of cases (but not for the case of the λ-transition of helium). However, the fact that internal parameters are not variables of state has consequences which must be taken into account in the development of the theory. Firstly, the comparison phase is now no longer characterized by macroscopic measurable quantities in the sense of § 2. A critical investigation of the question how far thermodynamic formalism is applicable is, therefore, not superfluous. We cannot enter into a detailed discussion of this question but merely state that Tisza and Klein have formulated a criterion which must be obeyed by the internal parameters introduced into the fundamental equation. The second consequence results from the relationship

$$\left(\frac{\partial U}{\partial \eta}\right)_x = 0 \tag{49.2}$$

which defines the equilibrium value of the internal parameter. Comparison with (20.9) shows that (49.2) means that the intensive parameter conjugate to η must at equilibrium be identically equal to zero. We showed this explicitly in §§ 14 and 37 using chemical affinity as our example. Internal parameters can, therefore, not be incorporated into the Legendre transformation of the fundamental equation as far as quasi-static processes are concerned. This limitation does not apply to the virtual displacements used for the derivation of equilibrium and stability conditions since these displacements by definition lead to non-equilibrium states. The 'intensive' parameters conjugate to the internal parameters here differ from zero (as can again be seen by using affinity as an example). Since the derivatives of these quantities with respect to the internal parameters generally differ from zero even at equilibrium, the internal parameters can now be wholly incorporated into the formalism of thermodynamics. We need not, for the present, pay any attention to their special position.

§ 50 Tisza's theory. II: Formulation of the stability conditions

We shall first turn the stability conditions into a form which takes into account the above-mentioned generalizations and is, at the same time, more convenient for the present problem than the formulations on §§ 40 and 41. We can here use the characteristic functions and extensive parameters in a form independent of the size of the system as was established in detail in § 40. We shall not, however, confine ourselves to mean molar quantities but leave the form of the reduction open; we shall use lower-case letters for the reduced quantities following a generalized form of the notation used in § 20. The symbol X_i for extensive parameters will now also include internal parameters.† Furthermore, we denote, otherwise than in eq. (21.1), the total number of independent variables (including internal parameters) of the fundamental equation by $r+1$. The fundamental equation can now be written in the reduced form

$$u = u(x_1, \dots, x_r). \tag{50.1‡}$$

The generalization of (20.41) gives for the intensive parameters

$$P_i = \frac{\partial U}{\partial X_i} = \frac{\partial u}{\partial x_i}. \tag{50.2}$$

In comparison with the terminology of elasticity theory we shall give the second derivatives of u the name *thermodynamic moduli*. These are always finite or zero. We can write them as

$$u_{ij} = \frac{\partial^2 u}{\partial x_i \, \partial x_j} = \left(\frac{\partial P_i}{\partial x_j}\right)_{x_{k \neq j}} = \left(\frac{\partial P_j}{\partial x_i}\right)_{x_{k \neq j}} = u_{ji}. \tag{50.3}$$

If the quadratic matrix formed from these quantities is denoted by $\boldsymbol{D}$, we have

$$\boldsymbol{D} \equiv [u_{ij}]. \tag{50.4}$$

If the matrix of the thermodynamic moduli is not singular, i.e. $D = 0$ for this determinant, we can define the reciprocal matrix by the equation

$$\boldsymbol{C} = \boldsymbol{D}^{-1} = \frac{\hat{\boldsymbol{D}}}{D}. \tag{50.5}$$

† The justification for this notation can be seen, for example, from eq. (36.5). The degree of long-range order defined by eq. (49.1) is obviously already a reduced variable in this context.

‡ The quantities x_i defined above must not be confused with mole fractions which do not occur in this section.

The matrix $\hat{\boldsymbol{D}}$ is obtained by first substituting for each element D_{ij} of the matrix $\boldsymbol{D}$ the co-factor of D_{ij} from the determinant D; rows and columns are now interchanged in the new matrix. This procedure would seem to make the calculation of the elements of $\boldsymbol{C}$ extremely difficult. This calculation is, in fact, very easy since, according to (47.5), D represents a Jacobi determinant whose reciprocal is obtained directly from (25.16)–(25.18). In this way we find that the elements of $\boldsymbol{C}$ are given by

$$v_{ij} = \left(\frac{\partial x_i}{\partial P_j}\right)_{P_{k\neq j}} = \left(\frac{\partial x_j}{\partial P_i}\right)_{P_{k\neq j}} = v_{ji}. \tag{50.6}$$

We shall again draw comparisons with elasticity theory and call the quantities v_{ij} *thermodynamic coefficients*.†

From § 21 we have for the thermodynamic potentials that‡

$$\psi^{(k)} = u - \sum_{i=1}^{k} P_i x_i, \tag{50.7}$$

$$\psi^{(k)} = \psi^{(k)}(P_1, \ldots, P_k, x_{k+1}, \ldots, x_r) \tag{50.8}$$

with the first derivatives

$$\frac{\partial \psi^{(k)}}{\partial P_i} = -x_i \quad (i = 1, \ldots, k), \tag{50.9}$$

$$\frac{\partial \psi^{(k)}}{\partial x_j} = P_j \quad (j = k+1, \ldots, r). \tag{50.10}$$

We can write the second derivatives of $\psi^{(k)}$ with respect to extensive parameters as

$$\psi_{jl}^{(k)} = \frac{\partial^2 \psi^{(k)}}{\partial x_j\, \partial x_l} = \left(\frac{\partial P_j}{\partial x_l}\right)_{P_i, x_{m\neq l}} = \left(\frac{\partial P_l}{\partial x_j}\right)_{P_i, x_{m\neq j}} = \psi_{lj}^{(k)}$$

$$(i = 1, \ldots, k;\ j, l = k+1, \ldots, r;\ m = k+1, \ldots, r;\ m \neq l, m \neq j \text{ respectively}). \tag{50.11}$$

It should be noted that the quantities defined by (50.11) are not identical with the thermodynamic moduli defined by (50.3), although, like the thermodynamic moduli, they are always finite or zero.

The generalized stability condition has a form completely analogous to that of (40.11). We have merely to introduce into the quadratic

† In English language publications u_{ij} are usually called *stiffness moduli*, the v_{ij} are called *compliance coefficients* while $\boldsymbol{D}$ and $\boldsymbol{C}$ are respectively called *stiffness matrix* and *compliance matrix*.

‡ The reasons for the slight change in notation compared with § 21 are easily recognized.

form the above-mentioned additional variables. Using (50.3) we thus obtain explicitly

$$\delta^2 u = \frac{1}{2} \sum_{i,j}^{r} u_{ij}\, \delta x_i\, \delta x_j > 0, \tag{50.12}$$

or, in matrix notation,†

$$\delta^2 u = \tfrac{1}{2} \widetilde{\delta \boldsymbol{x}} . \boldsymbol{D} . \delta \boldsymbol{x} > 0 \tag{50.13}$$

where $\delta \boldsymbol{x}$ is the column vector formed from the components δx_i and $\widetilde{\delta \boldsymbol{x}}$ is the corresponding transposed vector (row vector). Since, according to (47.5) the matrix $\boldsymbol{D}$ is symmetric it can always be changed into a diagonal matrix $\boldsymbol{\Lambda}$ by a congruent transformation. We shall denote the elements of $\boldsymbol{\Lambda}$ by λ_i. We have, therefore,

$$\boldsymbol{\Lambda} = \tilde{\boldsymbol{Q}} . \boldsymbol{D} . \boldsymbol{Q} = [\lambda_i \delta_{ij}], \tag{50.14}$$

where $\boldsymbol{Q}$ is the transformation matrix, $\tilde{\boldsymbol{Q}}$ its transposed matrix,‡ and δ_{ij} is the Kronecker delta.§

The quantities λ_i defined by (50.14) are not the eigenvalues of the matrix $\boldsymbol{D}$ since the congruent transformation (50.14) is not orthogonal. Since, furthermore, a congruent transformation is not unique, the values of the λ_i are also not unique. They depend on the initially chosen sequence of variables. According to Sylvester's law of the inertia of quadratic forms, however, the numbers of positive, negative, and vanishing λ_i are not changed by a permutation of the variables. Now, our thermodynamic problem is concerned only with these numbers and any sequence of the variables therefore leads to the same ultimate result. This can be easily verified by means of explicit examples. The lack of uniqueness is, therefore, of no significance for our purpose.‖

If we now introduce the new variables y_i by the transformation

$$\delta \boldsymbol{x} = \boldsymbol{Q} . \delta \boldsymbol{y}, \tag{50.15}$$

we obtain, from (50.13)–(50.15),

$$\delta^2 u = \tfrac{1}{2} \widetilde{\delta \boldsymbol{y}} . \boldsymbol{\Lambda} . \delta \boldsymbol{y} > 0, \tag{50.16}$$

† In connexion with the following discussion the reader is reminded that the product of two matrices $\boldsymbol{AB} = \boldsymbol{C}$ is a matrix with the elements $c_{ij} = \sum_k A_{ik} B_{kj}$.

‡ The matrix formed from a given matrix $\boldsymbol{A}$ by interchanging rows and columns is called the transpose of matrix $\boldsymbol{A}$ and is denoted by $\tilde{\boldsymbol{A}}$.

§ The Kronecker delta is defined by $\delta_{ij} = \begin{cases} 1 \text{ for } i = j \\ 0 \text{ for } i = j \end{cases}$.

‖ The orthogonal transformation thus offers no advantage as far as uniqueness is concerned but has the disadvantage that the eigenvalues have no direct physical significance.

or, explicitly,

$$\delta^2 u = \frac{1}{2} \sum_i \lambda_i (\delta y_i)^2 > 0. \tag{50.17}$$

The stability condition can, therefore, be fulfilled only if

$$\lambda_i > 0 \quad (i = 1, \ldots, r). \tag{50.18}$$

It is true, however, although we cannot give the proof here, that

$$\lambda_i = \frac{D_i}{D_{i-1}}, \tag{50.19}$$

where the D_i are the principal minors of the determinant D and we put $D_0 \equiv 1$. From (50.18) and (50.19) we get the equivalent form of the stability conditions

$$D_i > 0 \quad (i = 1, \ldots, r). \tag{50.20}$$

If we confine ourselves to the variables used in § 40 and, in particular, we understand u to be the mean molar internal energy, then (50.20) becomes identical with (40.12). Equation (50.19) gives the important relationship

$$\prod_{i=1}^{r} \lambda_i = D_r \equiv D. \tag{50.21}$$

We can deduce another useful form of the stability conditions by starting with eq. (50.2). From this and the definitions (50.3) and (50.4) we get the set of inhomogeneous linear equations

$$\delta \boldsymbol{P} = \boldsymbol{D} . \delta \boldsymbol{x}. \tag{50.22}$$

If we now put

$$\left. \begin{aligned} \delta P_1 = \delta P_2 = \ldots = \delta P_{k-1} = 0, \\ \delta x_{k+1} = \delta x_{k+2} = \ldots = \delta x_r = 0 \end{aligned} \right\} \tag{50.23}$$

eq. (50.22) becomes

$$\left. \begin{aligned} 0 &= u_{11}\,\delta x_1 + u_{12}\,\delta x_2 + \ldots + u_{1k}\,\delta x_k, \\ &\cdot \quad \cdot \quad \cdot \quad \cdot \quad \cdot \\ 0 &= u_{k-1,1}\,\delta x_1 + u_{k-1,2}\,\delta x_2 + \ldots + u_{k-1,k}\,\delta x_k, \\ \delta P_k &= u_{k1}\,\delta x_1 + u_{k2}\,\delta x_2 + \ldots + u_{kk}\,\delta x_k, \\ &\cdot \quad \cdot \quad \cdot \quad \cdot \quad \cdot \\ \delta P_r &= u_{r1}\,\delta x_1 + u_{r2}\,\delta x_2 + \ldots + u_{rk}\,\delta x_k. \end{aligned} \right\} \tag{50.24}$$

We can confine the argument to the first k equations, although we shall not give the proof for this. Application of Cramer's rule then gives

$$\delta x_k = \frac{D_{k-1}}{D_k}\,\delta P_k, \tag{50.25}$$

where the D_k are again the principal minors of the determinant D. Because of (50.23) we can write (50.25) as

$$\left(\frac{\partial P_k}{\partial x_k}\right)_{P_i,x_j} = \frac{D_k}{D_{k-1}}, \quad (i = 1,\ldots,k-1;\, j = k+1,\ldots,r). \tag{50.26}$$

Equations (50.19) and (50.26) together with the definition (50.11) therefore give

$$\lambda_k = \psi_{kk}^{(k-1)}. \tag{50.27}$$

The stability conditions may thus be written in the form

$$\psi_{kk}^{(k-1)} > 0 \quad (k = 1,\ldots,r). \tag{50.28}$$

§ 51 Tisza's theory. III: Critical points and higher-order transitions

We now define a critical point in the generalized sense discussed above† by the statement that it is the point where the quadratic form (50.13) becomes positive semidefinite. We then immediately get from (50.17) that one of the quantities λ_i vanishes at the critical point and, further, that $D = 0$ because of (50.21). We have already used the latter in § 46. We also conclude from this, in connexion with eq. (50.5), that all thermodynamic coefficients with an associated co-factor $|D_{ij}| \neq 0$ tend to infinity. It is therefore possible in principle that certain thermodynamic coefficients remain finite at the critical point. Thermodynamics can give no information about this if it happens accidentally (i.e. because of special properties of the system under consideration). We shall, however, now show that such behaviour can occur for systematic reasons, i.e. because of the thermodynamic structure of the critical point.

We shall start by asking: for which subscript s is $\lambda_s = 0$ at the critical point. Since the values of the λ_i depend on the original sequence of the variables, we shall assume the sequence to be such that s has the smallest possible value. We now distinguish between two cases: (a) the normal case $s = r$; (b) the exceptional case $s < r$.‡

In order to clarify the meaning of this division we shall now carry out the transformations (50.14), (50.15) explicitly (we previously only

† We shall use the expression 'critical point' in the generalized sense of Tisza's theory throughout this section.

‡ This distinction obviously makes sense only if the definition of the normal case is independent of the sequence of the variables. We shall show explicitly later, by means of an example, that this is so.

used a theorem of linear algebra). We therefore transform the quadratic form

$$\delta^2 u = \frac{1}{2}\sum_{i,j}^{r} u_{ij}\,\delta x_i\,\delta x_j \tag{51.1}$$

into a sum of squares by stepwise completion of the squares. We shall assume for this purpose that $u_{ii} \neq 0$ for all i.† Since we have

$$\left(\delta x_1 + \frac{1}{u_{11}}\sum_{j=2}^{r} u_{1j}\,\delta x_j\right)^2 = (\delta x_1)^2 + 2\frac{1}{u_{11}}\sum_{j=2}^{r} u_{1j}\,\delta x_1\,\delta x_j + \frac{1}{u_{11}^2}\left(\sum_{j=2}^{r} u_{1j}\,\delta x_j\right)^2 \tag{51.2}$$

we can write (51.1) as

$$\delta^2 u = \frac{1}{2}\left[u_{11}\left(\delta x_1 + \frac{1}{u_{11}}\sum_{j=2}^{r} u_{1j}\,\delta x_j\right)^2 - \frac{1}{u_{11}}\left(\sum_{j=2}^{r} u_{1j}\,\delta x_j\right)^2 + \sum_{i,j=2}^{r} u_{ij}\,\delta x_i\,\delta x_j\right]. \tag{51.3}$$

With the notation

$$\delta y_1 = \delta x_1 + \frac{1}{u_{11}}\sum_{j=2}^{r} u_{1j}\,\delta x_j \tag{51.4}$$

and

$$u_{ij}^{(1)} = u_{ij} - \frac{u_{1i}\,u_{1j}}{u_{11}} = \frac{1}{u_{11}}\begin{vmatrix} u_{11} & u_{1j} \\ u_{1i} & u_{ij} \end{vmatrix} \tag{51.5}$$

we obtain as the first step

$$\delta^2 u = \tfrac{1}{2}u_{11}(\delta y_1)^2 + \frac{1}{2}\sum_{i,j=2}^{r} u_{ij}^{(1)}\,\delta x_i\,\delta x_j. \tag{51.6}$$

The equation can easily be changed to a different form. In (50.23) we put $k = 2$ and add to the system (50.24) all the analogous systems of equations obtained by replacing δx_k by δx_j with $k+1 \leqslant j \leqslant r$. If we omit superfluous equations we obtain $r-1$ systems of equations of the form

$$\left.\begin{aligned} 0 &= u_{11}\,\delta x_1 + u_{1j}\,\delta x_j, \\ \delta P_i &= u_{i1}\,\delta x_1 + u_{ij}\,\delta x_j \quad (i,j = 2,\ldots,r) \end{aligned}\right\}. \tag{51.7}$$

This, with (50.11), (50.27), and (51.5), gives

$$u_{11} = \lambda_1, \quad u_{ij}^{(1)} = \psi_{ij}^{(1)} \tag{51.8}$$

† The mathematical condition for this is that the matrix $\boldsymbol{D}$ is of rank r which implies the above assumption.

and, with (51.6), we now get

$$\delta^2 u = \tfrac{1}{2}\lambda_1(\delta y_1)^2 + \frac{1}{2}\sum_{i,j=2}^{r} \psi_{ij}^{(1)}\,\delta x_i\,\delta x_j. \tag{51.9}$$

The second term of the right-hand side contains only $r-1$ variables. We now get, from the original supposition, that $\psi_{ii}^{(1)} \neq 0$ for $i = 2, \ldots, r$. The process can, therefore, be continued and the next step leads to the expression

$$\delta^2 u = \tfrac{1}{2}\lambda_1(\delta y_1)^2 + \tfrac{1}{2}\lambda_2(\delta y_2)^2 + \frac{1}{2}\sum_{i,j=3}^{r} \psi_{ij}^{(2)}\,\delta x_i\,\delta x_j \tag{51.10}$$

with

$$\delta y_2 = \delta x_2 + \frac{1}{\psi_{22}^{(1)}}\sum_{j=3}^{r} \psi_{2j}^{(1)}\,\delta x_j. \tag{51.11}$$

The quadratic form (51.1) is thus finally completely transformed into the sum of squares (50.17). Equations (51.4), (51.11), and their further analogues represent explicitly the reversal of the transformation (50.15). We can see from the general form of these equations that the last of them is simply

$$\delta y_r = \delta x_r. \tag{51.12}$$

We now drop the assumption we made initially and assume instead that

$$\psi_{ss}^{(s-1)} = 0. \tag{51.13}$$

Because of (51.12), this does not cause any difficulties in the transformation when $s = r$. The quadratic form is, however, now positive semidefinite as we can see clearly by putting

$$\delta y_1 = \delta y_2 = \ldots = \delta y_{r-1} = 0, \quad \delta y_r \neq 0.$$

This is the above-mentioned normal case.

The situation is different in the exceptional case $s < r$. The process of completion of squares now stops with the subscript $s-1$ as can be seen from eq. (51.11). We therefore have

$$\delta^2 u = \frac{1}{2}\sum_{i=1}^{s-1} \lambda_i(\delta y_i)^2 + \psi_{ss}^{(s-1)}(\delta x_s)^2 + \sum_{j=s+1}^{r} \psi_{sj}^{(s-1)}\,\delta x_s\,\delta x_j$$
$$+\frac{1}{2}\sum_{i,j=s+1}^{r} \psi_{ij}^{(s-1)}\,\delta x_i\,\delta x_j. \tag{51.14}$$

If we now put

$$\delta y_1 = \delta y_2 = \ldots = \delta y_{s-1} = 0, \quad \delta x_s \neq 0,$$

and

$$\delta x_{s+1} = \delta x_{s+2} = \ldots = \delta x_r = 0$$

we can see that at least the form (51.14) can assume the value zero because of (51.13); however, (51.14) is not positive semidefinite but indefinite, i.e. it can assume positive or negative values. We can show this to be so by assuming that $\psi_{sl}^{(s-1)} \neq 0$ $(l > s)$ and by putting

$$\delta y_1 = \delta y_2 = \ldots = \delta y_{s-1} = 0, \quad \delta x_s \neq 0, \quad \delta x_l \neq 0, \quad \delta x_j = 0$$

for $j \neq l$. With (51.13) we then get from (51.14) that

$$\delta^2 u = \psi_{sl}^{(s-1)}\, \delta x_s\, \delta x_l + \tfrac{1}{2}\psi_{ll}^{(s-1)}\, (\delta x_l)^2. \tag{51.15}$$

This expression can, however, be made positive or negative by a suitable choice of δx_s and δx_l and the assertion is thus proved. If, therefore, (51.13) applies, the quadratic form (51.14) can be positive semidefinite only if the condition

$$\psi_{sj}^{(s-1)} \equiv \left(\frac{\partial P_s}{\partial x_j}\right)_{P_1,\ldots,P_{s-1},x_s,x_{m\neq j}} = 0 \quad (j, m > s) \tag{51.16}$$

is fulfilled. With (51.13) and (51.16), eq. (51.14) reduces to the expression

$$\delta^2 u = \frac{1}{2}\sum_{i=1}^{s-1}\lambda_i(\delta y_i)^2 + \frac{1}{2}\sum_{i,j=s+1}^{r}\psi_{ij}^{(s-1)}\,\delta x_i\,\delta x_j. \tag{51.17}$$

In the reduced form the right-hand side no longer contains the variable δx_s. It is, therefore, positive definite and the completion of the squares can be continued. We must, however, remember that the variable x_s is no longer included in the sequence of Legendre transformations. In the second term of the right-hand side of (51.17) the quantities

$${}^{s}\psi_{kk}^{(k-1)} = \left(\frac{\partial P_k}{\partial x_k}\right)_{P_1,\ldots,P_{s-1},x_s,P_{s+1},\ldots,P_{k-1},x_{k+1},\ldots x_r} \tag{51.18}$$

derived from the thermodynamic potential

$${}^{s}\psi^{(k-1)} = u - \sum_{\substack{i=1\\ i\neq s}}^{k-1} P_i\, x_i \tag{51.19}$$

therefore replace the quantities $\psi_{kk}^{(k-1)}$. We shall not follow this through since we shall not require these details in our further discussion.

So far we have considered only second-order variations. It is easy to see that this corresponds to the derivation of the first equation for the critical point when applied to phase transitions. A further relationship for the critical point must, therefore, exist within the framework of the generalized definition of this section. This relationship is obtained from a consideration of third-order variations. We shall for this purpose investigate a special variation for which $\delta u = \delta^2 u = 0$. The general stability condition (18.2) can then be fulfilled if $\delta^3 u = 0$ also. We thus now imagine eq. (51.1) written with the inclusion of third-order terms and we again carry out the same transformation of the independent variables. The variation

$$\delta y_1 = \ldots = \delta y_{s-1} = 0, \quad \delta x_s \neq 0, \quad \delta x_{s+1} = \ldots = \delta x_r = 0$$

then corresponds strictly to the above definition and we obtain

$$\psi_{sss}^{(s-1)} \equiv \left(\frac{\partial^3 \psi^{(s-1)}}{\partial x_s^3}\right)_{P_1,\ldots,P_{s-1},x_{s+1},\ldots,x_r} = 0. \tag{51.20}$$

Thermodynamics can only make a conditional statement concerning the question of the behaviour of the variations (or derivatives) of higher order. If $\delta^{2i} u = 0$ for $i = 1, 2, \ldots, n$, then $\delta^{2n+1} u = 0$ also. The question itself can, however, only be answered by experiment or by statistical thermodynamics. No one has succeeded, so far, in solving this problem completely. The problem is, however, not very significant for the purposes of thermodynamics since statements about the fourth derivatives of thermodynamic potentials already lie at the limits of experiment, particularly in the vicinity of critical points.

We shall now summarize the relationships which define a critical point in Tisza's formulation. They are:

$$\psi_{ss}^{(s-1)} = 0, \quad \psi_{sss}^{(s-1)} = 0, \tag{51.21}$$

$$\psi_{kk}^{(k-1)} > 0 \quad (k < s), \tag{51.22}$$

$$\psi_{sj}^{(s-1)} = 0, \quad (j > s). \tag{51.23}$$

For the normal case ($s = r$), the conditions (51.21) and (51.22) correspond exactly to the formulations derived in § 46 for the special case of phase transitions. The condition (51.23) is then redundant. We have mentioned previously that all thermodynamic coefficients tend to infinity at the critical point. This statement can also be formulated in a different way. We derive from eq. (50.22) that

$$-\frac{\partial^2 \psi^{(r)}}{\partial P_i \, \partial P_j} = \left(\frac{\partial x_i}{\partial P_j}\right)_{P_l} = \frac{|D|_{ij}}{D}, \quad (i,j \leqslant r;\, j \neq l \leqslant r). \tag{51.24}$$

Since the matrix $\boldsymbol{D}$ is singular at the critical point, we have

$$\left|\left(\frac{\partial x_i}{\partial P_j}\right)_{P_l}\right| \to \infty. \tag{51.25†}$$

For the exceptional case, the extensive parameters are divided into two classes, $x_1, \ldots, x_s$ and $x_{s+1}, \ldots, x_r$, by the condition (51.23). They are independent in the sense that, in the immediate neighbourhood of the critical point, a change in the parameters of the second class at constant $P_1, \ldots, P_{s-1}$ does not influence the intensive parameter P_s. Since $\psi_{sj}^{(s-1)} = \psi_{js}^{(s-1)}$, we have the equivalent statement that a change in x_s under the same secondary conditions does not influence the intensive parameters $P_{s+1}, \ldots, P_r$. In order to investigate the behaviour of the second derivatives, we first consider a state far from the critical point and put

$$\left.\begin{aligned} &\delta P_j \neq 0, \quad \delta P_l = 0 \quad (j, l = 1, \ldots, s;\, j \neq l), \\ &\delta x_{s+1} = \ldots = \delta x_r = 0. \end{aligned}\right\} \tag{51.26}$$

If we resolve the system of equations (50.22) with respect to $\delta x_1, \ldots, \delta x_s$ under these conditions, we obtain

$$-\frac{\partial^2 \psi^{(s)}}{\partial P_i \, \partial P_j} = \left(\frac{\partial x_i}{\partial P_j}\right)_{P_l, x_m} = \frac{|D_s|_{ij}}{D_s} \quad (i, j \leqslant s;\; l \leqslant s;\; m > s). \tag{51.27}$$

The matrix $\boldsymbol{D}_s$ becomes singular at the critical point according to (50.19), (50.27), and (51.21). We therefore obtain

$$\left|\left(\frac{\partial x_i}{\partial P_j}\right)_{P_l, x_m}\right| \to \infty \quad (i, j \leqslant s). \tag{51.28‡}$$

The second derivatives with respect to the extensive parameters $\psi_{kl}^{(s)}$ $(k, l > s)$ can, as already mentioned, have no singularities. Second derivatives of mixed extensive and intensive parameters are without significance for stability problems as was shown in § 41; they must, therefore, likewise show 'normal' behaviour.§

According to the foregoing results we can say that the critical point is generated by the co-operation of the s extensive parameters of the first class; we shall from now on call them *critical parameters*. Tisza and Callen give transitions for which the entropy is one of the critical parameters the name *order–disorder transitions* whereas transitions

† The vertical lines do not here indicate a determinant but the absolute value.

‡ See footnote above.

§ Analogous statements apply to thermodynamic coefficients. We shall not prove this here.

for which the entropy belongs to the second class (i.e. a parameter irrelevant for the critical point) are called *displacive transitions*.

We now give a few simple illustrative examples. The most important applications can be discussed only when we reach Chapter IX since they require the introduction of additional work co-ordinates.

We first consider the critical point of vaporization of a one-component system. The fundamental equation for molar quantities is

$$u = u(s, v). \tag{51.29}$$

With the definitions (20.21a), (25.4)–(25.6) the matrix of the thermodynamic moduli now becomes

$$\boldsymbol{D} = \begin{bmatrix} \left(\dfrac{\partial T}{\partial s}\right)_v & \left(\dfrac{\partial T}{\partial v}\right)_s \\ +\left(\dfrac{\partial P}{\partial s}\right)_v & -\left(\dfrac{\partial P}{\partial v}\right)_s \end{bmatrix} = \begin{bmatrix} \dfrac{T}{C_V} & \dfrac{T\alpha}{\kappa C_V} \\ \dfrac{T\alpha}{\kappa C_V} & \dfrac{C_p}{C_V \kappa v} \end{bmatrix}. \tag{51.30}$$

For the matrix of the thermodynamic coefficients we obtain

$$\boldsymbol{C} = \boldsymbol{D}^{-1} = \begin{bmatrix} \left(\dfrac{\partial s}{\partial T}\right)_P & -\left(\dfrac{\partial s}{\partial P}\right)_T \\ \left(\dfrac{\partial v}{\partial T}\right)_P & \left(\dfrac{\partial v}{\partial P}\right)_T \end{bmatrix} = \begin{bmatrix} \dfrac{C_p}{T} & \alpha \\ \alpha & v\kappa \end{bmatrix}. \tag{51.31}$$

Transformation into the diagonal form according to eq. (50.27) gives

$$\lambda_1 = \frac{\partial^2 u}{\partial s^2} = \frac{T}{C_V}, \quad \lambda_2 = \frac{\partial^2 f}{\partial v^2} = -\left(\frac{\partial P}{\partial v}\right)_T, \tag{51.32}$$

where f is the molar Helmholtz free energy, s and v are the critical parameters. We are thus dealing with the normal case. At the critical point $\lambda_2 = 0$. From (51.21) we obtain explicitly

$$\left(\frac{\partial P}{\partial v}\right)_T = 0, \quad \left(\frac{\partial^2 P}{\partial v^2}\right)_T = 0 \tag{51.33}$$

in agreement with (46.1). All the thermodynamic coefficients tend to infinity at the critical point. We have, therefore, according to (51.31) that

$$C_P \to \infty, \quad \alpha \to \infty, \quad \kappa \to \infty. \tag{51.34}$$

We now change the sequence of variables and put $x_1 = v$, $x_2 = s$. The diagonal elements then have to be interchanged in (51.30) and (51.31). We thus find the elements of the diagonal matrix to be

$$\lambda_1 = \frac{\partial^2 u}{\partial v^2} = \frac{C_P}{C_V \kappa v}, \quad \lambda_2 = \frac{\partial^2 h}{\partial s^2} = \frac{T}{C_P}. \tag{51.35}$$

At the critical point we again have $\lambda_2 = 0$. We have thus shown explicitly that the definition of the normal case is independent of the sequence of the variables. The relationships (51.34) remain valid here and the first one of eqs. (51.33) follows directly. The derivation of the second eq. (51.33) is complicated and will not be carried out.

We shall choose as our second example the critical solution point of a binary liquid solution. We shall choose as our concentration variable the ratio of the mole numbers $c_1 \equiv n_1/n_2$. The critical parameters are then s, v, and c_1 and we are again dealing with the normal case. From (51.21) we again obtain

$$\left(\frac{\partial \mu_1}{\partial c_1}\right)_{T,P} = 0, \quad \left(\frac{\partial^2 \mu_1}{\partial c_1^2}\right)_{T,P} = 0. \tag{51.36}$$

The agreement of this relationship with (46.2) is easily established. The expression (51.22) gives, for the present case,

$$-\left(\frac{\partial P}{\partial v}\right)_{T,c_1} > 0. \tag{51.37}$$

This relationship states that the isothermal compressibility remains finite at the critical solution point.

The hyperstructure transition of β-brass mentioned in § 48 is a simple example of a state of affairs which leads to the definition of the exceptional case and to the introduction of internal parameters. We have initially the same variables of state as in the previous case, i.e. s, v, and c_1. The conditions (15.21) are, however, fulfilled neither for v nor for c_1 so that the entropy s is the only critical parameter among the variables of state. We are, therefore, dealing with the exceptional case. The entropy alone does not, however, permit a proper description of the circumstances. The reasons for this have already been discussed in § 49. We therefore introduce the degree of long-range order η defined by eq. (49.1) into the fundamental equation as an additional independent variable. We then have

$$u = u(s, v, c_1, \eta). \tag{51.38}$$

The critical parameters are now $x_1 = s$, $x_2 = x_s = \eta$. From (51.21) we have for the critical point that

$$\frac{\partial f}{\partial \eta} = 0, \quad \frac{\partial^2 f}{\partial \eta^2} = 0. \qquad (51.39)†$$

We find from (51.28) that C_V and, because of (25.15), C_P tend to infinity at the transition point.

† The second of eqs. (51.21) is not required since eq. (49.2) also applies for internal parameters.

CHAPTER VIII

Solids

§ 52 The strain tensor

We have so far used only one work co-ordinate, i.e. the volume, in explicit calculations. We did, however, come to the conclusion in § 14 that the introduction of further work co-ordinates is quite simple within the framework of thermodynamic formalism. We accordingly formulated, in § 15, the Second Law for any (finite) number of work co-ordinates. We mentioned, in § 49, some physical problems which cannot be dealt with unless additional work co-ordinates are introduced. In this and the following chapters we discuss some particularly important examples of such work co-ordinates and use them in the treatment of various physical problems.

According to § 14 a general definition of a work co-ordinate y_j is that the quantity $Y_j \, dy_j$ gives quasi-static work done on a closed system,† where Y_j denotes the generalized force conjugate to y_j. The structure of the fundamental equation discussed in § 20 requires that the y_j are extensive parameters and that the Y_j are intensive parameters. Apart from this, however, the explicit form of the quantities $Y_j \, dy_j$ must be deduced from the appropriate field of theoretical physics. The derivation of this form is not strictly a part of thermodynamics. For the special case of the mechanical work done on a fluid phase, however, this derivation is fairly trivial so that we simply wrote down the corresponding expression in § 3. The situation is less simple for the work co-ordinates discussed in this chapter. We shall, therefore, outline the derivation in each case as far as is necessary for an understanding of the thermodynamic considerations. Further details can be found in textbooks of theoretical physics.

We have mentioned previously that the expression $-P \, dV$ is generally valid only for fluid phases. The reason for this is that mechanical work done on a fluid phase is necessarily volume work

† From here on the expression 'work' is always used in the sense of this definition.

whereas a change of form of a solid body also requires work. This can be expressed more precisely by the statement that definite geometric interrelations exist between the material elements within a solid body and that their interrelations can be changed only by expending mechanical work. A thermodynamic consideration of solid bodies therefore requires the introduction of additional work co-ordinates even for the case of purely mechanical work (i.e. in the absence of external fields and when surface effects are ignored). The definition and properties of these additional work co-ordinates will be discussed in this section.

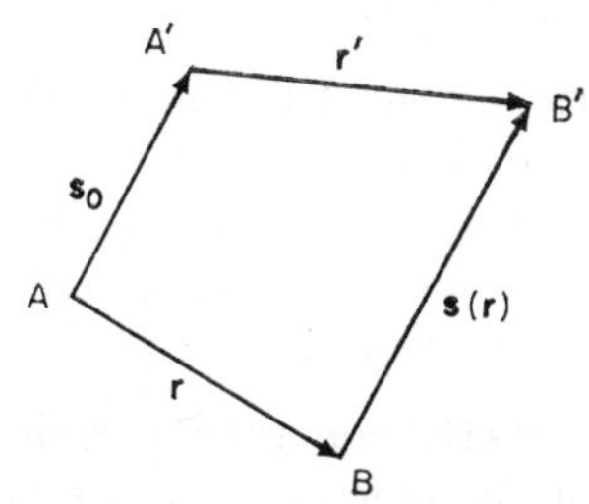

FIG. 45. Description of small deformations

We shall first investigate the mathematical description of small deformations. Let us examine a small region round a point A inside a deformable body.† We shall choose A as the origin of co-ordinates. Let B be a point in the neighbourhood of A within the region. We shall denote the position vector of B by $\boldsymbol{r}$. During a deformation, A undergoes a displacement $\boldsymbol{s}_0$ and becomes A′ while B undergoes a displacement $\boldsymbol{s}(\boldsymbol{r})$ dependent on the distance $\boldsymbol{r}$ and becomes B′. We denote the distance A′B′ by $\boldsymbol{r}'$. We can read off from Fig. 45 that

$$\boldsymbol{r}+\boldsymbol{s}(\boldsymbol{r}) = \boldsymbol{s}_0+\boldsymbol{r}'. \tag{52.1}$$

We now make the following assumptions:

(a) The region under consideration is so small that powers and products of the distances from A of all points B can be neglected.

(b) The displacement $\boldsymbol{s}(\boldsymbol{r})$ is a continuous function of the position co-ordinates.

(c) $\boldsymbol{s}(\boldsymbol{r})$ is so small and changes so slowly that products of the derivatives $\partial s_x/\partial x, \partial s_x/\partial y, \partial s_x/\partial z, \partial s_y/\partial x, \ldots$ with each other and with the components of $\boldsymbol{s}(\boldsymbol{r})$ can also be neglected.

† The following general discussion includes, as we shall see, fluid phases as a special case. We shall, therefore, not assume, for the present, that we are dealing only with solid bodies.

With these assumptions we can expand $\boldsymbol{s}(\boldsymbol{r})$ at the point A into a Taylor series and cut the series short after the linear term. In the components representation we then have

$$\left.\begin{aligned} s_x &= s_{0x}+\boldsymbol{r}.\operatorname{grad} s_x = s_{0x}+\left(\frac{\partial s_x}{\partial x}\right)_0 x+\left(\frac{\partial s_x}{\partial y}\right)_0 y+\left(\frac{\partial s_x}{\partial z}\right)_0 z, \\ s_y &= s_{0y}+\boldsymbol{r}.\operatorname{grad} s_y = s_{0y}+\left(\frac{\partial s_y}{\partial x}\right)_0 x+\left(\frac{\partial s_y}{\partial y}\right)_0 y+\left(\frac{\partial s_y}{\partial z}\right)_0 z, \\ s_z &= s_{0z}+\boldsymbol{r}.\operatorname{grad} s_z = s_{0z}+\left(\frac{\partial s_z}{\partial x}\right)_0 x+\left(\frac{\partial s_z}{\partial y}\right)_0 y+\left(\frac{\partial s_z}{\partial z}\right)_0 z. \end{aligned}\right\} \tag{52.2}$$

The partial derivatives on the right-hand side represent the nine components of a tensor called the vector gradient of $\boldsymbol{s}$ and denoted by Grad $\boldsymbol{s}$. In tensor notation eqs. (48.2) therefore take the simple form

$$\boldsymbol{s}(\boldsymbol{r}) = \boldsymbol{s}_0+\boldsymbol{r}.\operatorname{Grad}\boldsymbol{s}. \tag{52.3}$$

This equation represents the general expression for the displacement of the points of the volume element under consideration. According to Helmholtz' fundamental theorem of kinematics this displacement can be represented by a superposition of the following three movements:

(a) translation;
(b) rotation;
(c) dilation or contraction in three mutually perpendicular directions.

The movements (a) and (b) do not affect the internal energy of the body and are, therefore, of no significance for thermodynamics. The required work co-ordinates must therefore be obtained from the proof of Helmholtz' theorem. We use for this the fact that any tensor $\boldsymbol{T}$ can be expressed as the sum of a symmetric tensor $\boldsymbol{T}^s$ and an antisymmetric tensor $\boldsymbol{T}^a$.† We have, therefore, that

$$\boldsymbol{T} = \boldsymbol{T}^s+\boldsymbol{T}^a \tag{52.4}$$

with

$$\left.\begin{aligned} \boldsymbol{T}^s &= \tfrac{1}{2}(\boldsymbol{T}+\tilde{\boldsymbol{T}}), \\ \boldsymbol{T}^a &= \tfrac{1}{2}(\boldsymbol{T}-\tilde{\boldsymbol{T}}), \end{aligned}\right\} \tag{52.5}$$

where $\tilde{\boldsymbol{T}}$ is the 'transpose' of $\boldsymbol{T}$ (i.e. obtained from $\boldsymbol{T}$ by interchanging rows and columns). If we now denote by $\boldsymbol{R}$ the antisymmetric tensor

† For the components of a symmetric tensor we have that $T_{ij} = T_{ji}$ while for those of an antisymmetric tensor $T_{ij} = -T_{ji}$.

derived from Grad $\boldsymbol{s}$ according to (52.5) and by $\boldsymbol{S}$ the corresponding symmetric tensor, then eq. (52.3) can be written

$$\boldsymbol{s}(\boldsymbol{r}) = \boldsymbol{s}_0 + \boldsymbol{r}\,.\,\boldsymbol{R} + \boldsymbol{r}\,.\,\boldsymbol{S}. \tag{52.6}$$

We shall now show that the second term of the right-hand side represents a rotation. The velocity $\boldsymbol{v}$ of the points of a rigid body rotating at an angular velocity $\boldsymbol{\omega}$ is given by the equation

$$\boldsymbol{v} = \boldsymbol{\omega} \times \boldsymbol{r}. \tag{52.7}$$

This gives (since $\boldsymbol{\omega}$ is not a function of position)†

$$\mathrm{rot}\,\mathrm{v} = \boldsymbol{\omega}\,\mathrm{div}\,\boldsymbol{r} - \boldsymbol{\omega}\,.\,\mathrm{Grad}\,\boldsymbol{r} = 2\boldsymbol{\omega}. \tag{52.8}$$

The displacement caused by an infinitesimal rotation is therefore

$$\mathrm{d}\boldsymbol{s}_{\mathrm{R}} = \boldsymbol{v}\,\mathrm{d}t = \tfrac{1}{2}\,\mathrm{rot}\frac{\mathrm{d}\boldsymbol{s}}{\mathrm{d}t}\,\mathrm{d}t \times \boldsymbol{r} \tag{52.9}$$

or if, as before, we no longer use differentials for displacements assumed to be very small,

$$\boldsymbol{s}_{\mathrm{R}} = \tfrac{1}{2}\,\mathrm{rot}\,\boldsymbol{s} \times \boldsymbol{r}. \tag{52.10}$$

This equation in component notation is

$$\left.\begin{aligned} s_{\mathrm{R}x} &= \frac{1}{2}\left(\frac{\partial s_x}{\partial y} - \frac{\partial s_y}{\partial x}\right) y + \frac{1}{2}\left(\frac{\partial s_x}{\partial z} - \frac{\partial s_z}{\partial x}\right) z, \\ s_{\mathrm{R}y} &= \frac{1}{2}\left(\frac{\partial s_y}{\partial x} - \frac{\partial s_x}{\partial y}\right) x + \frac{1}{2}\left(\frac{\partial s_z}{\partial x} - \frac{\partial s_x}{\partial z}\right) z, \\ s_{\mathrm{R}z} &= \frac{1}{2}\left(\frac{\partial s_z}{\partial x} - \frac{\partial s_x}{\partial z}\right) x + \frac{1}{2}\left(\frac{\partial s_z}{\partial y} - \frac{\partial s_y}{\partial z}\right) y. \end{aligned}\right\} \tag{52.11}$$

The coefficients on the right-hand side are easily seen to be exactly the components of the antisymmetric tensor $\boldsymbol{R}$. This shows that the second term of eq. (52.6) represents a rotation. Since the first term corresponds to a translation, the last term must represent a displacement caused purely by a deformation. We shall now prove this explicitly.

We shall from now on ignore translation and rotation; we can use (52.1) to write eq. (52.6) in the abbreviated form

$$\boldsymbol{r}' = \boldsymbol{r} + \boldsymbol{r}\,.\,\boldsymbol{S}. \tag{52.12}$$

† It is generally true that

$$\mathrm{rot}\,(\boldsymbol{a} \times \boldsymbol{b}) = \boldsymbol{b}\cdot\mathrm{Grad}\,\boldsymbol{a} - \boldsymbol{a}\cdot\mathrm{Grad}\,\boldsymbol{b} + \boldsymbol{a}\,\mathrm{div}\,\boldsymbol{b} - \boldsymbol{b}\,\mathrm{div}\,\boldsymbol{a}.$$

The symmetric tensor $\boldsymbol{S}$ which is explicitly given by

$$\boldsymbol{S} = \begin{bmatrix} \frac{\partial s_x}{\partial x} & \frac{1}{2}\left(\frac{\partial s_x}{\partial y}+\frac{\partial s_y}{\partial x}\right) & \frac{1}{2}\left(\frac{\partial s_x}{\partial z}+\frac{\partial s_z}{\partial x}\right) \\ \frac{1}{2}\left(\frac{\partial s_x}{\partial y}+\frac{\partial s_y}{\partial x}\right) & \frac{\partial s_y}{\partial y} & \frac{1}{2}\left(\frac{\partial s_y}{\partial z}+\frac{\partial s_z}{\partial y}\right) \\ \frac{1}{2}\left(\frac{\partial s_x}{\partial z}+\frac{\partial s_z}{\partial x}\right) & \frac{1}{2}\left(\frac{\partial s_y}{\partial z}+\frac{\partial s_z}{\partial y}\right) & \frac{\partial s_z}{\partial z} \end{bmatrix} \quad (52.13)$$

is called the *strain tensor*.

In order to investigate eq. (52.12) we first write the displacement $\boldsymbol{s}_{\mathrm{D}}$ which corresponds to the last term of (52.6) in a form analogous to (52.11), i.e.

$$\left.\begin{aligned} s_{\mathrm{D}x} &= S_{xx}x+S_{xy}y+S_{xz}z,\\ s_{\mathrm{D}y} &= S_{yx}x+S_{yy}y+S_{yz}z,\\ s_{\mathrm{D}z} &= S_{zx}x+S_{zy}y+S_{zz}z. \end{aligned}\right\} \quad (52.14)$$

From a scalar multiplication of eq. (52.12) by $\boldsymbol{r}$ we obtain

$$(\boldsymbol{r}'-\boldsymbol{r}).\boldsymbol{r} = \boldsymbol{s}_{\mathrm{D}}.\boldsymbol{r} = S_{xx}x^2+S_{yy}y^2+S_{zz}z^2+2S_{xy}xy+2S_{yz}yz + 2S_{zx}zx \equiv f(x,y,z). \quad (52.15)$$

For $f(x,y,z) = \text{const.}$, (52.15) is the equation of a second-order surface called the *strain ellipsoid*.† The components of the displacement $\boldsymbol{s}_{\mathrm{D}}$ are now obtained from (52.15) in the simple form

$$s_{\mathrm{D}x} = \frac{1}{2}\frac{\partial f}{\partial x}, \quad s_{\mathrm{D}y} = \frac{1}{2}\frac{\partial f}{\partial y}, \quad s_{\mathrm{D}z} = \frac{1}{2}\frac{\partial f}{\partial z}. \quad (52.16)$$

We now transform the strain ellipsoid into principal axes and denote the new co-ordinates by X_1, X_2, X_3. Equation (52.15) then becomes

$$S_{\mathrm{I}}X_1^2+S_{\mathrm{II}}X_2^2+S_{\mathrm{III}}X_3^2 = F(X_1,X_2,X_3). \quad (52.17)$$

Since eq. (52.16) is independent of the choice of co-ordinates we now have

$$\left.\begin{aligned} s_{\mathrm{D}X_1} &= \frac{1}{2}\frac{\partial F}{\partial X_1} = S_{\mathrm{I}}X_1,\\ s_{\mathrm{D}X_2} &= \frac{1}{2}\frac{\partial F}{\partial X_2} = S_{\mathrm{II}}X_2,\\ s_{\mathrm{D}X_3} &= \frac{1}{2}\frac{\partial F}{\partial X_3} = S_{\mathrm{III}}X_3. \end{aligned}\right\} \quad (52.18)$$

† This name is generally used although the surface is not necessarily an ellipsoid.

These equations are quite simple to interpret. If we assume that the point B in Fig. 45 (p. 217) has the Cartesian co-ordinates X_i, (52.18) tells us that B will change into the point B′ which has the co-ordinates

$$X_i + s_{DX_i} = X_i(1+S_i). \tag{52.19}$$

We have thus shown that the third term in eq. (52.6) represents dilation or contraction in three mutually perpendicular directions. Helmholtz' fundamental theorem is thus completely proved. The quantities $S_{\mathrm{I}}, S_{\mathrm{II}}, S_{\mathrm{III}}$ are called *linear dilations*.

The relative volume change θ associated with a deformation is easily calculated in the principal axes representation. If we denote the original volume by ΔV and the volume in the deformed state by $\Delta V'$ we can give the definition

$$\theta = \frac{\Delta V' - \Delta V}{\Delta V}. \tag{52.20}$$

Suppose that the volume element is a rectangular parallelepiped with sides parallel to the principal axes and with one corner at the origin. If a_i denotes the lengths of the edges in the original state and a_i' the lengths of the edges in the deformed state we have

$$\Delta V = a_1 a_2 a_3, \quad \Delta V' = a_1' a_2' a_3'. \tag{52.21}$$

Equation (52.19) gives

$$a_i' = a_i(1+S_i). \tag{52.22}$$

We thus find

$$\Delta V' = \Delta V(1+S_{\mathrm{I}})(1+S_{\mathrm{II}})(1+S_{\mathrm{III}}) \tag{52.23}$$

and, if we neglect higher-order terms as before,

$$\theta = S_{\mathrm{I}} + S_{\mathrm{II}} + S_{\mathrm{III}}. \tag{52.24}$$

But the sum of the diagonal elements of a tensor is invariant, i.e. it is independent of the choice of the system of co-ordinates. It is, therefore, generally true that

$$\theta = \frac{\partial s_x}{\partial x} + \frac{\partial s_y}{\partial y} + \frac{\partial s_z}{\partial z} \equiv \operatorname{div} \boldsymbol{s}. \tag{52.25}$$

We finally consider the physical meaning of the tensor components S_{ij} for any system of co-ordinates. For simplicity we start with the unit vectors of our system of co-ordinates. We shall denote these vectors by $\boldsymbol{i}, \boldsymbol{j}, \boldsymbol{k}$. From (52.12) we then get

$$\boldsymbol{r}_i' = \boldsymbol{i} + S_{xx}\boldsymbol{i} + S_{xy}\boldsymbol{j} + S_{xz}\boldsymbol{k}. \tag{52.26}$$

The unit vector $\boldsymbol{j}$ is correspondingly changed to the vector

$$\boldsymbol{r}_j = \boldsymbol{j} + S_{yx}\boldsymbol{i} + S_{yy}\boldsymbol{j} + S_{yz}\boldsymbol{k}. \tag{52.27}$$

The relative change in length of the vector $\boldsymbol{i}$ is thus

$$\frac{l_i' - l_i}{l_i} = \frac{r_i' - i}{i} = \frac{\sqrt{(\boldsymbol{r}_i' . \boldsymbol{r}_i')} - 1}{1}. \tag{52.28}$$

If we introduce (52.26) and expand the root we obtain

$$\frac{l_i' - l_i}{l_i} = \frac{\partial s_x}{\partial x} \equiv S_{xx}, \tag{52.29}$$

where higher-order terms have been neglected. The diagonal elements of the strain tensor thus represent the relative changes in length of lines which originally followed the direction of the axes of co-ordinates.

The vectors $\boldsymbol{i}$ and $\boldsymbol{j}$ which are mutually perpendicular in the original state enclose an angle $\frac{1}{2}\pi - \delta_{ij}$ in the deformed state. Since δ_{ij} is assumed to be a very small quantity we have

$$\cos\left(\tfrac{1}{2}\pi - \delta_{ij}\right) = \sin\delta_{ij} = \frac{\boldsymbol{r}_i' . \boldsymbol{r}_j'}{r_i' r_j'} \approx \delta_{ij}. \tag{52.30}$$

If we introduce the expressions (52.26) and (52.27) and again neglect high-order terms, we find that

$$\delta_{ij} = \frac{\partial s_x}{\partial y} + \frac{\partial s_y}{\partial x} = 2S_{xy}. \tag{52.31}$$

The off-diagonal elements of the strain tensor thus represent half the angular change undergone during the deformation by two lines which originally lay in the direction of the co-ordinate axes.

§ 53 The stress tensor

Let us now consider the forces acting on a volume element of the deformable body. We must distinguish between forces acting on the volume element as a whole (gravity, inertia) and forces that are due to neighbouring volume elements and act only on the surface of the volume element under consideration. The latter forces are, therefore, not proportional to the volume of the volume element but to its surface. The force per unit area is called the stress $\boldsymbol{T}$. For a given surface element the stress generally has one component in a direction normal to the surface (normal stress, pressure or tension stress) and another component perpendicular to the first (tangential stress, shear stress). Let us suppose that we know, at an appropriate point, the stresses in the normal directions corresponding to the axes of a

Cartesian co-ordinate system; we shall show that we then also know how $\boldsymbol{T}$ depends on the direction of the normal at that point. We choose a small tetrahedron as our volume element. Three of the tetrahedral surfaces, $\mathrm{d}f_x, \mathrm{d}f_y, \mathrm{d}f_z$ are placed in co-ordinate planes in such a way that their outgoing normals are in the directions of the negative co-ordinate axes (Fig. 46). The stresses applied to these three surfaces

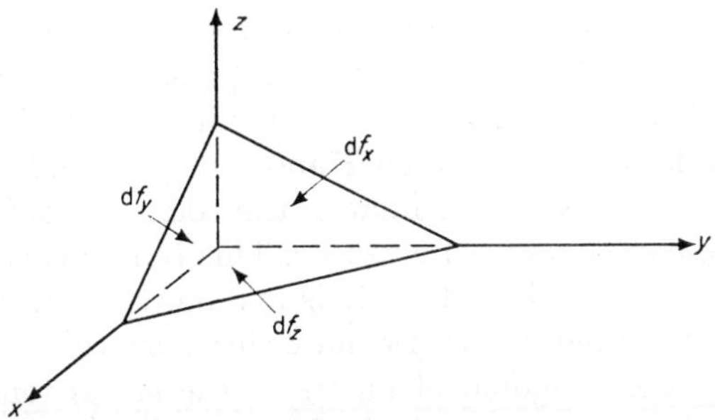

FIG. 46. Calculation of stress

are denoted by $\boldsymbol{T}_1, \boldsymbol{T}_2, \boldsymbol{T}_3$. Let $\boldsymbol{n}$ be the normal to the fourth tetrahedral plane, $\mathrm{d}f$, and let the stress applied to this surface be $\boldsymbol{T}$. Forces proportional to the volume may be neglected in the formulation of the mechanical equilibrium condition for the tetrahedron since they disappear when the surface/volume ratio becomes very large with decreasing volume. We therefore obtain

$$\boldsymbol{T}\,\mathrm{d}f + \boldsymbol{T}_1\,\mathrm{d}f_x + \boldsymbol{T}_2\,\mathrm{d}f_y + \boldsymbol{T}_3\,\mathrm{d}f_z = 0. \tag{53.1}$$

If, however, we take into account the sign of the surfaces, we have

$$\boldsymbol{n}\,\mathrm{d}f = -\boldsymbol{i}\,\mathrm{d}f_x - \boldsymbol{j}\,\mathrm{d}f_y - \boldsymbol{k}\,\mathrm{d}f_z \tag{53.2}$$

and, therefore,

$$\mathrm{d}f_x = -\mathrm{d}f\cos\alpha, \quad \mathrm{d}f_y = -\mathrm{d}f\cos\beta, \quad \mathrm{d}f_z = -\mathrm{d}f\cos\gamma \tag{53.3}$$

where α, β, γ are the angles between the normal $\boldsymbol{n}$ and the positive co-ordinate axes. Equation (53.1) thus becomes

$$\boldsymbol{T} = \boldsymbol{T}_1\cos\alpha + \boldsymbol{T}_2\cos\beta + \boldsymbol{T}_3\cos\gamma. \tag{53.4}$$

If we denote the j-component of $\boldsymbol{T}_i$ by T_{ij}, we can write (53.4) in component representation:

$$\left.\begin{aligned} T_x &= T_{11}\cos\alpha + T_{21}\cos\beta + T_{31}\cos\gamma, \\ T_y &= T_{12}\cos\alpha + T_{22}\cos\beta + T_{32}\cos\gamma, \\ T_z &= T_{13}\cos\alpha + T_{23}\cos\beta + T_{33}\cos\gamma. \end{aligned}\right\} \tag{53.5}$$

Each normal direction $\boldsymbol{n}$ is thus now associated with a stress vector $\boldsymbol{T}$.

Let us now consider a cube whose edge has length unity and whose sides are parallel to the co-ordinate planes. For the sides parallel to the y–z-plane, we have

$$\cos\alpha = 1, \quad \cos\beta = \cos\gamma = 0 \tag{53.6}$$

and, therefore,

$$T_x = T_{11}, \quad T_y = T_{12}, \quad T_z = T_{13}. \tag{53.7}$$

Corresponding relationships apply to the other sides. Coefficients with both subscripts the same thus represent *normal stresses* and coefficients with mixed subscripts represent *tangential stresses.*

Since the relationship between $\boldsymbol{T}$ and $\boldsymbol{n}$ must be independent of the choice of the system of co-ordinates, the coefficients T_{ij} in eq. (53.5) represent the components of a tensor. This tensor is called the *stress tensor*. We shall now show that it is symmetric and can, therefore, again be represented by a second-order surface called the *stress ellipsoid*. The non-diagonal elements of the stress tensor disappear with a transformation to principal axes. The remaining diagonal elements are called *normal stresses* and we denote them by $T_{\mathrm{I}}, T_{\mathrm{II}}, T_{\mathrm{III}}$. In the principal axes representation, eqs. (53.5) take the simple form

$$T_x = T_{\mathrm{I}}\cos\alpha, \quad T_y = T_{\mathrm{II}}\cos\beta, \quad T_z = T_{\mathrm{III}}\cos\gamma. \tag{53.8}$$

The set of equations (53.5) can be quite generally represented in tensor notation as the simple expression

$$\boldsymbol{T} = \boldsymbol{n}\,.\,\mathsf{T}. \tag{53.9}$$

We still need the equilibrium conditions for a deformable body of finite volume. We now have to include in our considerations [in contrast with eq. (53.1)] the forces proportional to the volume; we shall summarize them in a vector $\boldsymbol{F}$ per unit volume. It is necessary and sufficient for the mechanical equilibrium of our volume that both the resultant force and the resultant moment of the forces vanish. It must be true, therefore, that

$$\int \boldsymbol{F}\,\mathrm{d}V + \int \boldsymbol{T}\,\mathrm{d}f = 0, \tag{53.10}$$

$$\int \boldsymbol{r}\times\boldsymbol{F}\,\mathrm{d}V + \int \boldsymbol{r}\times\boldsymbol{T}\,\mathrm{d}f = 0. \tag{53.11}$$

The surface integrals are to be taken over the surface of the volume under consideration since the contributions of internal dividing surfaces cancel.†

† Each internal dividing surface belongs to two adjacent volume elements and therefore makes two equal and opposite contributions.

We shall look first at the x-component of eq. (53.10). The second integral can be transformed if we remember that, because of (53.5), we can write

$$T_x \, df = (T_{11} \cos \alpha + T_{21} \cos \beta + T_{31} \cos \gamma) \, df$$
$$= \boldsymbol{T}_1^* . \boldsymbol{n} \, df = \boldsymbol{T}_1^* . d\boldsymbol{f}, \tag{53.12}$$

where $\boldsymbol{T}_1^*$ is a new vector with components T_{11}, T_{21}, T_{31}, and $d\boldsymbol{f}$ is the vector surface element. According to Gauss' theorem† we then get

$$\int T_x \, df = \int \boldsymbol{T}_1^* . d\boldsymbol{f} = \int \operatorname{div} \boldsymbol{T}_1^* \, dV. \tag{53.13}$$

Analogous transformations can be applied to the y- and z-components. Since the equilibrium condition must apply to every volume, we obtain

$$\left.\begin{aligned} F_x + \operatorname{div} \boldsymbol{T}_1^* &= F_x + \frac{\partial T_{11}}{\partial x} + \frac{\partial T_{21}}{\partial y} + \frac{\partial T_{31}}{\partial z} = 0, \\ F_y + \operatorname{div} \boldsymbol{T}_2^* &= F_y + \frac{\partial T_{12}}{\partial x} + \frac{\partial T_{22}}{\partial y} + \frac{\partial T_{32}}{\partial z} = 0, \\ F_z + \operatorname{div} \boldsymbol{T}_3^* &= F_z + \frac{\partial T_{13}}{\partial x} + \frac{\partial T_{23}}{\partial y} + \frac{\partial T_{33}}{\partial z} = 0, \end{aligned}\right\} \tag{53.14}$$

or, in tensor notation,

$$\boldsymbol{F} + \operatorname{Div} \boldsymbol{T} = 0. \tag{53.15}$$

The operation Div, explained by eqs. (53.14) and (53.15), is called *vector divergence.*

The x-component of eq. (53.11) is

$$\int (yF_z - zF_y) \, dV + \int (yT_z - zT_y) \, df = 0. \tag{53.16}$$

The second integral may again be changed to a triple integral in the same way and gives

$$\int (yT_z - zT_y) \, df = \int (\operatorname{div} y \boldsymbol{T}_3^* - \operatorname{div} z \boldsymbol{T}_2^*) \, dV. \tag{53.17}$$

Since the condition (53.16) must also apply to any volume, we find that

$$yF_z - zF_y + \operatorname{div} y \boldsymbol{T}_3^* - \operatorname{div} z \boldsymbol{T}_2^* = 0 \tag{53.18}$$

† Gauss' theorem states that

$$\int \boldsymbol{a} \cdot d\boldsymbol{f} = \int \operatorname{div} \boldsymbol{a} \, dV$$

where the left-hand integral stretches over a closed surface and the right-hand integral over the volume enclosed by the surface.

or, since

$$\operatorname{div} y\boldsymbol{T}_3^* = y \operatorname{div} \boldsymbol{T}_3^* + \boldsymbol{T}_3^* . j = y \operatorname{div} \boldsymbol{T}_3^* + T_{23}, \tag{53.19}$$

that

$$y(F_z + \operatorname{div} \boldsymbol{T}_3^*) - z(F_y + \operatorname{div} \boldsymbol{T}_2^*) - T_{32} + T_{23} = 0. \tag{53.20}$$

The two expressions in parentheses disappear because of (53.14). If we carry out analogous calculations for the y- and z-components, we get

$$T_{12} = T_{21}, \quad T_{23} = T_{32}, \quad T_{13} = T_{31}. \tag{53.21}$$

We have thus shown that the stress tensor, too, is symmetric at equilibrium (and we are interested only in equilibrium).†

§ 54 The fundamental equation and thermal equations of state

We are now in a position to derive the explicit expression for the quasi-static work done in an infinitesimal deformation. We consider a cubic volume element $\Delta V = \Delta x\, \Delta y\, \Delta z$ and calculate first the work done by the stresses applied to the two sides parallel to the y–z-plane. We shall denote these two amounts of work by $\mathrm{d}W_x$ and $\mathrm{d}W_{x+\Delta x}$. They have opposite signs so that we can write

$$\mathrm{d}W_x = -(T_{xx}\,\mathrm{d}s_x + T_{xy}\,\mathrm{d}s_y + T_{xz}\,\mathrm{d}s_z)\,\Delta y\,\Delta z, \tag{54.1}$$

and work done on the parallel side is then

$$\mathrm{d}W_{x+\Delta x} = (T_{xx}\,\mathrm{d}s_x + T_{xy}\,\mathrm{d}s_y + T_{xz}\,\mathrm{d}s_z)\,\Delta y\,\Delta z + \frac{\partial}{\partial x}(T_{xx}\,\mathrm{d}s_x + T_{xy}\,\mathrm{d}s_y + T_{xz}\,\mathrm{d}s_z)\,\Delta x\,\Delta y\,\Delta z. \tag{54.2}$$

In order to obtain the total work done, we have to add (54.1) to (54.2), add the work done by the external force $\boldsymbol{F}$ in the x-direction, and finally add the analogous expressions for the y- and z-direction. We thus find that

$$\mathrm{d}W = \left[\frac{\partial}{\partial x}(T_{xx}\,\mathrm{d}s_x + T_{xy}\,\mathrm{d}s_y + T_{xz}\,\mathrm{d}s_z) + F_x\,\mathrm{d}s_x + \frac{\partial}{\partial y}(T_{yx}\,\mathrm{d}s_x + T_{yy}\,\mathrm{d}s_y + T_{yz}\,\mathrm{d}s_z) + F_y\,\mathrm{d}s_y + \frac{\partial}{\partial z}(T_{zx}\,\mathrm{d}s_x + T_{zy}\,\mathrm{d}s_y + T_{zz}\,\mathrm{d}s_z) + F_z\,\mathrm{d}s_z\right]\Delta V. \tag{54.3}$$

† The solution of the mechanical problem still requires the definition of the boundary conditions at the surface of the volume. Discussion of this subject will be found in textbooks of theoretical physics.

After the differentiations have been done, the terms $F_i \, \mathrm{d}s_i$ disappear because of the equilibrium condition (53.14). Furthermore, any terms containing spatial derivatives of the components of the stress tensor also vanish. If we now remember that the operations d, $\partial/\partial x$, $\partial/\partial y$, $\partial/\partial z$ are interchangeable we obtain, because of the symmetry of the stress tensor, that

$$\mathrm{d}W = \left[T_{xx}\,\mathrm{d}\frac{\partial s_x}{\partial x} + T_{yy}\,\mathrm{d}\frac{\partial s_y}{\partial y} + T_{zz}\,\mathrm{d}\frac{\partial s_z}{\partial z} + T_{xy}\,\mathrm{d}\left(\frac{\partial s_x}{\partial y} + \frac{\partial s_y}{\partial x}\right)\right.$$
$$\left. + T_{yz}\,\mathrm{d}\left(\frac{\partial s_y}{\partial z} + \frac{\partial s_z}{\partial y}\right) + T_{zx}\,\mathrm{d}\left(\frac{\partial s_z}{\partial x} + \frac{\partial s_x}{\partial z}\right)\right] \Delta \mathrm{V} \quad (54.4)$$

or, if we introduce the components of the strain tensor defined by eq. (52.13), that

$$\mathrm{d}W = [T_{xx}\,\mathrm{d}S_{xx} + T_{yy}\,\mathrm{d}S_{yy} + T_{zz}\,\mathrm{d}S_{zz} + 2T_{xy}\,\mathrm{d}S_{xy}$$
$$+ 2T_{yz}\,\mathrm{d}S_{yz} + 2T_{zx}\,\mathrm{d}S_{zx}]. \quad (54.5)$$

This equation is, so far, valid only for an infinitesimal volume element ΔV. In general, the components of the stress tensor as well as those of the strain tensor must be regarded as field quantities, i.e. as functions of the position co-ordinates. The thermodynamic treatment of this general case was done by Gibbs but is relatively complicated. We shall, therefore, confine ourselves from now on to homogeneous systems for which stress and strain components are position independent properties of the system. The validity of (54.5) for any finite volume V_0 is thus confirmed.

Because of their symmetry the stress and the strain tensor have only six independent components each. We can, therefore, now discontinue the cumbersome double subscript notation. We simply number the components according to the scheme

$$\left.\begin{aligned}
S_{xx} &\to S_1, & T_{xx} &\to T_1,\\
S_{yy} &\to S_2, & T_{yy} &\to T_2,\\
S_{zz} &\to S_3, & T_{zz} &\to T_3,\\
2S_{yz} &\to S_4, & T_{yz} &\to T_4,\\
2S_{zx} &\to S_5, & T_{zx} &\to T_5,\\
2S_{xy} &\to S_6, & T_{xy} &\to T_6.
\end{aligned}\right\} \quad (54.6)$$

The strain components S_1, S_2, S_3 thus represent fractional changes of length in the directions of the co-ordinate axes while S_4, S_5, S_6 are angular changes. The stress components T_1, T_2, T_3 are normal stresses and the components T_4, T_5, T_6 are tangential stresses. We thus obtain the equation

$$\mathrm{d}W = \sum_{j=1}^{6} T_j V_0 \,\mathrm{d}S_j \tag{54.7}$$

for the quasi-static work done on a deformable body.

We have so far made no use of the special properties of solid bodies. We shall now show that eq. (54.7) is a true generalization of the earlier formulations and that it contains eq. (3.5) for fluid phases as a special case. The characteristic mechanical property of a fluid phase is that at equilibrium all tangential stresses vanish. It follows that the stress ellipsoid is a sphere and the normal stresses are equal to each other. We therefore have

$$T_1 = T_2 = T_3, \quad T_4 = 0, \quad T_5 = 0, \quad T_6 = 0. \tag{54.8}$$

In hydrostatics, pressure is defined as a negative stress. With (52.20) and (52.25) we therefore obtain

$$\mathrm{d}W = -P\,\mathrm{d}V \quad \text{(fluid phases)}. \tag{54.9}$$

In order to get a basis for a thermodynamic treatment of solids we have to replace the term $-P\,\mathrm{d}V$ in the fundamental equation by the more general expression (54.7). We thus find that

$$\mathrm{d}U = T\,\mathrm{d}S + \sum_{j=1}^{6} T_j V_0 \,\mathrm{d}S_j + \sum \mu_i \,\mathrm{d}n_i. \tag{54.10}$$

We can see that the quantities $V_0 S_j$ represent extensive parameters and the quantities T_j represent intensive parameters. The integrated form of eq. (54.10) is, therefore,

$$U = U(S, V_0 S_1, V_0 S_2, V_0 S_3, V_0 S_4, V_0 S_5, V_0 S_6, n_1, \ldots, n_m). \tag{54.11}$$

Further use of eqs. (54.10) and (54.11) is usually made according to the methods developed in the previous chapters. Even the relationships written in terms of generalized quantities of state can be adopted directly. The problem of the present section is, in fact, solved with the statement of the fundamental equation since the plan of this book excludes a systematic treatment of special classes of substances. We shall, however, use a few examples to show how equations derived earlier for solids are modified. In this connexion, we shall also define a few concepts important in the thermodynamics of solids.

The generalization of the systematic equations of state (20.22)–(20.27) is trivial and we shall not deal with it. Much more important is the analogue of the thermal equation of state (20.29) which is represented by six equations of the form

$$T_j = T_j(T, V_0 S_1, \ldots, V_0 S_6) \quad (j = 1, 2, \ldots, 6). \tag{5.12}$$

Experience shows that the relationships between stress components and strain components is linear for sufficiently small deformations. We have, therefore, that

$$T_j = \sum_{k=1}^{6} c_{jk} S_k \quad (j = 1, 2, \ldots, 6) \tag{54.13}$$

or, if we combine the stress and strain components into vectors each with six components,

$$T = \mathbf{c} . S \tag{54.14†}$$

(*Hooke's law*). The quantities c_{jk} depend only on temperature and are called *isothermal elastic stiffness coefficients*. We shall return to them below. Hooke's law practically always applies in the only region relevant to thermodynamics, i.e. that of completely reversible deformations.

Application of Euler's theorem shows the explicit form of eq. (54.11) to be

$$U = TS + \sum_{j=1}^{6} T_j V_0 S_j + \sum_{i=1}^{m} \mu_i n_i \tag{54.15}$$

and this gives the generalized Gibbs–Duhem equation

$$S\,\mathrm{d}T + \sum_{j=1}^{6} V_0 S_j \,\mathrm{d}T_j + \sum_{i=1}^{m} n_i \,\mathrm{d}\mu_i = 0. \tag{54.16}$$

§ 55 Thermodynamic potentials and Maxwell's relations

An increase in the number of extensive parameters necessarily causes a very great increase in the number of possible thermodynamic potentials, since one or more stress components can be introduced in various combinations as independent variables. Such functions can be useful in the treatment of particular problems‡ but cannot be

† If, in the deformation, entropy is kept constant instead of temperature, a formally similar relationship is valid. The coefficients occurring in this relationship are the *adiabatic* (or *isentropic*) *elastic stiffness coefficients*. Cf. the analogous definitions of compressibility in §§ 25 and 40.

‡ An example can be found at the end of § 63.

discussed here. In § 21 we introduced special designations for particularly important thermodynamic potentials. We shall use these, here and in the following chapters, for functions in which the totality of work co-ordinates is treated in the same way as was the quantity V in earlier sections. We merely point out that the *Helmholtz free energy* is now defined by the equations

$$F = U - TS, \tag{55.1}$$

$$\mathrm{d}F = -S\,\mathrm{d}T + \sum_{j=1}^{6} T_j V_0\,\mathrm{d}S_j + \sum_{i=1}^{m} \mu_i\,\mathrm{d}n_i, \tag{55.2}$$

$$\left(\frac{\partial F}{\partial T}\right)_{S_j,n_i} = -S, \quad \left(\frac{\partial F}{\partial (V_0 S_j)}\right)_{T,S_{k\neq j},n_i} = T_j, \quad \left(\frac{\partial F}{\partial n_i}\right)_{T,S_j,n_{l\neq j}} = \mu_i, \tag{55.3}$$

while the definition of the *Gibbs free energy* is

$$G = U - TS - \sum_{j=1}^{6} T_j V_0 S_j, \tag{55.4}$$

$$\mathrm{d}G = -S\,\mathrm{d}T - \sum_{j=1}^{6} V_0 S_j\,\mathrm{d}T_j + \sum_{i=1}^{m} \mu_i\,\mathrm{d}n_i, \tag{55.5}$$

$$\left(\frac{\partial G}{\partial T}\right)_{T_j,n_i} = -S, \quad \left(\frac{\partial G}{\partial T_j}\right)_{T,T_{k\neq j},n_i} = -V_0 S_j, \quad \left(\frac{\partial G}{\partial n_i}\right)_{T,T_j,n_{l\neq i}} = \mu_i. \tag{55.6}$$

We derived the formulation of the equilibrium conditions with the aid of thermodynamic potentials in § 23, using generalized quantities of state. This method can be adopted here without modification. The Gibbs–Helmholtz equations in the general form (24.3), (24.4), (24.10b) and, similarly, Maxwell's relations (24.11)–(24.13) retain their validity here. Among the many special cases obtained from them without difficulty we note only the relationship

$$\left(\frac{\partial T_k}{\partial S_j}\right)_{T,S_k,n_i} = \left(\frac{\partial T_j}{\partial S_k}\right)_{T,S_j,n_i} \tag{55.7}$$

derived from the Helmholtz free energy and the analogous equation

$$\left(\frac{\partial S_k}{\partial T_j}\right)_{T,T_k,n_i} = \left(\frac{\partial S_j}{\partial T_k}\right)_{T,T_j,n_i} \tag{55.8}$$

which appertains to the Gibbs free energy. If we work on the basis of the internal energy and the suitably generalized enthalpy, we obtain formulae in which the entropy is kept constant in place of the temperature.

Among the second derivatives of the thermodynamic potentials the quantities

$$c_{jk} = \left(\frac{\partial T_j}{\partial S_k}\right)_{T,S_j,n_i} \tag{55.9}$$

derived from the Helmholtz free energy are particularly important. It is easy to see that they are identical with the isothermal elastic stiffness coefficients defined by eq. (54.13). The quantities

$$\kappa_{jk} = \left(\frac{\partial S_j}{\partial T_k}\right)_{T,T_j,n_i} \tag{55.10}$$

derived from the Gibbs free energy are called *isothermal elastic compliance coefficients*. There is a simple connexion between elastic stiffness coefficients and elastic compliance coefficients. If we interpret the κ_{jk} (as we did previously the c_{jk}) as elements of a six-row square matrix $\boldsymbol{\kappa}$, eqs. (55.9) and (55.10) immediately give

$$\boldsymbol{\kappa} = \mathbf{c}^{-1}. \tag{55.11}$$

Equation (54.14) can, therefore, also be written in the form of the strain–stress relation

$$\boldsymbol{S} = \boldsymbol{\kappa}\,.\,\boldsymbol{T}. \tag{55.12}$$

The matrix of the isothermal elastic compliance coefficients is the analogue of the isothermal compressibility defined for fluid phases by eq. (25.6).

§ 56 Symmetry properties of solids

Because of Maxwell's relations (55.7) and (55.8), the matrix of the stiffness coefficients and the matrix of the compliance coefficients are both symmetric. We thus have only 21 independent elements in each case. This number can be reduced still further by a consideration of the spatial symmetry properties of the system since eqs. (54.14) and (55.12) must be invariant with respect to co-ordinate transformations appropriate to the symmetry properties of the system. We shall illustrate the principle of this reduction by means of two examples.

The number 21 is appropriate for a *triclinic crystal* which has no symmetry properties. The three crystal axes have different lengths and arbitrary directions. We shall take as our first example a *monoclinic crystal* in which one crystal axis is perpendicular to the other two and thus represents a twofold axis of symmetry. If we choose this axis as the z-axis of a system of rectangular co-ordinates, then the co-ordinate transformation corresponding to the symmetry property (revolution of 180° round the z-axis) is

$$x \to -x, \quad y \to -y, \quad z \to z. \tag{56.1}$$

The components of the displacement are transformed correspondingly according to

$$s_x \to -s_x, \quad s_y \to -s_y, \quad s_z \to s_z. \tag{56.2}$$

The definitions (52.13) and (54.6) then give for the transformation of the strain components that

$$\left.\begin{aligned} S_1 \to S_1, \quad S_2 \to S_2, \quad S_3 \to S_3, \quad S_6 \to S_6, \\ S_4 \to -S_4, \quad S_5 \to -S_5, \end{aligned}\right\} \tag{56.3}$$

and, according to (54.6), for the transformation of the stress components that

$$\left.\begin{aligned} T_1 \to T_1, \quad T_2 \to T_2, \quad T_3 \to T_3, \quad T_6 \to T_6, \\ T_4 \to -T_4, \quad T_5 \to -T_5. \end{aligned}\right\} \tag{56.4}$$

Since the S_j and T_j transform in the same way, the invariance requirement can be fulfilled only if the stiffness coefficients are themselves invariant with respect to the transformation (56.1). We have thus, for example, that

$$c_{14} = \frac{\partial T_1}{\partial S_4} = \frac{\partial T_1}{\partial(-S_4)} = -c_{14} = 0 \tag{56.5}$$

and, generally, that

$$c_{14} = c_{24} = c_{34} = c_{64} = c_{15} = c_{25} = c_{35} = c_{65} = 0 \text{ (monoclinic crystal).} \tag{56.6}$$

A monoclinic crystal thus has only 13 independent elastic stiffness coefficients.

We choose as our second example an *isotropic solid* whose properties are independent of direction, i.e. invariant with respect to any rotation of the system of co-ordinates. We consider a volume element in the form of a rectangular parallelepiped whose edges are parallel to the principal axes of the stress ellipsoid. Only normal stresses now act on the sides of the volume element, i.e. the normal stresses $T_{\text{I}}, T_{\text{II}}, T_{\text{III}}$, while the tangential stresses vanish. The deformation thus consists of pure dilations in the directions of the principal axes of the stress ellipsoid and no angular changes occur. The principal axis of stress and strain ellipsoids therefore coincide.

For the special case under consideration we can write (54.13) as

$$\left.\begin{aligned} T_{\mathrm{I}} &= c_{11}S_{\mathrm{I}}+c_{12}S_{\mathrm{II}}+c_{13}S_{\mathrm{III}}, \\ T_{\mathrm{II}} &= c_{21}S_{\mathrm{I}}+c_{22}S_{\mathrm{II}}+c_{23}S_{\mathrm{III}}, \\ T_{\mathrm{III}} &= c_{31}S_{\mathrm{I}}+c_{32}S_{\mathrm{II}}+c_{33}S_{\mathrm{III}}, \end{aligned}\right\} \tag{56.7}$$

where $S_{\mathrm{I}}, S_{\mathrm{II}}, S_{\mathrm{III}}$ are the linear dilations. Because of the already-proved symmetry of the coefficient matrix we have

$$c_{12} = c_{21}, \quad c_{13} = c_{31}, \quad c_{23} = c_{32}. \tag{56.8}$$

A result of the assumed isotropy is that the directions of the principal axes are equivalent and, therefore,

$$c_{11} = c_{22} = c_{33}. \tag{56.9}$$

Furthermore, the two axis directions perpendicular to a normal stress must be equivalent and, therefore,

$$c_{12} = c_{13} = c_{23}. \tag{56.10}$$

Comparison of (56.8)–(56.10) with (56.7) shows that the latter set of equations contains only two independent variables. We can, therefore, write

$$T_i = (c_{11}-c_{12})S_i+c_{12}(S_{\mathrm{I}}+S_{\mathrm{II}}+S_{\mathrm{III}}) \tag{56.11}$$

or, with new coefficients,

$$T_i = 2\mu_{\mathrm{L}}S_i+\lambda_{\mathrm{L}}(S_{\mathrm{I}}+S_{\mathrm{II}}+S_{\mathrm{III}}) \quad (i = \mathrm{I, II, III}). \tag{56.12}$$

In order to generalize this result to include a volume element of any orientation, we simply have to transform the components of the stress and the strain tensors from the principal axes representation to any arbitrary system of co-ordinates. We denote by α_i the direction of the new x-axis with respect to the principal axes, by β_i that of the new y-axis, and by γ_i that of the new z-axis. We then have

$$\sum_{i=1}^{3}\alpha_i^2 = 1, \quad \sum_{i=1}^{3}\alpha_i\beta_i = 0 \tag{56.13}$$

and corresponding expressions for the β_i and γ_i. We then have, for the tensor components,† that

$$\left.\begin{aligned} T_{xx} &= \alpha_1^2 T_{\rm I} + \alpha_2^2 T_{\rm II} + \alpha_3^2 T_{\rm III}, \\ T_{xy} &= \alpha_1 \beta_1 T_{\rm I} + \alpha_2 \beta_2 T_{\rm II} + \alpha_3 \beta_3 T_{\rm III}, \end{aligned}\right\} \tag{56.14}$$

$$\left.\begin{aligned} S_{xx} &= \alpha_1^2 S_{\rm I} + \alpha_2^2 S_{\rm II} + \alpha_3^2 T_{\rm III}, \\ S_{xy} &= \alpha_1 \beta_1 S_{\rm I} + \alpha_2 \beta_2 S_{\rm II} + \alpha_3 \beta_3 T_{\rm III}, \end{aligned}\right\} \tag{56.15}$$

and corresponding expressions, with cyclic exchanges, for the rest of the components. If we now multiply eq. (56.12) by α_i^2, add, and take into account (56.13)–(56.15), we obtain

$$T_{xx} = 2\mu_{\rm L} S_{xx} + \lambda_{\rm L}(S_{xx} + S_{yy} + S_{zz}). \tag{56.16}$$

If we multiply (56.12) by $\alpha_i \beta_i$, we obtain in an analogous way

$$T_{xy} = 2\mu_{\rm L} S_{xy}. \tag{56.17}$$

The remaining components of the stress tensor are found by simple cyclic exchange of the subscripts. We have thus shown that isotropic solids have only two independent elastic moduli. The quantities $\mu_{\rm L}$ and $\lambda_{\rm L}$ which we have just introduced are known as *Lamé constants.* In their place, Young's modulus $E_{\rm Y}$ and *Poisson's ratio* ν are often used. They are defined by the equations

$$2\mu_{\rm L} = \frac{E_{\rm Y}}{1+\nu}, \quad \lambda_{\rm L} = \frac{\nu E_{\rm Y}}{(1+\nu)(1-2\nu)}. \tag{56.18}$$

Their advantage over the Lamé constants is that they have a concrete physical meaning.

We now give a clear summary of our results by writing out the stiffness matrix explicitly and giving, in addition, the matrix for a cubic crystal.

Triclinic crystal:

$$\mathbf{c} = \begin{bmatrix} c_{11} & c_{12} & c_{13} & c_{14} & c_{15} & c_{16} \\ c_{12} & c_{22} & c_{23} & c_{24} & c_{25} & c_{26} \\ c_{13} & c_{23} & c_{33} & c_{34} & c_{35} & c_{36} \\ c_{14} & c_{24} & c_{34} & c_{44} & c_{45} & c_{46} \\ c_{15} & c_{25} & c_{35} & c_{45} & c_{55} & c_{56} \\ c_{16} & c_{26} & c_{36} & c_{46} & c_{56} & c_{66} \end{bmatrix}; \tag{56.19}$$

† It is convenient to reintroduce here the double subscript notation for tensor components.

monoclinic crystal:

$$\boldsymbol{c} = \begin{bmatrix} c_{11} & c_{12} & c_{13} & 0 & 0 & c_{16} \\ c_{12} & c_{22} & c_{23} & 0 & 0 & c_{26} \\ c_{13} & c_{23} & c_{33} & 0 & 0 & c_{36} \\ 0 & 0 & 0 & c_{44} & c_{45} & 0 \\ 0 & 0 & 0 & c_{45} & c_{55} & 0 \\ c_{16} & c_{26} & c_{36} & 0 & 0 & c_{66} \end{bmatrix}; \tag{56.20}$$

cubic crystal:

$$\boldsymbol{c} = \begin{bmatrix} c_{11} & c_{12} & c_{12} & 0 & 0 & 0 \\ c_{12} & c_{11} & c_{12} & 0 & 0 & 0 \\ c_{12} & c_{12} & c_{11} & 0 & 0 & 0 \\ 0 & 0 & 0 & c_{44} & 0 & 0 \\ 0 & 0 & 0 & 0 & c_{44} & 0 \\ 0 & 0 & 0 & 0 & 0 & c_{44} \end{bmatrix}; \tag{56.21}$$

isotropic solid:

$$\boldsymbol{c} = \begin{bmatrix} c_{11} & c_{12} & c_{12} & 0 & 0 & 0 \\ c_{12} & c_{11} & c_{12} & 0 & 0 & 0 \\ c_{12} & c_{12} & c_{11} & 0 & 0 & 0 \\ 0 & 0 & 0 & \frac{1}{2}(c_{11}-c_{12}) & 0 & 0 \\ 0 & 0 & 0 & 0 & \frac{1}{2}(c_{11}-c_{12}) & 0 \\ 0 & 0 & 0 & 0 & 0 & \frac{1}{2}(c_{11}-c_{12}) \end{bmatrix}. \tag{56.22}$$

If we introduce the Lamé constants into (56.22) and form the reciprocal matrix, we obtain, using (56.18), the compliance matrix in terms of the elastic constants:

isotropic solid:

$$\boldsymbol{\kappa} = \begin{bmatrix} 1/E_{\mathrm{Y}} & -\nu/E_{\mathrm{Y}} & -\nu/E_{\mathrm{Y}} & 0 & 0 & 0 \\ -\nu/E_{\mathrm{Y}} & 1/E_{\mathrm{Y}} & -\nu/E_{\mathrm{Y}} & 0 & 0 & 0 \\ -\nu/E_{\mathrm{Y}} & -\nu/E_{\mathrm{Y}} & 1/E_{\mathrm{Y}} & 0 & 0 & 0 \\ 0 & 0 & 0 & 2\dfrac{1+\nu}{E_{\mathrm{Y}}} & 0 & 0 \\ 0 & 0 & 0 & 0 & 2\dfrac{1+\nu}{E_{\mathrm{Y}}} & 0 \\ 0 & 0 & 0 & 0 & 0 & 2\dfrac{1+\nu}{E_{\mathrm{Y}}} \end{bmatrix}. \tag{56.23}$$

The matrix (56.23) immediately gives the concrete meaning of the quantities E_Y and ν. We choose the x-axis as our vertical axis and imagine a vertical rod of uniform cross section q supported at the upper end and weighted at the lower end by a mass p. The only stress component which does not vanish is then the normal stress $T_1 = p/q$. Equations (55.12) and (56.23) then give

$$S_1 = \frac{1}{E_Y} T_1, \tag{56.24}$$

$$S_2 = S_3 = -\frac{\nu}{E_Y} T_1. \tag{56.25}$$

Because of (52.29), eq. (56.24) states that $1/E_Y$ is the fractional change in length (stretch) per unit stress. According to (56.24) and (56.25), we have that $\nu = -S_2/S_1$; ν is therefore the ratio of the fractional lateral contraction to the fractional stretch in the x-direction.

Comparison of (56.21) and (56.22) shows that the elastic behaviour of cubic crystals is less simple than that of isotropic solids since the former have three, the latter only two independent elastic moduli. The most important kinds of isotropic solids are amorphous substances, such as glass and many synthetic polymers, and polycrystalline (i.e. made up of countless tiny crystallites orientated in a completely disordered way) substances such as marble, most metals used in technology, and metallic alloys. In both cases all 'texture' (e.g. a preferred orientation of the crystallites) must be absent.

The treatment of heterogeneous equilibria on the basis of the general theory developed in this section was carried out by Gibbs. It is naturally much more complicated than the case of fluid phases discussed in Chapter IV. We cannot go into detail here and can only note that, in particular, the fulfilment of the requirements for the validity of the phase rule (§ 29) is not at all obvious.

The application of the stability conditions in the general form of the inequalities (41.12)–(41.15) or (50.28), however, presents no difficulties. It leads to the conclusion that each of the principal minors of the determinant of elastic coefficients must be positive.

We now consider briefly the case where a hydrostatic pressure P acts on the solid from all directions but uni-axial normal stresses and tangential stresses are absent. The stress ellipsoid is a sphere under these conditions. Equations (54.8) and (54.9) apply and we arrive at the form of the fundamental equation laid down previously for fluid phases. Gibbs showed that, under these conditions, the theory of heterogeneous equilibria discussed in Chapter IV is also applicable to solid phases. Furthermore, the isothermal compressibility is given by the general expression

$$\kappa = -\frac{1}{V}\left(\frac{\partial V}{\partial P}\right)_T = \frac{|\theta|}{P} \tag{56.26}$$

[where θ is given by (52.20)] and, using (55.12), by the elastic compliance or stiffness coefficients. We can easily verify that the special stability conditions formulated above imply the stability condition (41.18) for the isothermal compressibility.

Finally, we mention a useful approximation which we shall use in Chapter IX. Experience shows that the entropy is usually (but not always)† practically independent of the deformation. If this is so, integration of (55.2) and (54.13) at constant temperature and constant mole numbers gives

$$F = F_0(T, n_1, \ldots, n_m) + \tfrac{1}{2}V_0 \sum_{j,k=1}^{6} c_{jk} S_j S_k, \tag{56.27}$$

where F_0, the Helmholtz free energy in the undeformed state, depends only on the temperature and the mole numbers.

† The entropy is not independent of the deformation, in particular, for the so-called elastomers (e.g. rubber) where a deformation results in large entropy changes.

CHAPTER IX

Systems in an electric field

§ 57 Electrostatic work

The structure of classical thermodynamics imposes two limitations on the treatment of systems in electric fields if we exclude chemical reactions.† Firstly only systems in an electrostatic field are to be considered since the discussion is limited to equilibria and quasi-static processes. Electric and magnetic phenomena can, therefore, be treated separately. Secondly, conductors are excluded altogether since in them a time-invariant electric field can be maintained only as a stationary state of a dissipative process (i.e. conduction of a current). We have thus defined the subject of this section as the thermodynamics of non-conductors (dielectrics) in an electrostatic field. The derivation of the explicit expressions for the quasi-static work done on the system [eq. (14.1)] and for the fundamental equation [eq. (21.3)] is, however, less simple here than in the case of a solid where both are obtained directly from the formalism of the mechanics of continuous media. We shall, therefore, discuss some of the difficulties in greater detail.

We consider first quite generally a dielectric surrounding a conductor. Let us suppose that the surface of the conductor is the site of the charges causing the field.‡ We denote the total charge by e, its surface density by σ, and the potential of the conductor (with respect to infinity) by ϕ. The electric field strength in the dielectric is then

$$\boldsymbol{E} = -\operatorname{grad} \phi, \tag{57.1}$$

while the surface charge density is related to the dielectric displacement $\boldsymbol{D}$§ by the equation

$$D_n = -4\pi\sigma, \tag{57.2}$$

† These will be discussed in Chapter XI.

‡ The lines of force end at the site of the opposite charges. This site must be assumed to be at infinity.

§ The vector $\boldsymbol{D}$ is often called the electric induction.

where D_n is the component of $\boldsymbol{D}$ in a direction (outwards, i.e. towards the conductor) normal to the surface of the dielectric. The total charge on the conductor is, therefore,

$$e = -\frac{1}{4\pi}\int D_n\,\mathrm{d}f = -\frac{1}{4\pi}\int \boldsymbol{D}\,.\,\mathrm{d}\boldsymbol{f}, \tag{57.3}$$

where $\mathrm{d}f$ is a surface element and $\mathrm{d}\boldsymbol{f}$ a vector surface element. The vectors $\boldsymbol{E}$ and $\boldsymbol{D}$ are related by

$$\boldsymbol{D} = \boldsymbol{\epsilon}\,.\,\boldsymbol{E}, \tag{57.4}$$

when $\boldsymbol{\epsilon}$ is called the *dielectric tensor*. For isotropic substances and cubic crystals,† eq. (57.4) reduces to the simpler form

$$\boldsymbol{D} = \varepsilon \boldsymbol{E}, \tag{57.5}$$

where ε is the *dielectric constant* of the substance. Both $\boldsymbol{\epsilon}$ and ε are, however, functions of thermodynamic quantities of state. If the dielectric is not homogeneous, $\boldsymbol{\epsilon}$ and ε thus represent field quantities dependent on positional co-ordinates.‡ With the aid of the vectors $\boldsymbol{E}$ and $\boldsymbol{D}$ the basic equations for the electrostatic field in the dielectric can be written

$$\mathrm{rot}\,\boldsymbol{E} = 0, \quad \mathrm{div}\,\boldsymbol{D} = 0. \tag{57.6}$$

We now calculate the work δW^* necessary to bring an infinitesimal charge δe from infinity on to the conductor. By definition we have

$$\delta W^* = \phi\delta e. \tag{57.7}$$

Since the surface of the conductor is at a uniform potential (equipotential surface), we obtain by means of (57.3)

$$\delta W^* = -\frac{1}{4\pi}\int \phi\delta\boldsymbol{D}\,.\,\mathrm{d}\boldsymbol{f}. \tag{57.8}$$

If we change the integral taken over the surface into one taken over the volume by using Gauss' theorem,§ we find that

$$\delta W^* = -\frac{1}{4\pi}\int \mathrm{div}\,(\phi\delta\boldsymbol{D})\,\mathrm{d}V, \tag{57.9}$$

† Isotropic substances and cubic crystals thus do not differ in their dielectric behaviour, in contrast with their elastic behaviour (§ 56).

‡ $\boldsymbol{\epsilon}$ and ε must, in general, be regarded as functions of the electric field strength as well.

§ Cf. § 53.

where the integral extends over the whole of space outside the conductor. Because of (57.1) and (57.6) we have

$$\operatorname{div} \phi \delta \boldsymbol{D} = \phi \operatorname{div} \delta \boldsymbol{D} + \delta \boldsymbol{D} . \operatorname{grad} \phi = -\boldsymbol{E} . \delta \boldsymbol{D}. \tag{57.10}$$

This with (57.9) gives

$$\delta W^* = \frac{1}{4\pi} \int \boldsymbol{E} . \delta \boldsymbol{D} \, \mathrm{d} V. \tag{57.11}$$

Equation (57.11) is a completely general result obtained in electrostatics and is the basis for the whole of the following discussion.

We now consider a condenser consisting of two flat parallel plates 1 and 2 with charges $-e$ and e. Let the constant surface densities on these plates be $-\sigma$ and σ. We denote the area of each plate by a, the distance between the plates by l and assume that $a \gg l^2$. The distortion of the fields at the edge ('edge effects') can then be neglected. If we assume the condenser to be *in vacuo* we have inside the condenser a homogeneous electric field perpendicular to the surface of the plates

$$\boldsymbol{E}_0 = -\operatorname{grad} \phi_0, \quad \boldsymbol{E}_0 = \Delta \phi_0 / l, \tag{57.12}$$

where

$$\Delta \phi_0 = \int_1^2 \boldsymbol{E}_0 . \mathrm{d} \boldsymbol{r} \tag{57.13}$$

is the potential difference between the plates. Furthermore, the expression

$$E_{0n} = E_0 = 4\pi\sigma, \tag{57.14}$$

analogous to (57.2), applies. If we bring a charge $-\delta e$ from infinity on to plate 1 and a charge δe on to plate 2, the same amount of work δW_0 is required as is needed to transfer a charge δe from plate 1 to plate 2. By definition we have, therefore, that

$$\delta W_0^* = \Delta \phi \delta e. \tag{57.15}$$

A calculation analogous to the previous one then gives

$$\delta W_0^* = \frac{1}{4\pi} \int \boldsymbol{E}_0 . \delta \boldsymbol{E}_0 \, \mathrm{d} V \tag{57.16}$$

or, if we neglect edge effects and put $al \equiv V_c$, that

$$\delta W_0^* = \frac{1}{4\pi} \boldsymbol{E}_0 . \delta \boldsymbol{E}_0 V_c. \tag{57.17}$$

Let us now suppose that a dielectric of volume V is placed between the plates of the condenser, that the part of the volume V_c not occupied by the dielectric is V', and that V' is evacuated. The potential difference between the plates is now decreased from $\Delta\phi_0$ to $\Delta\phi$ and the electric field inside the condenser is no longer homogeneous. According to (57.11) the work done for an infinitesimal change in the charge on the plates is now

$$\delta W^* = \frac{1}{4\pi}\int_{V_c} \boldsymbol{E}.\delta\boldsymbol{D}\,\mathrm{d}V \tag{57.18}$$

since $\boldsymbol{E} = 0$ outside V_c if we ignore edge effects.

The expression (57.18) obviously does not represent the quasi-static work done on the dielectric with a change in the charge on the plates since it must also include the work done with the change in the field in the evacuated volume V'. This field is, in general, distorted by the presence of the dielectric, in contrast with the case of eq. (57.12); we shall denote this field by $\boldsymbol{E}^{(a)}$ and write

$$\delta W^* = \frac{1}{4\pi}\int_{V'} \boldsymbol{E}^{(a)}.\delta\boldsymbol{E}^{(a)}\,\mathrm{d}V + \frac{1}{4\pi}\int_{V} \boldsymbol{E}.\delta\boldsymbol{D}\,\mathrm{d}V. \tag{57.19}$$

As far as thermodynamics is concerned, only the second term on the right-hand side is relevant. However, this also cannot be identified with the work done on the dielectric. If we assume V also to be *in vacuo*, the second term does not vanish but we again obtain eq. (57.16) in the form

$$\delta W_0^* = \frac{1}{4\pi}\int_{V'} \boldsymbol{E}_0.\delta\boldsymbol{E}_0\,\mathrm{d}V + \frac{1}{4\pi}\int_{V} \boldsymbol{E}_0.\delta\boldsymbol{E}_0\,\mathrm{d}V. \tag{57.20}$$

A part of the second term of (57.19) must thus be interpreted as a change in the energy of the electrostatic field. This part reduces to the second term of (57.20) *in vacuo*. The expression

$$\boldsymbol{D} = \boldsymbol{E} + 4\pi\boldsymbol{p} \tag{57.21}$$

is well known in electrostatics and allows us to separate out the work done on the material system. $\boldsymbol{p}$ is the *polarization per unit volume* (polarization density), i.e. the dipole moment per unit volume. The relationship between polarization density and field strength is given by

$$\boldsymbol{p} = \boldsymbol{\chi}.\boldsymbol{E} \tag{57.22}$$

where $\boldsymbol{\chi}$ is the *tensor electrical susceptibility.* From (57.4), (57.21), and (57.22) we obtain

$$\boldsymbol{\chi} = \frac{1}{4\pi}(\boldsymbol{\epsilon} - \boldsymbol{U}), \tag{57.23}$$

where $\boldsymbol{U}$ is the unit tensor. For isotropic systems and cubic crystals, (57.22) and (57.23) reduce to the expressions

$$\boldsymbol{p} = \chi \boldsymbol{E} \tag{57.24}$$

and

$$\chi = \frac{\varepsilon - 1}{4\pi}, \tag{57.25}$$

where χ is the (scalar) *electrical susceptibility.* Introduction of (57.21) into (57.19) now gives

$$\delta W^* = \frac{1}{4\pi}\int_{V'} \boldsymbol{E}^{(a)}.\delta \boldsymbol{E}^{(a)} \, \mathrm{d}V + \frac{1}{4\pi}\int_V \boldsymbol{E}.\delta \boldsymbol{E} \, \mathrm{d}V + \int_V \boldsymbol{E}.\delta \boldsymbol{p} \, \mathrm{d}V. \tag{57.26}$$

Comparison with (57.20) shows that the second term on the right-hand side is the energy change of the electrostatic field in the dielectric while the last term represents the quasi-static work done on the material system.

§ 58 The fundamental equation for a dielectric in an electric field

Equation (57.26) contains no particular assumptions concerning the electric field in the dielectric. To evaluate the integral, $\boldsymbol{E}$ must be calculated in each case from the equations of electrostatics. The last term of (57.26) in the form given above must, therefore, be introduced into the equations of thermodynamics without further assumptions. This formulation has (as in the analogous case of elastic solids) the advantage of greater generality but requires the introduction of field quantities into thermodynamics. This extension lies, as we have mentioned previously, outside the framework of our discussion. We therefore now investigate the conditions under which the last term of (57.26) may be brought into the form $Y_j \, \mathrm{d}y_j$ corresponding to the definition (14.1). We see immediately that this is always possible if the electric field inside the dielectric is homogeneous and the dielectric constant (or the dielectric tensor) is independent of the position co-ordinates. The problem thus reduces to the question: under what conditions are these two requirements fulfilled?

We assume, as before, that the electric field $\boldsymbol{E}_0$ in the absence of the dielectric is homogeneous. An inhomogeneous field inside the dielectric can then be due to two causes, namely

(a) inhomogeneity of the dielectric substance,

(b) the shape of the dielectric substance.

Case (a) is also the only possible cause of a position dependence of ε or $\boldsymbol{\epsilon}$.

Re (a) An originally homogeneous body brought into an electric field generally becomes inhomogeneous. A completely rigorous treatment will, therefore, have to start from eq. (57.26). Inhomogeneities caused by the field may, however, be neglected if the field strength is kept low so that the expansion of the thermodynamic functions in powers of the field strength can terminate with the square term. In order to prove this, we denote local variations from the mean density by $\delta\rho$ and the associated change in the internal energy by δU. We then have

$$\delta U = \left(\frac{\partial U}{\partial \rho}\right)_0 \int_V \delta\rho \,\mathrm{d}V + \frac{1}{2}\left(\frac{\partial^2 U}{\partial \rho^2}\right)_0 \int_V (\delta\rho)^2 \,\mathrm{d}V + \dots, \quad (58.1)$$

where the derivatives are to be taken for the homogeneous system. The first term on the right-hand side vanishes since application of the electric field does not change the mass of the system. We shall show in § 60 that $\delta\rho \sim E^2$ from which follows that

$$\delta U \sim E^4. \quad (58.2)$$

The effect of the inhomogeneity on the thermodynamic functions is thus a fourth-order effect and completely negligible according to our supposition.

Re (b) The effect of the shape of the dielectric on the field $\boldsymbol{E}$ is based on the fact that the boundary conditions at the dielectric-vacuum boundary have to be taken into consideration in the solution of the field equations (57.6). Suppose that we have a homogeneous field $\boldsymbol{E}_0$ *in vacuo*. We now place in this field a homogeneous dielectric ellipsoid. It can be shown in electrostatics (cf. e.g. Landau and Lifshitz)† that the field inside this ellipsoid is also homogeneous. If the ellipsoid is oriented so that a major axis falls in the direction of $\boldsymbol{E}_0$, then $\boldsymbol{E}_0$

† L. D. Landau and E. M. Lifshitz, *Electrodynamics of Continuous Media*, Oxford, 1960.

and $\boldsymbol{E}$ are parallel. This condition is always fulfilled for a sphere. For an isotropic dielectric we thus have the simple expressions

$$\boldsymbol{E} = \frac{3}{\varepsilon+2}\boldsymbol{E}_0 \tag{58.3}$$

and

$$\boldsymbol{p} = \frac{3}{4\pi}\frac{\varepsilon-1}{\varepsilon+1}\boldsymbol{E}_0. \tag{58.4}$$

For an infinitely long cylinder oriented in the direction of the field we have simply that

$$\boldsymbol{E} = \boldsymbol{E}_0. \tag{58.5}$$

On the basis of these ideas we shall make the following assumptions for the following discussion:†

(a) The field $\boldsymbol{E}_0$ in the condenser *in vacuo* (the 'applied field') is homogeneous and perpendicular to the surface of the plates.

(b) The field $\boldsymbol{E}_0$ is weak. Terms higher than the square term can therefore be ignored in the expansion of the thermodynamic functions as a power series in $\boldsymbol{E}$. This assumption implies that ε and χ do not depend on the field strength.‡

(c) The dielectric substance brought into the field is originally homogeneous. Its shape is an ellipsoid with one of the major axes in the direction of the field $\boldsymbol{E}_0$.

The quasi-static work done on the system (i.e. on the dielectric) with an infinitesimal change in the charge on the plates is then given by (57.26) as

$$\mathrm{d}W = \boldsymbol{E}.\mathrm{d}(\boldsymbol{p}.V) = \boldsymbol{E}.\mathrm{d}\boldsymbol{P}, \tag{58.6}$$

where $\boldsymbol{P} = \boldsymbol{p}V$ is the *total polarization* of the dielectric.

We still have to answer the question whether $\boldsymbol{E}$ and $\boldsymbol{P}$ have the properties of quantities of state discussed in § 20. The answer for $\boldsymbol{P}$ is obtained immediately if we remember that $\boldsymbol{P}$ is by definition an extensive parameter. The considerations for $\boldsymbol{E}$ are less simple since the boundary conditions are the deciding factor when two partial systems ′ and ″ are in contact. The boundary conditions state that the normal components§ of $\boldsymbol{E}$ at the boundary surface of two dielectrics

† These assumptions are obviously not necessary for the application of thermodynamics to our problems, but they result in a particularly lucid presentation.

‡ This will become obvious with the later explicit expressions.

§ I.e. normal to the boundary surface.

with dielectric constants ε' and ε'' are given by

$$\varepsilon' E'_n = \varepsilon'' E''_n \tag{58.7}$$

while the tangential components are given by

$$E'_{\text{tg}} = E''_{\text{tg}}. \tag{58.8}$$

If we require the boundary surface to be everywhere in the direction of $\boldsymbol{E}$ we have $E'_n = E''_n = 0$ and we are left with the condition (58.8) which agrees with the definition (20.7) of intensive parameters. It is hardly surprising that the boundary surface is here subject to a condition since the simple example of a gravitational field has already shown that the position of the boundary surface cannot be freely chosen in the presence of an external field. We shall, however, not discuss this problem in any greater detail. If the two partial systems consist of the same dielectric substance and, therefore, $\varepsilon' = \varepsilon''$, we obtain from (58.7) and (58.8) simply that $\boldsymbol{E}' = \boldsymbol{E}''$ which again agrees with (20.7).

The explicit fundamental equation is now obtained by introducing (58.6) into (21.3) and is, for fluid phases,

$$\mathrm{d}U = T\,\mathrm{d}S - P\,\mathrm{d}V + \boldsymbol{E}.\mathrm{d}\boldsymbol{P} + \sum_{i=1}^{m} \mu_i\,\mathrm{d}n_i. \tag{58.9}$$

Use of (54.7) gives for solid substances

$$\mathrm{d}U = T\,\mathrm{d}S + V_0 \sum_{j=1}^{6} T_j\,\mathrm{d}S_j + \boldsymbol{E}.\mathrm{d}\boldsymbol{P} + \sum_{i=1}^{m} \mu_i\,\mathrm{d}n_i. \tag{58.10}$$

For a detailed discussion of these equations the reader is referred to earlier considerations (particularly Chapter III) whose generalization should offer no difficulties. We shall only discuss certain aspects connected with the presence of the electric field. As long as no special properties of solids are involved, we shall confine our discussion to the expressions for fluid phases since their generalization is trivial. In order to make clear the modifications of the thermodynamic functions caused by the electric field, we denote, from now on, by the subscript 0 the quantity defined for $\boldsymbol{E} = 0$ with all other conditions unchanged.

Application of Euler's theorem (§ 19) to eq. (58.9) gives

$$U = TS - PV + \boldsymbol{E}.\boldsymbol{P} + \sum_{i=1}^{m} \mu_i n_i. \tag{58.11}$$

Equation (58.10) correspondingly gives

$$U = TS + V_0 \sum_{j=1}^{6} T_j S_j + \boldsymbol{E}.\boldsymbol{P} + \sum_{i=1}^{m} \mu_i n_i. \tag{58.12}$$

§ 59 Thermodynamic potentials

In the presence of the electric field we denote the Helmholtz free energy by the thermodynamic potential

$$F = U - TS \tag{59.1}$$

whose differential is the equation

$$\mathrm{d}F = -S\,\mathrm{d}T - P\,\mathrm{d}V + \boldsymbol{E}.\mathrm{d}\boldsymbol{P} + \sum_{i=1}^{m} \mu_i\,\mathrm{d}n_i. \tag{59.2}$$

Correspondingly, the Gibbs free energy is defined by the thermodynamic potential

$$G = U - TS + PV - \boldsymbol{E}.\boldsymbol{P}, \tag{59.3}$$

whose differential is

$$\mathrm{d}G = -S\,\mathrm{d}T + V\,\mathrm{d}P - \boldsymbol{P}.\mathrm{d}\boldsymbol{E} + \sum_{i=1}^{m} \mu_i\,\mathrm{d}n_i. \tag{59.4}$$

Equation (21.8) once more gives the relationship

$$G = \sum_{i=1}^{m} \mu_i n_i \tag{59.5}$$

which is formally identical with (21.40). This shows immediately that the chemical potentials are modified by the electric field. In order to derive an explicit expression for this modification we start with Maxwell's relation

$$\left(\frac{\partial \mu_i}{\partial \boldsymbol{E}}\right)_{T,P,n} = -\left(\frac{\partial \boldsymbol{P}}{\partial n_i}\right)_{T,P,\mathbf{E},n_{j\neq i}} \tag{59.6}$$

which follows from (59.4). Equation (59.6) with (57.24) becomes

$$\left(\frac{\partial \mu_i}{\partial \boldsymbol{E}}\right)_{T,P,n} = -\left(\frac{\partial (V\chi)}{\partial n_i}\right)_{T,P,\mathbf{E},n_{j\neq i}} \mathbf{E}. \tag{59.7}$$

If we define the partial molar susceptibility χ_i by the equation

$$\chi_i = \left(\frac{\partial (V\chi)}{\partial n_i}\right)_{T,P,\mathbf{E},n_{j\neq i}} \tag{59.8}$$

we obtain from (59.7) by integration from 0 to $\boldsymbol{E}$ at constant T, P, and n_i (all i)

$$\mu_i = \mu_{0i} - \tfrac{1}{2}\chi_i\,\mathbf{E}^2. \tag{59.9}$$

Analogous expressions can be derived for other partial molar quantities.†

The interaction of the electric field with the dielectric leads to a number of effects some of which have technical applications. Some of these effects will now be discussed but only from a thermodynamic point of view.

§ 60 Electrostriction

We again start by considering a fluid phase. From eq. (59.4) we get Maxwell's relation

$$\left(\frac{\partial V}{\partial \mathbf{E}}\right)_{T,P,n} = -\left(\frac{\partial \boldsymbol{P}}{\partial P}\right)_{T,\mathbf{E},n} \tag{60.1}$$

or, with (57.24),

$$\left(\frac{\partial V}{\partial \mathbf{E}}\right)_{T,P,n} = -\left(\frac{\partial (V\chi)}{\partial P}\right)_{T,\mathbf{E},n} \mathbf{E}. \tag{60.2}$$

Integration between 0 and $\boldsymbol{E}$ gives

$$V - V_0 = -\frac{1}{2}\left(\frac{\partial (V\chi)}{\partial P}\right)_{T,\mathbf{E},n} \boldsymbol{E}^2. \tag{60.3}$$

This change of volume which occurs when an electric field is applied is called *electrostriction*. This effect can be positive or negative. It is proportional to the square of the field strength and is, therefore, not affected by reversing the direction of the field. The effect is not reversible, i.e. compression or expansion causes neither the generation of a field inside the dielectric nor, according to (57.24), any polarization.

Electrostriction in solids is a complicated process since the volume change is generally accompanied by a deformation. We choose as a simple example the case of an isotropic dielectric sphere. Since the effects are small we can, as an approximation, consider the volume change and the deformation separately. If, therefore, we neglect the deformation, the volume change is again given by eq. (60.3). The deformation at constant volume can be calculated if it is described by means of suitable parameters; the Helmholtz free energy is then expressed as a function of these parameters and the general equilibrium condition (23.10) is used. The choice of parameters must

† The definition of partial molar quantities [eq. (26.6)] must here be augmented by the condition that the field strength $\mathbf{E}$ also remains constant for the differentiation with respect to n_i.

naturally rest on assumptions concerning the general character of the deformation. We note that the deformation must be rotationally symmetrical with respect to the direction of the field, the direction being chosen as the x-axis. Furthermore, the whole of the deformed surface must be finite. We shall not consider this subject further (it has little relevance to the real object of our discussion) but start by supposing that the sphere is deformed to an ellipsoid of rotation (spheroid) whose rotational axis coincides with the direction of the field. If we denote the original radius of the sphere by R, the semi-axis of the spheroid in the x-direction by a, and the second semi-axis by b, the deformation is described by the quantity

$$d = \frac{a-b}{R}. \tag{60.4}$$

An originally isotropic body generally becomes anisotropic when it is deformed.† In the deformed state, therefore, the dielectric tensor $\boldsymbol{\epsilon}$ introduced in eq. (57.4) takes the place of the scalar dielectric constant. The components of the dielectric tensor depend on the components of the strain tensor. This dependence must be found by experiment but may be taken to be linear for sufficiently small deformations. We can, therefore, make the statement‡

$$\varepsilon_{ij} = \varepsilon_0\,\delta_{ij} + \alpha S_{ij} + \beta(S_{xx} + S_{yy} + S_{zz})\,\delta_{ij}, \tag{60.5}$$

which agrees formally with eqs. (56.16) and (56.17); ε_0 is the scalar dielectric constant of the undeformed body and α and β are scalar constants. The components of the tensor electric susceptibility are then

$$\chi_{ij} = \frac{1}{4\pi}[\varepsilon_0\,\delta_{ij} + \alpha S_{ij} + \beta(S_{xx} + S_{yy} + S_{zz})\,\delta_{ij} - 1]. \tag{60.6}$$

The Helmholtz free energy is given by (58.12) and (59.1) as

$$F = V_0 \sum_{j=1}^{6} T_j S_j + \boldsymbol{E}.\boldsymbol{P} + \sum_{i=1}^{m} \mu_i n_i. \tag{60.7}$$

The application of the correspondingly generalized equilibrium condition (23.10) is, however, not trivial and we shall discuss it in some detail.

† A well-known example is found in the textures occurring during the working of metals (rolling, drawing).

‡ We again use the double subscript notation for the strain components since it makes matters particularly clear.

If we start from the general formulation (23.5) we find for the Helmholtz free energy that†

$$(\delta F)_T = 0 \tag{60.8}$$

with the secondary conditions

$$\delta S_j = 0 \quad (j = 1, 2, ..., 6), \tag{60.9}$$

$$\delta P_x = 0, \quad \delta P_y = 0, \quad \delta P_z = 0, \tag{60.10}$$

$$\delta n_i = 0 \quad (i = 1, 2, ..., m). \tag{60.11}$$

These secondary conditions are obviously absurd here, since they admit of no statement relating to our problem, although eq. (60.8) is fulfilled by (60.7) together with (60.9)–(60.11). This seeming contradiction is explained by the fact that in the present case the variations of the extensive parameters in eq. (60.7) are not independent of each other. This situation is completely analogous to that met in connexion with chemical reactions where additional conditions for the variations of the mole numbers are introduced. We can, therefore, in principle use the procedure used in § 33.

We first have to formulate secondary conditions to replace the conditions (60.9) and (60.10). They emerge from the assumed geometry of the deformation, the condition of constant volume, and the connexion between polarization and deformation given by eq. (60.6) together with (57.22). The geometry of the deformation implies

$$S_1 \neq 0, \quad S_2 = S_3 \neq 0, \quad S_4 = S_5 = S_6 = 0. \tag{60.12}$$

The condition of constant volume may, according to (52.25), be written

$$S_1 + S_2 + S_3 = 0 \tag{60.13}$$

Furthermore, we have, according to (52.29) and (60.4), that

$$S_1 - S_2 = \frac{a-b}{R} = d \tag{60.14}$$

and, according to (60.12)–(60.14), that

$$S_1 = \tfrac{2}{3}(S_1 - S_2) = \tfrac{2}{3}d. \tag{60.15}$$

Only the x-component of the field strength differs from zero because of the choice of the system of co-ordinates. That the x-component of the polarization is the only one that differs from zero then follows

† We can use (23.5) here only with the equilibrium sign.

from (57.22), (60.6), (60.12), and (60.13). We obtain

$$\left.\begin{aligned} P_x &= \frac{V}{4\pi}(\varepsilon_0 + \alpha S_1 - 1)E_x, \\ P_y &= 0, \quad P_z = 0. \end{aligned}\right\} \tag{60.16}$$

The relationships (60.12), (60.13), and (60.16) here play the role of the reaction equation (33.3). It is easy to see that

$$\left.\begin{aligned} \delta S_4 &= \delta S_5 = \delta S_6 = 0, \\ \delta P_y &= \delta P_z = 0, \\ \delta n_i &= 0 \quad (i = 1, 2, \ldots, m) \end{aligned}\right\} \tag{60.17}$$

are the only secondary conditions of (60.9)–(60.11) that remain valid without any change. For the extensive parameters S_1, S_2, S_3, and P_x, however, the secondary conditions

$$\left.\begin{aligned} &\delta S_2 - \delta S_3 = 0, \quad \delta S_1 + \delta S_2 + \delta S_3 = 0, \\ &\delta P_x - \frac{V}{4\pi}\alpha E_x \delta S_1 = 0 \end{aligned}\right\} \tag{60.18}$$

obtained from (60.12), (60.13), and (60.14) apply. These conditions constitute the analogue of eq. (33.9). The extremum problem (60.8) is reduced to the expression

$$V(T_1\,\delta S_1 + T_2\,\delta S_2 + T_3\,\delta S_3) + E_x\,\delta P_x = 0, \tag{60.19}$$

by the unchanged secondary conditions (60.17). This expression is analogous to (33.10). If we now use (60.18) to eliminate δS_2, δS_3, and δP_x we obtain

$$V(T_1 - T_2)\,\delta S_1 - \frac{V}{4\pi}\alpha E_x^2\,\delta S_1 = 0 \tag{60.20}$$

or, using (60.15),

$$(T_1 - T_2)\,\delta d - \frac{1}{4\pi}\alpha E_x^2\,\delta d = 0. \tag{60.21}$$

From here on the argument proceeds differently from the case of chemical equilibria since we are not interested in an equilibrium condition for internal parameters but in the equilibrium value of the quantity d. We therefore eliminate the stress component by means of eqs. (56.16) and (60.13). Using (60.14) we obtain

$$T_1 - T_2 = 2\mu_L(S_1 - S_2) = 2\mu_L\,d \tag{60.22}$$

where μ_{L} is the first Lamé constant. Equation (60.21) then becomes

$$2\mu_{\mathrm{L}}\, d\, \delta d - \frac{1}{4\pi}\alpha E_x^2 \delta d = 0 \tag{60.23}$$

or, since the variation δd is arbitrary,

$$d = \frac{a-b}{R} = \frac{\alpha}{8\pi\mu_{\mathrm{L}}}\boldsymbol{E}^2. \tag{60.24}$$

The deformation is thus likewise proportional to the square of the field strength.

The above argument may, as in the analogous case of chemical reactions (§ 36), also be followed through by representing the Helmholtz free energy as a function of the internal parameter d and then determining the equilibrium value of d from the extremum condition (60.8) with the secondary conditions (60.17).

§ 61 The electrocaloric effect

For simplicity we consider a fluid one-component system and use the thermodynamic potential†

$$\mathrm{d}H = T\,\mathrm{d}S + V\,\mathrm{d}P - \boldsymbol{P}.\mathrm{d}\boldsymbol{E} + \mu\,\mathrm{d}n \tag{61.1}$$

which gives the Maxwell relation

$$\left(\frac{\partial T}{\partial \boldsymbol{E}}\right)_{S,P,n} = -\left(\frac{\partial \boldsymbol{P}}{\partial S}\right)_{\boldsymbol{E},P,n}. \tag{61.2}$$

Using (25.8) and (57.24) we obtain

$$\left(\frac{\partial \boldsymbol{P}}{\partial S}\right)_{\boldsymbol{E},P,n} = \left(\frac{\partial (V\chi)}{\partial S}\right)_{\boldsymbol{E},P,n}\boldsymbol{E} = \left(\frac{\partial (V\chi)}{\partial T}\right)_{\boldsymbol{E},P,n}\left(\frac{\partial T}{\partial S}\right)_{\boldsymbol{E},P,n}\boldsymbol{E}. \tag{61.3}$$

If we introduce molar quantities and take into account (25.4), we finally find from (61.2) that

$$\left(\frac{\partial T}{\partial \boldsymbol{E}}\right)_{S,P} = -\frac{T}{C_{P,E}}\left(\frac{\partial (v\chi)}{\partial T}\right)_{\boldsymbol{E},P}\boldsymbol{E}, \tag{61.4}$$

where $C_{P,E}$ is the molar heat capacity at constant pressure and constant field strength. A quasi-static adiabatic change in the field strength thus leads to a temperature change whose sign is determined by the temperature dependence of the molar electric susceptibility. This phenomenon is called the *electrocaloric effect*. Experience shows

† The symbol H is justified by the convention mentioned in § 55.

that the quantity $\partial(v\chi)/\partial T$ is generally negative. A quasi-static adiabatic diminution in the field strength (and thus in the polarization) at constant pressure thus causes cooling. The effect is, however, extremely weak and, unlike its magnetic analogue (§ 66), only of theoretical interest.

§ 62 Piezoelectricity

Besides electrostriction discussed in § 60 there is a further electromechanical coupling effect which, however, occurs only in certain crystals and also differs considerably in other respects from electrostriction. In the discussion of electrostriction we assumed that the components of the dielectric tensor depend on the components of the strain tensor. We can obviously introduce the components of the stress tensor in an analogous way as independent variables. Both statements imply, however, that the polarization vanishes with the field strength. J. and P. Curie found, however, that in some cases an electric polarization is caused by the application of mechanical stresses even in the absence of an external electric field. This phenomenon is called the *piezoelectric effect* or, briefly, *piezoelectricity*. We can, therefore, make the statement†

$$p_i = \sum_{j=1}^{3} \chi_{ij} E_j + \sum_{k=1}^{6} \gamma_{ik} T_k \tag{62.1}$$

or, in the compact notation of eqs. (54.14) and (57.22),‡

$$\boldsymbol{p} = \boldsymbol{\chi} . \boldsymbol{E} + \boldsymbol{\gamma} . \boldsymbol{T}. \tag{62.2}$$

The components of the tensor $\boldsymbol{\gamma}$ are called *piezoelectric coefficients*. In order to obtain an explicit expression for the Gibbs free energy we introduce (55.12) and (62.2) into the equation

$$\mathrm{d}G = -S\,\mathrm{d}T - V_0 \sum_{j=1}^{6} S_j\,\mathrm{d}T_j - \boldsymbol{P} . \mathrm{d}\boldsymbol{E} + \mu\,\mathrm{d}n. \tag{62.3}$$

We now integrate, at T = const., n = const., $\boldsymbol{E} = 0$ for the stress components between 0 and T_j and use for this the approximation mentioned at the end of § 56. We then integrate, at T = const., n = const., T_j = const. (all j) for the field strength between 0 and $\boldsymbol{E}$.

† For certain classes of crystals an inhomogeneous term must be added to the right-hand side of (62.1). We shall ignore this for the present and return to it in § 63.

‡ It must be remembered that the vector $\boldsymbol{E}$ has three components and the vector $\boldsymbol{T}$ six in this notation. The quantity $\boldsymbol{\gamma}$ therefore appears as a rectangular matrix with three rows and six columns whereas we are really dealing with a third-degree tensor (with triple subscript components).

We shall assume in this connexion that the entropy is independent of the field strength; this is an unobjectionable approximation accord-to § 61.† We then obtain

$$G = G_0 - \tfrac{1}{2}V_0\,\boldsymbol{T}.\boldsymbol{\kappa}.\boldsymbol{T} - \tfrac{1}{2}V_0\,\boldsymbol{E}.\boldsymbol{\chi}.\boldsymbol{E} - V_0(\boldsymbol{\gamma}.\boldsymbol{T}).\boldsymbol{E} \tag{62.4}$$

which gives for the strain components

$$-\frac{1}{V_0}\left(\frac{\partial G}{\partial T_j}\right)_{T,\mathbf{E},n} = S_j = \sum_{k=1}^{6} \kappa_{jk}T_k + \sum_{l=1}^{3} \gamma_{jl}E_l. \tag{62.5}$$

Equations (62.1) and (62.5) contain the essential properties which distinguish piezoelectricity from electrostriction. Equation (62.5) shows that piezoelectricity is reversible in contrast to electrostriction. Even in the absence of external stresses an electric field (or the polarization caused by it) produces a deformation of the dielectric. Furthermore, the strain components depend linearly on the field strength in the piezoelectric effect according to (62.5), while they depend on the square of the field strength in electrostriction according to (60.24).

In the most general case the piezoelectric tensor $\boldsymbol{\gamma}$ contains 18 independent elements. This number is again reduced to a greater or lesser extent by the symmetry properties of the crystal. The condition that $\boldsymbol{\gamma}$ must be invariant with respect to symmetry transformations of the crystal determines which piezoelectric coefficients differ from zero, which are independent and, finally, which crystals can actually show the piezoelectric effect. We see immediately that a piezoelectric substance cannot have a centre of symmetry.‡ Isotropic substances in particular thus cannot be piezoelectric. Closer investigation shows that only 20 of the 32 possible crystal classes can show the piezoelectric effect. We can exemplify the reduction of $\boldsymbol{\gamma}$ by writing down eq. (62.5) for quartz. A suitable choice of co-ordinate system gives for vanishing stress components

$$\left.\begin{aligned} S_1 = \gamma_{11}E_x, \quad S_2 = -\gamma_{11}E_x, \quad S_3 = 0, \\ S_4 = \gamma_{14}E_x, \quad S_5 = -\gamma_{14}E_y, \quad S_6 = -2\gamma_{11}E_y. \end{aligned}\right\} \tag{62.6}$$

We are therefore left with only two independent piezoelectric coefficients.

† Electrostriction is a higher-order effect (see below) and may be ignored here.

‡ The mathematical reason for this depends on the fact that the components of a third-order tensor change their sign with an inversion (i.e. a reversal of the sign of all co-ordinates).

The calculation of the electric field in a piezoelectric substance requires the simultaneous solution of eqs. (53.15) and (57.6) for suitable boundary conditions. The problem is, therefore, usually extraordinarily complicated. It again becomes much simpler, however, for the case of an ellipsoid with no external forces acting on its surface. This again results in a homogeneous field within the ellipsoid, a homogeneous deformation and $T_j = 0$ for all j. The piezoelectric effect is extensively applied in electrical engineering for the frequency stabilization of oscillatory circuits, quartz being the substance generally used. Details can be found in appropriate textbooks.

§ 63 Ferroelectricity

Equation (57.22) does not strictly formally represent the most general relationship between polarization density and field strength. An inhomogeneous term may be added and we then get

$$\boldsymbol{p} = \boldsymbol{p}_0 + \boldsymbol{\chi} \cdot \boldsymbol{E}. \tag{63.1}$$

The vector $\boldsymbol{p}_0$ represents the polarization density at vanishing field strength, i.e. the spontaneous polarization per unit volume. $\boldsymbol{p}_0$ is thus a property of the dielectric, in which there must be a definite preferred direction which is conserved for all symmetry transformations of the substance. This condition is fulfilled only for symmetry groups consisting of a single axis of symmetry and planes of symmetry through this axis. Only 10 of the 32 crystal classes have these properties and form a sub-group among piezoelectric crystals called *pyroelectric crystals*. Crystals with a centre of symmetry cannot be pyroelectric. The dipole moment due to spontaneous polarization is generally not measurable since it is compensated by free charges which have reached the surface in the form of an internal current† or from the outside. This compensation is, however, a relatively slow process and the dipole moment is observed if the state of polarization of the crystal is changed by heating or cooling. This phenomenon is known as the *pyroelectric effect*.

The direction of spontaneous polarization in an ordinary pyroelectric crystal cannot be reversed by an external field since the relevant potential barriers are too high. There is, however, a sub-group among pyroelectric crystals which is characterized by the following properties:

(a) The pyroelectric structure exists only within a certain temperature range and arises *via* a thermodynamic transition.

† The electrical conductance of dielectric crystals is generally very small but not quite zero.

(b) The direction of spontaneous polarization can be reversed by an external field. The dependence of the polarization on the external field is characterized by a hysteresis loop.

Crystals possessing these properties are called *ferroelectric crystals* because of the analogy with ferromagnetism. The first observations attributable to the special properties of ferroelectrics were made by Pockels on Rochelle salt as early as 1894. The analogy with ferromagnetism was, however, realized much later, by Valasek in 1922. Ferroelectricity has since become an extensive part of solid-state physics. We shall, therefore, here confine our discussion to a few general remarks and a short thermodynamic consideration of transition points on the basis of the theory developed in § 51. We shall mainly follow the analysis by Tisza.

The occurrence of crystals in so-called 'twinned modifications' is widespread. Twinned crystals arise from a common point as mirror images or by rotation through 180° with respect to one another. Twinned modifications under hydrostatic pressure and in the absence of external fields have identical stability properties.† They can, therefore, co-exist in a crystal without any limitation and are then known in physics as *domains*. These domains in pyroelectric crystals correspond to regions of homogeneous spontaneous polarizations in opposite directions. The reversal of direction of the resulting dipole moment is due to nucleation and growth of the regions oriented in the direction of the field. The domains are not, however, phases within the meaning of § 2; their existence is, therefore, without influence on the number of thermodynamic degrees of freedom.

In certain cases the twinned modifications may be regarded as being due to equal and opposite displacements in a lattice which is symmetric with respect to the symmetry transformation connecting the twins. The difference between the domains can then decrease with a change in temperature until they eventually become identical at a critical temperature T_c. The undistorted lattice is then stable above T_c.

The above ideas outline a simple scheme for a ferroelectric transition. T_c is called the *ferroelectric Curie point*. T_c is a critical point in the sense of Tisza's theory and the domains then play the role of 'phases'. This mechanism is remarkable in that only minimal thermal effects are to be expected. We might also expect a ferroelectric transition to be an order–disorder transition like the magnetic and hyperstructure transitions we met in § 49. In that case the specific heat

† Certain stress components or components of the field strength can favour one of the two modifications.

should show a λ-point. These considerations indicate that ferroelectric transitions may occur by means of very different mechanisms and may therefore be expected to show differences in their thermodynamic properties. The details of the molecular mechanisms are obviously very much more complicated and still largely unexplained.

We shall now consider a few typical ferroelectrics and choose Rochelle salt (sodium potassium tartrate tetrahydrate, $NaKC_4H_4O_6 . 4H_2O$) as our first example. Rochelle salt is ferroelectric between 255 K and 297 K. It thus has a lower Curie point T_c^l and an upper Curie point T_c^u. Rochelle salt is orthorhombic for $T > T_c^u$ and $T < T_c^l$. It is monoclinic in the ferroelectric state and the spontaneous polarization is in the direction of the monoclinic axis.

In order to apply the theory developed in § 51 we first write the equation for the piezoelectric effect in the form

$$T_i = \sum_{j=1}^{6} c_{ij} S_j + \sum_{k=1}^{3} \delta_{ik} p_k, \tag{63.2}$$

where the c_{ij} are the elastic stiffness coefficients defined by eq. (54.13) and the δ_{ik} are the reverse piezoelectric coefficients.†

If we introduce (63.2) into (58.10) and then integrate with assumptions analogous to those of § 62 we obtain by means of (57.22) the fundamental equation per unit volume in the form

$$u = u_0 + \tfrac{1}{2}\boldsymbol{S}.\boldsymbol{c}.\boldsymbol{S} + \tfrac{1}{2}\boldsymbol{p}.\boldsymbol{\chi}^{-1}.\boldsymbol{p} + (\boldsymbol{\delta}.\boldsymbol{p}).\boldsymbol{S}. \tag{63.3}$$

We still need the explicit forms of the matrices $\boldsymbol{c}$ and $\boldsymbol{\delta}$ for orthorhombic crystals. These are

$$\boldsymbol{c} = \begin{bmatrix} c_{11} & c_{12} & c_{13} & 0 & 0 & 0 \\ c_{12} & c_{22} & c_{23} & 0 & 0 & 0 \\ c_{13} & c_{23} & c_{33} & 0 & 0 & 0 \\ 0 & 0 & 0 & c_{44} & 0 & 0 \\ 0 & 0 & 0 & 0 & c_{55} & 0 \\ 0 & 0 & 0 & 0 & 0 & c_{66} \end{bmatrix}, \tag{63.4}$$

$$\boldsymbol{\delta} = \begin{bmatrix} 0 & 0 & 0 & \delta_{14} & 0 & 0 \\ 0 & 0 & 0 & 0 & \delta_{25} & 0 \\ 0 & 0 & 0 & 0 & 0 & \delta_{36} \end{bmatrix}. \tag{63.5}$$

† We do not require the explicit relationship between the γ_{ik} and the δ_{ik}.

We choose the monoclinic axis as x-axis, i.e. the direction of spontaneous polarization in the ferroelectric state.

Although the critical parameters obviously have to be obtained from experimental data, we shall proceed in the opposite way since this is easier to follow. We therefore suppose that there are two critical parameters: the polarization density p_x in the direction of the monoclinic axis and the strain component S_4. We shall show that the conclusions obtained according to Tisza's theory then coincide with experimental fact.

We are here dealing with the exceptional case in the terminology of § 51. We therefore need consider only the critical variables. For small deviations from the orthorhombic state we find from (63.3)–(63.5) that

$$\frac{\partial u}{\partial S_4} = T_4 = c_{44} S_4 + \delta_{14} p_x, \tag{63.6}$$

$$\frac{\partial u}{\partial p_x} = E_x = \delta_{14} S_4 + (\chi^{-1})_{xx} p_x. \tag{63.7}$$

The coefficients occurring in these equations are the thermodynamic moduli defined by (50.3). The inverse relationships containing the thermodynamic coefficients are obtained directly from (62.4) and are

$$S_4 = \kappa_{44} T_4 + \gamma_{14} E_x, \tag{63.8}$$

$$p_x = \gamma_{14} T_4 + \chi_{xx} E_x. \tag{63.9}$$

We transform the matrix of the thermodynamic moduli into the diagonal form and obtain, according to (50.27),

$$\lambda_1 = \frac{\partial^2 u}{\partial S_4^2} = \left(\frac{\partial T_4}{\partial S_4}\right)_{p_x} = c_{44}, \tag{63.10}$$

$$\lambda_2 = \frac{\partial^2 \psi^{(1)}(T_4)}{\partial p_x^2} = \left(\frac{\partial E_x}{\partial p_x}\right)_{T_4} = \frac{1}{c_{44}} \begin{vmatrix} c_{44} & \delta_{14} \\ \delta_{14} & (\chi^{-1})_{xx} \end{vmatrix} = (\chi^{-1})_{xx} - \frac{\delta_{14}^2}{c_{44}}, \tag{63.11}$$

where $\psi^{(1)}(T_4)$ is the simple Legendre transform of the fundamental equation with respect to S_4, and the right-hand side of (63.11) follows from (51.8) and (51.5). The ferroelectric Curie point is defined by $\lambda_2 = 0$. According to (51.28) together with (63.8) and (63.9), this gives†

$$\kappa_{44} \to \infty, \quad \gamma_{14} \to \infty, \quad \chi_{xx} \to \infty. \tag{63.12}$$

Figures 47–49 show that a very steep maximum is observed at the Curie point in all three cases. It is immediately plausible that the statements (63.12) can be verified experimentally only in this sense. A more detailed discussion of this problem would lead us into a consideration of experimental techniques and is thus beyond the scope of this book.

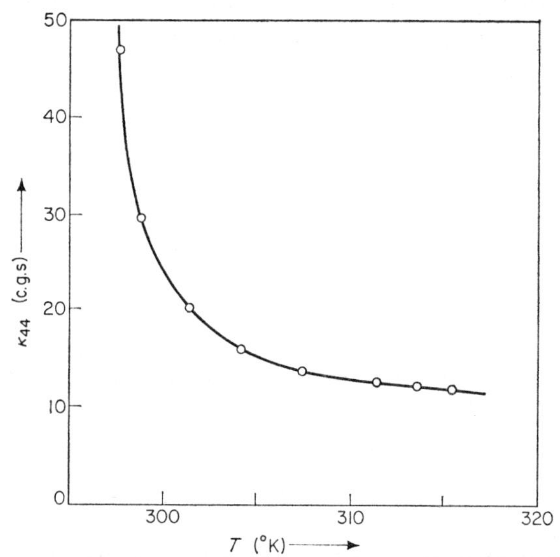

FIG. 47. The isothermal elastic compliance coefficient κ_{44} in the region of the upper Curie point of Rochelle salt [experimental data from H. Mueller, *Phys. Rev.*, **57**, 829 (1940)]

According to the theory the thermodynamic coefficients which do not appear in eqs. (63.8) and (63.9) must not show singularities at the Curie point. This too is confirmed by experiment. In particular, the specific heat does not show a λ-anomaly.‡

† This is, however, valid only for the free crystal. For the clamped crystal, for which the strain component S_4 is fixed, (63.7) and (63.11) give for the Curie point

$$\chi_{xx} = \left(\frac{\partial p_x}{\partial E_x}\right)_S = \frac{c_{44}}{\delta_{14}^2}.$$

‡ The specific heat has a very slight maximum (less than 1%) at the upper Curie point. The maximum is at the limit of experimental accuracy.

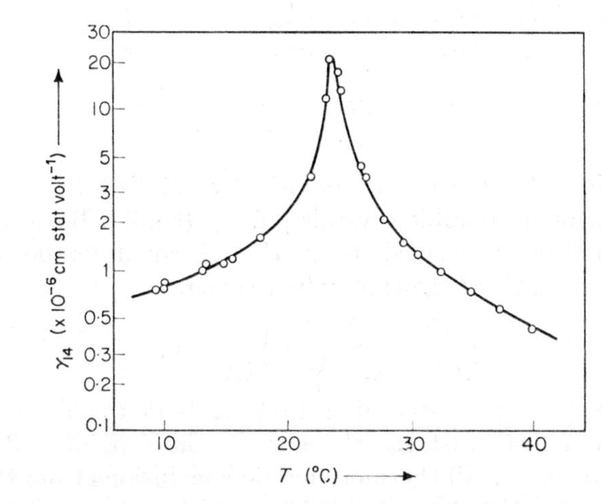

FIG. 48. The piezoelectric coefficient γ_{14} in the region of the Curie point of Rochelle salt

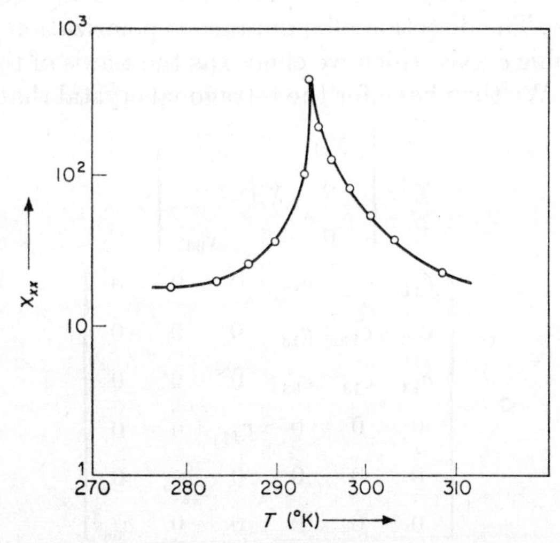

FIG. 49. The electric susceptibility χ_{xx} in the region of the upper Curie point of Rochelle salt [experimental data from J. Hablützel, *Helv. Phys. Acta*, **12**, 489 (1939)]

Finally, the condition (51.16) must be fulfilled. The explicit expression is

$$\frac{\partial E_x}{\partial p_y} = \frac{\partial E_x}{\partial p_z} = \frac{\partial E_x}{\partial S_1} = \frac{\partial E_x}{\partial S_2} = \frac{\partial E_x}{\partial S_3} = \frac{\partial E_x}{\partial S_5} = \frac{\partial E_x}{\partial S_6} = 0, \tag{63.13}$$

$$\frac{\partial E_x}{\partial s} = 0. \tag{63.14}$$

The conditions (63.13) are automatically fulfilled because of the symmetry of orthorhombic crystals [cf. eq. (63.5)]. By using one of Maxwell's relations derived from the thermodynamic potential $\psi^{(1)}(T_4)$, the condition (63.14) may be written

$$\left(\frac{\partial E}{\partial s}\right)_{T_4,p_x,} = \left(\frac{\partial T}{\partial p_x}\right)_{s,T_4} = 0. \tag{63.15}$$

This, however, means according to § 61, that the square of the electrocaloric effect must vanish near the Curie point and this has been found to be so. All thermodynamic conclusions from the initial assumptions have thus been verified by experiment.

We choose as our second example potassium dihydrogen phosphate which has only one ferroelectric Curie point, $T_c = 123$ K. The crystal is tetragonal for $T > T_c$ whereas the ferroelectric crystal ($T < T_c$) is orthorhombic. The direction of spontaneous polarization is along the crystallographic c-axis which we choose as the z-axis of the system of co-ordinates. We then have for the tetragonal crystal that

$$\boldsymbol{\chi} = \begin{bmatrix} \chi_{11} & 0 & 0 \\ 0 & \chi_{11} & 0 \\ 0 & 0 & \chi_{33} \end{bmatrix}, \tag{63.16}$$

$$\mathbf{c} = \begin{bmatrix} c_{11} & c_{12} & c_{13} & 0 & 0 & 0 \\ c_{12} & c_{11} & c_{13} & 0 & 0 & 0 \\ c_{13} & c_{13} & c_{33} & 0 & 0 & 0 \\ 0 & 0 & 0 & c_{44} & 0 & 0 \\ 0 & 0 & 0 & 0 & c_{55} & 0 \\ 0 & 0 & 0 & 0 & 0 & c_{66} \end{bmatrix}, \tag{63.17}$$

$$\boldsymbol{\delta} = \begin{bmatrix} 0 & 0 & 0 & \delta_{14} & 0 & 0 \\ 0 & 0 & 0 & 0 & \delta_{14} & 0 \\ 0 & 0 & 0 & 0 & 0 & \delta_{36} \end{bmatrix}. \qquad 63.18)$$

We now continue in the same way as before, starting with the assumption that the critical parameters are s, S_6, and p_z. We are, therefore, again considering the exceptional case. The equations of state are†

$$\delta T = \frac{T}{c_{0V}} \delta s + c_{66} S_6 + \delta_{36} p_z, \tag{63.19}$$

$$T_6 = \frac{T}{c_{0V}} \delta s + c_{66} S_6 + \delta_{36} p_z, \tag{63.20}$$

$$E_z = \frac{T}{c_{0V}} \delta s + \delta_{36} S_6 + (\chi^{-1})_{33} p_z. \tag{63.21}$$

The inverse relationships containing the thermodynamic coefficients are

$$\delta s = \frac{c_P}{T} \delta T + \kappa_{66} T_6 + \gamma_{36} E_z, \tag{63.22}$$

$$S_6 = \frac{c_P}{T} \delta T + \kappa_{66} T_6 + \gamma_{36} E_z, \tag{63.23}$$

$$p_z = \frac{c_P}{T} \delta T + \gamma_{36} T_6 + \chi_{33} E_z. \tag{63.24}$$

From (50.27) we obtain

$$\lambda_1 = \frac{\partial^2 u}{\partial s^2} = \left(\frac{\partial T}{\partial s}\right)_{T_6, p_z} = \frac{T}{c_V}, \tag{63.25}$$

$$\lambda_2 = \frac{\partial^2 \psi^{(1)}(T)}{\partial S_6^2} = \left(\frac{\partial T_6}{\partial S_6}\right)_{T, p_z} = c_{66}^{T,p}, \tag{63.26}$$

$$\lambda_3 = \frac{\partial^2 \psi^{(2)}(T, T_6)}{\partial p_z^2} = \left(\frac{\partial E_z}{\partial p_z}\right)_{T, T_6} = \frac{1}{c_{66}^{T,p}} \begin{vmatrix} c_{66}^{T,p} & \delta_{36} \\ \delta_{36} & (\chi^{-1})_{33} \end{vmatrix} \tag{63.27}$$

$$= (\chi^{-1})_{33} - \frac{\delta_{36}^2}{c_{66}^{T,p}}.$$

The Curie point is defined by $\lambda_3 = 0$. This, according to (51.28) together with (63.22)–(63.24), gives

$$c_P \to \infty, \quad \kappa_{66} \to \infty, \quad \gamma_{36} \to \infty, \quad \chi_{33} \to \infty. \tag{63.28}$$

† The reader is reminded that the fundamental equation (63.3) refers to unit volume and that the subscript 0 denotes the undeformed state at zero field strength.

Figures 50–53 show that these statements are confirmed by experiment. The thermodynamic coefficients which do not appear in eqs. (63.22)–(63.24), however, show no singularities at the Curie point.

The condition (51.16) for this case is

$$\frac{\partial E_z}{\partial_{p_x}} = \frac{\partial E_z}{\partial_{p_y}} = \frac{\partial E_z}{\partial S_1} = \frac{\partial E_z}{\partial S_2} = \frac{\partial E_z}{\partial S_3} = \frac{\partial E_z}{\partial S_4} = \frac{\partial E_z}{\partial S_5} = 0. \qquad (63.29)$$

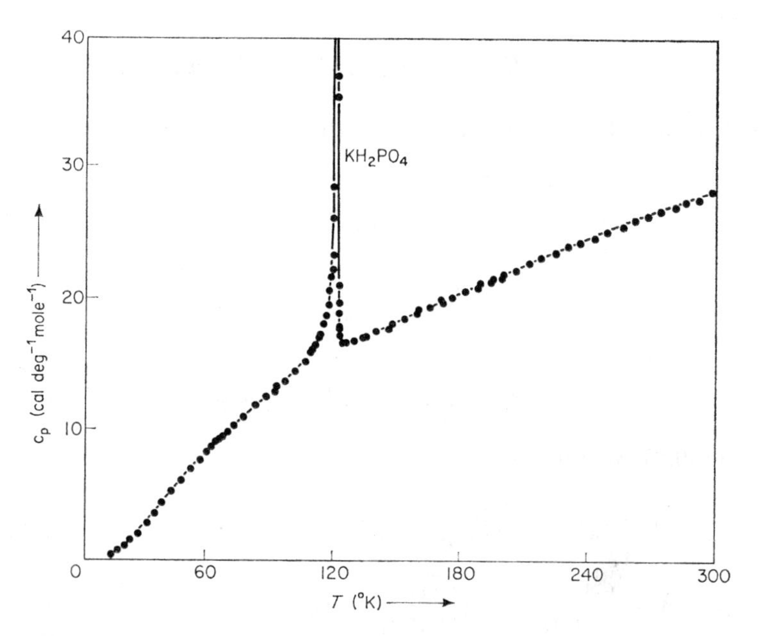

FIG. 50. The specific heat of KH_2PO_4 in the region of the Curie point

Equations (63.16)–(63.18) show that these conditions are automatically fulfilled because of the symmetry of tetragonal crystals. A condition analogous to eq. (63.14) does not occur here. KH_2PO_4 should, therefore, show a clear electrocaloric effect near the Curie point and this is confirmed by Fig. 54.

In the terminology of § 51, the ferroelectric Curie point of Rochelle salt constitutes a displacive transition, that of KH_2PO_4 an order–disorder transition. The two types correspond to the two schemes outlined above for the occurrence of a ferroelectric transition.

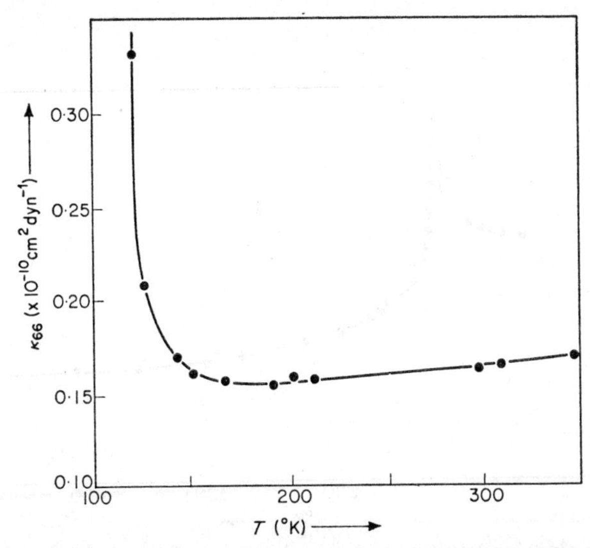

FIG. 51. The isothermal elastic compliance coefficient κ_{66} in the region of the Curie point of KH_2PO_4 [experimental data from W. P. Mason, *Phys. Rev.*, **69**, 173 (1946)]

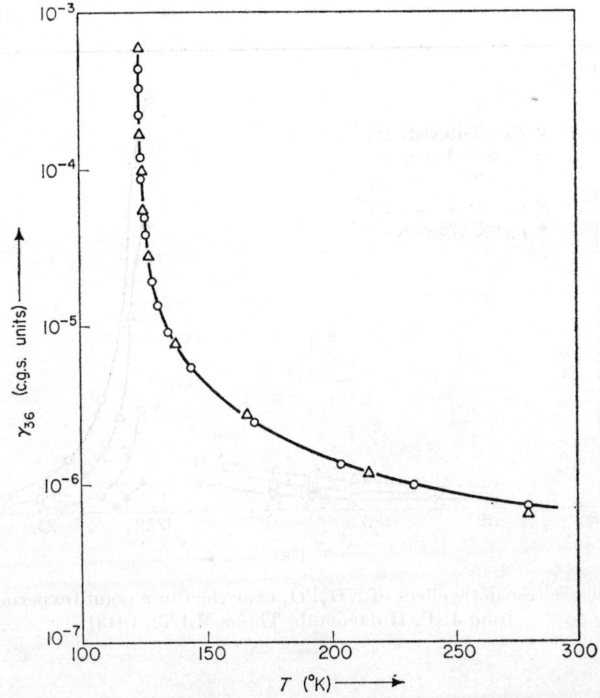

FIG. 52. The piezoelectric coefficient γ_{36} in the region of the Curie point of KH_2PO_4

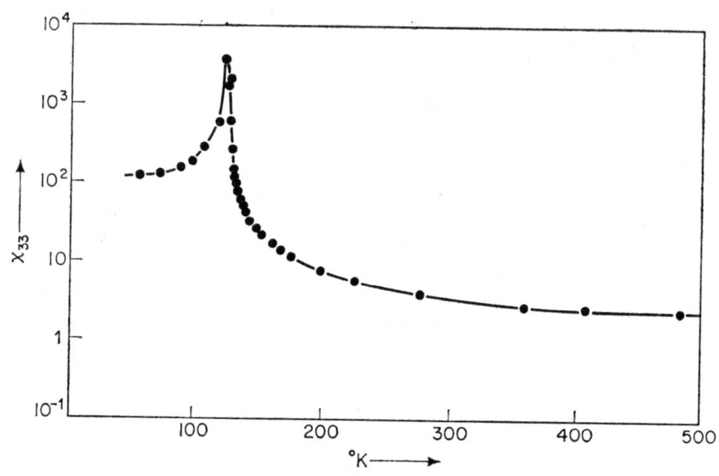

FIG. 53. The electric susceptibility χ_{33} in the region of the Curie point of KH_2PO_4 [experimental data from G. Busch, *Helv. Phys. Acta*, **11**, 269 (1938)]

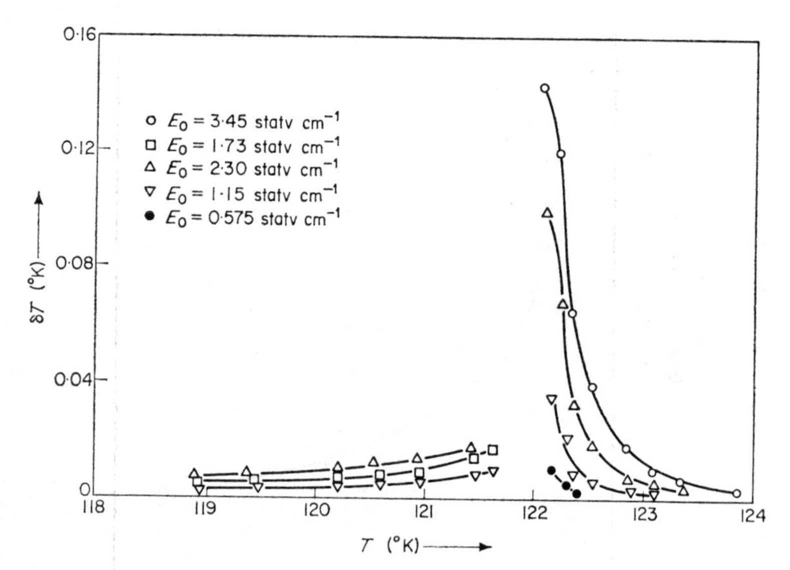

FIG. 54. The electrocaloric effect of KH_2PO_4 near the Curie point [experimental data from J. F. Hutzenlaub, Thesis M.I.T., 1943]

CHAPTER X

Systems in a magnetic field

§ 64 Magnetostatic work

The thermodynamic treatment of systems in a magnetic field shows many and extensive analogies with the treatment of systems in an electric field. We shall, therefore, confine our discussion mainly to those problems whose treatment cannot simply be taken from § 57.

The differences between the two cases (as far as they are relevant here) are based mainly on the fact that magnetic 'charges' do not exist. It is therefore not possible to define magnetic work in a form analogous to (57.7). We thus have to develop a suitable definition.

We consider a conductor carrying a steady electric current of current density $\boldsymbol{j}$. The current generates a magnetic field whose field strength is $\boldsymbol{H}$ and whose magnetic induction is $\boldsymbol{B}$.† These two quantities are related by

$$\boldsymbol{B} = \boldsymbol{\mu} . \boldsymbol{H}, \tag{64.1}$$

which, for isotropic substances, reduces to the equation

$$\boldsymbol{B} = \mu . \boldsymbol{H}. \tag{64.2}$$

The quantity μ is called the *magnetic permeability*. Under the stated conditions the field equations (as far as we need them here) are

$$\operatorname{div} \boldsymbol{B} = 0, \quad \operatorname{rot} \boldsymbol{H} = \frac{4\pi}{c} \boldsymbol{j}, \tag{64.3}$$

$$\frac{1}{c} \frac{\partial \boldsymbol{B}}{\partial t} + \operatorname{rot} \boldsymbol{E} = 0, \tag{64.4}$$

where c is the velocity of light *in vacuo* and t is the time.

† This notation corresponds to the international conventions [Document U.I.P. 11 (S.U.N. 65–3)] of IUPAP. Some authors have recently called $\boldsymbol{B}$ the magnetic field strength.

The calculation of the electrical work in § 57 is based on the fact that a change in the electric field strength requires a transport of charge in the electric field resulting in the performance of work. No work is done, however, when electric charge is transported in a magnetic field since the force exerted by the magnetic field on the charge acts perpendicularly to the direction of flow. A change in the distribution of static charge therefore has no effect on the magnetic field. We must, therefore, base our ideas on the fact that the magnetic field is generated by the electric current and the source of the current generally performs work even when the current is steady; this work will appear as Joule heating. This work used up in a dissipative process may, however, in principle be made as small as we like (superconduction) and has no connexion with the existence of the magnetic field. We may, therefore, ignore this work in connexion with the present discussion. These considerations have re-established the analogy with the electric case since the maintenance of an electrostatic field in principle requires no work to be done. A change in the magnetic field is caused by a change in current density and hence basically by a change in the electromotive force of the current source. The change in the magnetic field induces an electric field and thus an electromotive force in the conductor opposite to the original e.m.f. The current source has to do work to overcome this induced e.m.f. This work δW^* is the work necessary for a change in the magnetic field. This work is, therefore, equal and opposite to the work done by the induced electric field on the electric current, and that work must be expressed in terms of magnetic field quantities.

The work done in time δt by the induced field $\boldsymbol{E}$ is

$$-\delta W^* = \delta t \int \boldsymbol{j}.\boldsymbol{E}\,\mathrm{d}V. \tag{64.5}$$

With (64.3) this becomes

$$\delta W^* = -\delta t \frac{c}{4\pi} \int \boldsymbol{E}.\mathrm{rot}\,\boldsymbol{H}\,\mathrm{d}V \tag{64.6}$$

or†

$$\delta W^* = \delta t \frac{c}{4\pi} \int \mathrm{div}\,(\boldsymbol{E}\times\boldsymbol{H})\,\mathrm{d}V - \delta t \frac{c}{4\pi} \int \boldsymbol{H}.\mathrm{rot}\,\boldsymbol{E}\,\mathrm{d}V. \tag{64.7}$$

† We have that

$$\mathrm{div}\,\mathbf{a}\times\mathbf{b} = \mathbf{b}\,\mathrm{rot}\,\mathbf{a} - \mathbf{a}\,\mathrm{rot}\,\mathbf{b}.$$

We may assume without loss of generality that field strengths vanish at infinity. If, therefore, we transform the first integral by means of Gauss' theorem† into an integral over an infinitely distant surface, this integral must also vanish. If we introduce (64.4) into the second integral and write

$$\delta \boldsymbol{B} = \frac{\partial \boldsymbol{B}}{\partial t}\,\delta t \tag{64.8}$$

we finally find that

$$\delta W^* = \frac{1}{4\pi}\int \boldsymbol{H}.\delta\boldsymbol{B}\,\mathrm{d}V. \tag{64.9}$$

§ 65 The fundamental equation for a system in a magnetic field. Thermodynamic potentials

Further developments are so similar to those of §§ 57 and 58 that we shall only indicate briefly the most important points. The analogue of the plate condenser is a long, homogeneous, cylindrical coil. If we neglect boundary effects, we have, *in vacuo*, a homogeneous magnetic field $\boldsymbol{H}_0$ in the direction of the coil axis inside the coil while outside the coil $\boldsymbol{H}_0 = 0$. If a material body of volume V is introduced into this field, the field is modified by the *magnetic polarization* (*magnetization*) of the body. The same formal problems then occur as in the case of an electrostatic field. We can, therefore, simply adopt our previous results. The work to be done in connexion with a change in the field is also modified by the presence of the body. As far as thermodynamics is concerned only the work done on the body itself is of interest. In order to separate this work we use the expression

$$\boldsymbol{B} = \boldsymbol{H} + 4\pi\boldsymbol{m}, \tag{65.1}$$

where $\boldsymbol{m}$ is the *magnetic moment per unit volume* (*magnetization density*). The magnetization density is connected with the field strength by a relationship which, for an isotropic substance, is

$$\boldsymbol{m} = \chi\boldsymbol{H} \tag{65.2}$$

where χ is called the *magnetic susceptibility*.‡ From (64.2), (65.1), and (65.2) we get

$$\chi = \frac{\mu - 1}{4\pi}. \tag{65.3}$$

† Cf. § 53.

‡ The usual symbol for the magnetic susceptibility is χ_{m}. Since we are considering exclusively magnetic phenomena in this chapter we shall omit the subscript.

We now make the following assumptions analogous to those of § 58:

(a) The field $\boldsymbol{H}_0$ *in vacuo* inside the coil (the 'applied field') is homogeneous and is in the direction of the coil axis.

(b) The field $\boldsymbol{H}_0$ is weak. Terms higher than the square term can therefore be ignored in the expansion of the thermodynamic functions as a power series in $\boldsymbol{H}$. This assumption implies that μ and χ are independent of the field strength.

(c) The substance introduced into the field is originally homogeneous. It has the shape of an ellipsoid with one of the main axes in the direction of the field $\boldsymbol{H}_0$.

An analogous application of the considerations of § 58 to eqs. (64.9) and (65.1) then gives the quasi-static work done on the system with an infinitesimal change in current density as

$$\mathrm{d}W = \boldsymbol{H}.\mathrm{d}(\boldsymbol{m}.V) = \boldsymbol{H}.\mathrm{d}\boldsymbol{M}, \tag{65.4}$$

where $\boldsymbol{M} = \boldsymbol{m}V$ is the *total magnetization* of the substance.

The explicit fundamental equation is obtained by introducing (65.4) into (21.3). Since we shall not be considering anisotropy effects, we write it in the form

$$\mathrm{d}U = T\,\mathrm{d}S - P\,\mathrm{d}V + \boldsymbol{H}.\mathrm{d}\boldsymbol{M} + \sum_{i=1}^{m} \mu_i\,\mathrm{d}n_i \tag{65.5}$$

or, using Euler's theorem,

$$U = TS - PV + \boldsymbol{H}.\boldsymbol{M} + \sum_{i=1}^{m} \mu_i n_i. \tag{65.6}$$

We now introduce some thermodynamic potentials which are often used in the present context. We define the *enthalpy* by the equation

$$H = U + PV - \boldsymbol{H}.\boldsymbol{M} \tag{65.7}$$

or, in differential form,

$$\mathrm{d}H = T\,\mathrm{d}S + V\,\mathrm{d}P - \boldsymbol{M}.\mathrm{d}\boldsymbol{H} + \sum_{i=1}^{m} \mu_i\,\mathrm{d}n_i. \tag{65.8}$$

The *Helmholtz free energy* is

$$F = U - TS \tag{65.9}$$

or

$$\mathrm{d}F = -S\,\mathrm{d}T - P\,\mathrm{d}V + \boldsymbol{H}.\mathrm{d}\boldsymbol{M} + \sum_{i=1}^{m} \mu_i\,\mathrm{d}n_i. \tag{65.10}$$

Finally, we denote the *Gibbs free energy* by the function

$$G = U - TS + PV - \boldsymbol{H}.\boldsymbol{M} \qquad (65.11)$$

or

$$\mathrm{d}G = -S\,\mathrm{d}T + V\,\mathrm{d}P - \boldsymbol{M}.\mathrm{d}\boldsymbol{H} + \sum_{i=1}^{m} \mu_i\,\mathrm{d}n_i, \qquad (65.12)$$

where we again have that

$$G = \sum_{i=1}^{m} \mu_i n_i. \qquad (65.13)$$

Despite the formal analogies between the most important equations, the magnetic properties of substances differ in various ways from their dielectric ones. The magnetic permeability μ may be greater or less than one while we always have $\varepsilon > 1$ for the dielectric constant. The magnetic susceptibility χ can correspondingly be positive or negative. Substances with $\chi < 0$ are called *diamagnetic* and substances with $\chi > 0$ are *paramagnetic*. The electric susceptibilities of liquids and solids are generally between 10^{-1} and 10 whereas the absolute values of the magnetic susceptibilities are usually of an order of magnitude $\leqslant 10^{-4}$ with the exception of those of ferromagnetic substances. Absolute values of diamagnetic susceptibilities are much smaller than paramagnetic ones. The difference between the 'local field' $\boldsymbol{H}$ and the 'applied field' $\boldsymbol{H}_0$ can, therefore, often be ignored.

A further difference is that the magnetic symmetry properties of anisotropic substances are not necessarily the same as their crystallographic symmetry properties.

Finally, magnetic transitions differ quite considerably from electrical ones. Although the transition from the paramagnetic to the ferromagnetic state shows many analogies with the ferroelectric transition discussed in § 63, the former differs from the latter, on the one hand, in that the close connexion with the crystal structure is absent and, on the other, in that the ferromagnetic transition is necessarily an order–disorder transition and the entropy is thus always a critical parameter. This means that the ferromagnetic, unlike the ferroelectric, Curie point is always accompanied by a λ-anomaly of the specific heat. Quite apart from this, many metals and alloys undergo a transition for which no electrical analogue exists; this transition is associated with the phenomenon of superconduction.

We shall confine further discussion to two effects which are thermodynamically of particular interest and have considerable general significance.

§ 66 The magnetocaloric effect

We consider a one-component system and start from eq. (65.8) which gives the Maxwell relation

$$\left(\frac{\partial T}{\partial \boldsymbol{H}}\right)_{S,P,n} = -\left(\frac{\partial \boldsymbol{M}}{\partial S}\right)_{\boldsymbol{H},P,n}. \tag{66.1}$$

By means of (25.8) and (65.2) we obtain

$$\left(\frac{\partial \boldsymbol{M}}{\partial S}\right)_{\boldsymbol{H},P,n} = \left(\frac{\partial (V\chi)}{\partial S}\right)_{\boldsymbol{H},P,n} \boldsymbol{H} = \left(\frac{\partial (V\chi)}{\partial T}\right)_{\boldsymbol{H},P,n} \left(\frac{\partial T}{\partial S}\right)_{\boldsymbol{H},P,n} \boldsymbol{H}. \tag{66.2}$$

If we introduce molar quantities and apply (25.4), eq. (66.1) finally becomes

$$\left(\frac{\partial T}{\partial \boldsymbol{H}}\right)_{S,P} = -\frac{T}{C_{P,H}} \left(\frac{\partial (v\chi)}{\partial T}\right)_{\boldsymbol{H},P} \boldsymbol{H}, \tag{66.3}$$

where $C_{P,H}$ is the molar heat capacity at constant pressure and constant magnetic field strength. A quasi-static adiabatic change in field strength at constant pressure thus causes a temperature change whose sign is determined by the way in which the molar magnetic susceptibility $v\chi \equiv \bar{\chi}$ depends on the temperature. This phenomenon is called the *magnetocaloric effect* and is formally completely analogous to the electrocaloric effect (§ 61). Unlike the latter, however, the magnetocaloric effect is of great practical importance since it is the basis of the most important method of generating extremely low temperatures ($T < 1$ K). The method is called *adiabatic demagnetization* and consists in principle of the following steps:

(a) A paramagnetic salt is put into a vessel filled with helium at low pressure. The specimen is cooled to about 1 K by means of an external bath of liquid helium.

(b) After thermal equilibrium has been established the salt is magnetized by the application of a field of about 10^5 Oersted. Thermal equilibrium is maintained by removing the liberated heat by means of the external helium bath.

(c) The sample is thermally insulated by evacuating the container.

(d) The magnetic field is reduced to zero (or to a small finite value). The salt is cooled by means of the magnetocaloric effect.

The greatest attainable cooling is found by integrating eq. (66.3). We shall not carry out this calculation since the main features are obtained much more simply and clearly from a graphical representation. We start with eq. (65.12) which gives the Maxwell relation

$$\left(\frac{\partial S}{\partial \boldsymbol{H}}\right)_{T,P} = \left(\frac{\partial \boldsymbol{M}}{\partial T}\right)_{\boldsymbol{H},P} \tag{66.4}$$

which gives for molar quantities, using (65.2),

$$\left(\frac{\partial s}{\partial \boldsymbol{H}}\right)_{T,P} = \left(\frac{\partial (v\chi)}{\partial T}\right)_{\boldsymbol{H},P} \boldsymbol{H}. \tag{66.5}$$

The paramagnetic susceptibility within certain temperature regions is often described sufficiently accurately by the Curie–Weiss law

$$v\chi \equiv \bar{\chi} = \frac{C}{T-\theta}, \tag{66.6}$$

where C and θ are constants. With (66.5) this becomes

$$\left(\frac{\partial s}{\partial \boldsymbol{H}}\right)_{T,P} = -\frac{C}{(T-\theta)^2} \boldsymbol{H}. \tag{66.7}$$

Integration at constant T and P gives

$$s(T,H) = s(T,0) - \frac{1}{2}\frac{C}{(T-\theta)^2} \boldsymbol{H}^2. \tag{66.8}$$

Isothermal magnetization thus decreases the entropy; the decrease in entropy increases with decreasing temperature. This statement (like the Curie–Weiss law) cannot apply to the lowest temperatures since the entropy change must vanish at the absolute zero. We can, therefore, draw the schematic diagram shown in Fig. 55. This shows the molar entropy as a function of temperature with the field strength as a parameter. Steps (b) and (d) of the adiabatic demagnetization are represented by the lines marked with arrows. Two curves are shown for $\boldsymbol{H} = 0$; No. 1 refers to a substance unsuitable for cooling whereas No. 2 represents schematically the behaviour of paramagnetic salts actually used in the process. The figure shows immediately that strong cooling is due mainly to two factors. Firstly, the substance used must have a strongly rising $s(T,0)$ curve in the region of the required temperature T_2, i.e. the specific heat must have a steep maximum. Secondly, the entropy at $T_1 \approx 1$ K must be changed by a technically realizable magnetic field in such a way that it has a value within the steep region of the curve at $\boldsymbol{H} = 0$. These conditions are fulfilled by certain rare earth salts (e.g. gadolinium sulphate, cerium magnesium nitrate) and by certain alums (e.g. iron ammonium sulphate, chromium methylammonium sulphate). These salts are the usual ones used in adiabatic demagnetization. Temperatures of the order of 10^{-3} K have been reached in this way. The temperature scale in this region is initially an empirical one defined by the Curie–Weiss law. The conversion into absolute temperature is done by the method described in § 12.

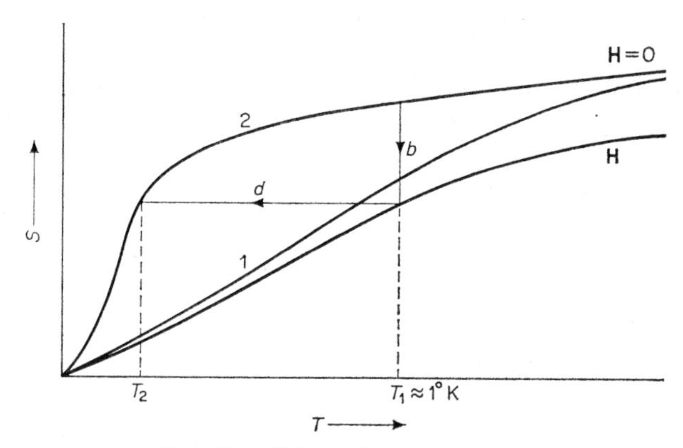

Fig. 55. Adiabatic demagnetization

§ 67 Superconduction

Many metals, metallic compounds, and alloys at low temperatures (generally between 0·1 and 20 K) show a transition which is macroscopically characterized by the following manifestations:

(a) The electrical resistance of the sample falls to an immeasurably small value at the transition point T_c. It is probable that this value is exactly zero.

(b) When a magnetic field of increasing strength is applied to the sample, the original resistance reappears at a critical field strength $\boldsymbol{H}_c$.

(c) The magnetic induction $\boldsymbol{B}$ vanishes inside the superconducting substance (Meissner effect).

The name superconduction is nowadays applied to the totality of these effects.†

† We shall confine our discussion in this section to the so-called superconductors of the first kind, i.e. mainly pure metals. The reader is referred to the specialist literature for certain peculiarities of superconductors of the second kind.

Since the detailed nature of the effects depends on the shape of the sample and its orientation in the magnetic field we shall assume that the sample is a long cylinder oriented in the direction of the field. As has already been mentioned, we then have $\boldsymbol{H}_0 = \boldsymbol{H}$ and the effect (b) appears as a discontinuous change at the field strength $\boldsymbol{H}_c$.

Experience shows that the appearance of superconduction at finite field strength constitutes a first-order transition. We therefore have to consider the co-existence of two phases, the superconducting phase (superscript s) and the 'normal' phase (superscript n).†

We shall confine our discussion to one-component systems. The general equilibrium conditions are obtained from the fundamental equation in the form of eq. (65.5) by the method of § 27. We find that

$$T^{(s)} = T^{(n)}, \quad P^{(s)} = P^{(n)}, \quad \boldsymbol{H}^{(s)} = \boldsymbol{H}^{(n)} \tag{67.1}$$

and

$$\mu^{(s)} = \mu^{(n)}. \tag{67.2}$$

An argument completely analogous to that of § 29 gives the phase rule here as

$$f = m + 3 - \sigma. \tag{67.3}$$

The system thus has two thermodynamic degrees of freedom, i.e. it is bivariant. Despite this, phase reactions are possible without limitation since the formal considerations of § 30 do not depend on the number of work co-ordinates and the condition for case (a) ($m < \sigma$) is satisfied here. To obtain a more comprehensible argument, however, it is useful to keep one variable constant and thus to reduce the problem formally to one concerning a univariant system.

We consider first the most important case, i.e. $P = \text{const.}$ Since

$$\mathrm{d}\mu = -s\,\mathrm{d}T + v\,\mathrm{d}P - v\boldsymbol{m}.\mathrm{d}\boldsymbol{H}, \tag{67.4}$$

we find from (67.1) and (67.2) that the equation for the co-existence curve in the $\boldsymbol{H}$–T-plane is

$$\frac{\mathrm{d}\boldsymbol{H}_c}{\mathrm{d}T_c} = -\frac{s^{(n)} - s^{(s)}}{[(v\boldsymbol{m})^{(n)} - (v\boldsymbol{m})^{(s)}]}. \tag{67.5}$$

Because of the Meissner effect we have, according to (65.1), that

$$\boldsymbol{m}^{(s)} = -\boldsymbol{H}/4\pi \tag{67.6}$$

for the superconducting phase. The very small paramagnetic magnetization can be completely ignored compared with this

† The co-existence of the two phases is capable of direct experimental proof.

quantity. The entropy difference therefore is

$$s^{(n)} - s^{(s)} = -\frac{v^{(s)}\,\boldsymbol{H}_c}{4\pi}\frac{d\boldsymbol{H}_c}{dT_c}. \tag{67.7}$$

This shows the molar heat of transition to be

$$L = T_c(s^{(n)} - s^{(s)}) = -\frac{v^{(s)}\,T_c\,\boldsymbol{H}_c}{4\pi}\frac{d\boldsymbol{H}_c}{dT_c}. \tag{67.8}$$

Equation (67.7) shows that the course of the co-existence curve is determined almost exclusively by the entropy difference between the normal phase and the superconducting phase. According to the Nernst heat theorem (§ 38) we should have

$$\lim_{T\to 0} s^{(n)} = \lim_{T\to 0} s^{(s)} \tag{67.9}$$

and thus, according to (67.7),

$$\lim_{T\to 0}\frac{d\boldsymbol{H}_c}{dT_c} = 0. \tag{67.10}$$

Figures 56 and 57 show that this is in very reasonable agreement with experimental data although a strict experimental proof can hardly be claimed. We may conclude from experimental data (Figs. 56 and 57) together with eq. (67.7) or from statistical theory that the superconducting phase always has a smaller entropy than the co-existing normal phase. Statistical theory in combination with (67.7) gives the result that the gradient of the co-existence curve is always negative and this is confirmed by experiment. Thermodynamics can naturally give no information about the limiting value $\lim_{H_c\to 0}(d\boldsymbol{H}_c/dT_c)$. This limiting value, however, unlike the limiting value (67.10), is experimentally easily accessible and we can see from Figs. 56 and 57 that it is always finite. This result is of particular interest for the discussion of specific heats which is our next subject.

We start by differentiating eq. (67.7) along the co-existence curve.† This operation which we simply write as an ordinary derivative gives

$$\frac{ds^{(n)}}{dT_c} = \left(\frac{\partial s^{(n)}}{\partial T}\right)_{\boldsymbol{H},P} + \left(\frac{\partial s^{(n)}}{\partial \boldsymbol{H}}\right)_{T,P}\frac{d\boldsymbol{H}_c}{dT_c} \tag{67.11}$$

† This operation is of more general significance. For liquid helium, for example, only the specific heat at the equilibrium vapour pressure can be obtained by experiment.

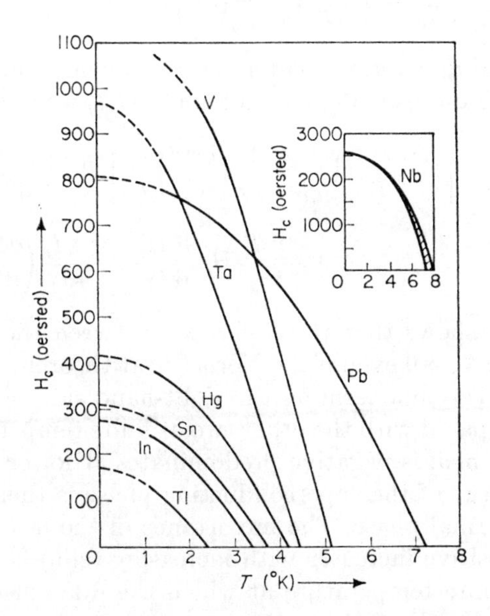

FIG. 56. Co-existence curves for superconduction

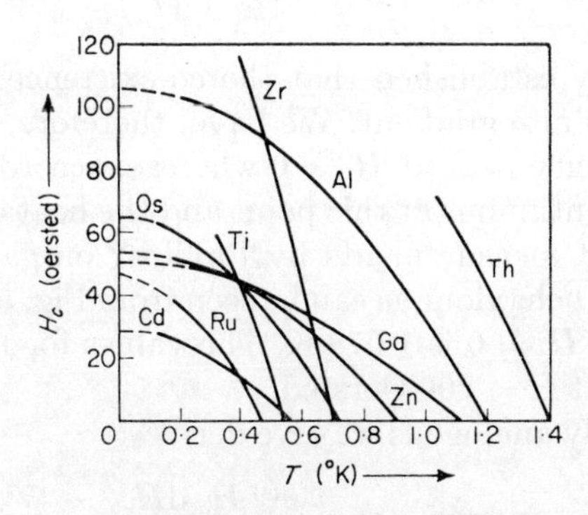

FIG. 57. Co-existence curves for superconduction

or, with (66.4),

$$\frac{\mathrm{d}s^{(\mathrm{n})}}{\mathrm{d}T_\mathrm{c}} = \left(\frac{\partial s^{(\mathrm{n})}}{\partial T}\right)_{\boldsymbol{H},P} + \left(\frac{\partial(v\boldsymbol{m})^{(\mathrm{n})}}{\partial T}\right)_{\boldsymbol{H},P} \frac{\mathrm{d}\boldsymbol{H}_\mathrm{c}}{\mathrm{d}T_\mathrm{c}} \tag{67.12}$$

and corresponding expressions for the superconducting phase. If we multiply by T_c and use (65.2), (67.6), and (67.7) we then obtain

$$C^{(\mathrm{s})}_{P,\boldsymbol{H}} - C^{(\mathrm{n})}_{P,\boldsymbol{H}} = \left[\left(\frac{\partial(v\chi)^{(\mathrm{n})}}{\partial T}\right)_{\boldsymbol{H},P} - \frac{1}{4\pi}\left(\frac{\partial v^{(\mathrm{s})}}{\partial T}\right)_{\boldsymbol{H},P}\right] T_\mathrm{c} \boldsymbol{H}_\mathrm{c} \frac{\mathrm{d}\boldsymbol{H}_\mathrm{c}}{\mathrm{d}T_\mathrm{c}}$$
$$+ \frac{v^{(\mathrm{s})} T_\mathrm{c}}{4\pi} \boldsymbol{H}_\mathrm{c} \frac{\mathrm{d}^2 \boldsymbol{H}_\mathrm{c}}{\mathrm{d}T^2} + \frac{v^{(\mathrm{s})} T_\mathrm{c}}{4\pi}\left(\frac{\mathrm{d}\boldsymbol{H}_\mathrm{c}}{\mathrm{d}T_\mathrm{c}}\right)^2. \tag{67.13}$$

This equation shows that the difference between the specific heats vanishes when $T_\mathrm{c} \to 0$ even if the Nernst heat theorem [eq. (67.10)] is not obeyed.† The first term on the right-hand side can generally be neglected compared with the other two at finite temperatures T_c. The second term which is negative predominates at lower temperatures. The specific heat of the superconducting phase is then smaller than that of the normal phase. The importance of the last term which is necessarily positive increases with increasing temperature. There is, therefore, a finite temperature at which the difference between the specific heats vanishes. Above this temperature the superconducting phase has the greater specific heat. Finally, we have, for $\boldsymbol{H}_\mathrm{c} \to 0$, that

$$C^{(\mathrm{s})}_{P,\boldsymbol{H}} - C^{(\mathrm{n})}_{P,\boldsymbol{H}} = \frac{v^{(\mathrm{s})} T_\mathrm{c}}{4\pi}\left(\frac{\mathrm{d}\boldsymbol{H}_\mathrm{c}}{\mathrm{d}T_\mathrm{c}}\right)^2_{\boldsymbol{H}_\mathrm{c}=0}. \tag{67.14}$$

We have already established that the co-existence curve meets the abscissa with a finite gradient. We have, therefore, a finite discontinuity in the specific heat at $\boldsymbol{H}_\mathrm{c} = 0$ whereas, according to eq. (67.7), the entropy is continuous at this point and the heat of transition thus vanishes. Recent measurements by Phillips‡ on gallium are a good example for this behaviour as can be seen from Fig. 58. The transition temperature for $\boldsymbol{H}_\mathrm{c} = 0$ is 1·078 K. The values for the normal phase were measured at $H = 200$ Oersted.

In a completely analogous way we derive

$$v^{(\mathrm{n})} - v^{(\mathrm{s})} = \frac{v^{(\mathrm{s})} \boldsymbol{H}_\mathrm{c}}{4\pi} \frac{\mathrm{d}\boldsymbol{H}_\mathrm{c}}{\mathrm{d}P_\mathrm{c}} \tag{67.15}$$

† The only assumption necessary is that $\mathrm{d}\boldsymbol{H}_\mathrm{c}/\mathrm{d}T_\mathrm{c}$ does not become infinite as $T \to 0$.

‡ N. E. Phillips, *Phys. Rev.*, **134A**, 385 (1964).

for $T = $ const. This is the differential equation for the co-existence curve in the P–$\boldsymbol{H}$-plane. For $\boldsymbol{H} = $ const. we obtain the ordinary form of the Clausius–Clapeyron equation

$$\frac{\mathrm{d}P_c}{\mathrm{d}T_c} = \frac{L}{T_c(v^{(n)} - v^{(s)})}. \tag{67.16}$$

These relationships are, as already mentioned, less important than eq. (67.7) since the effects at pressures attainable with ordinary laboratory apparatus are at the limit of measurability, e.g. the quantity $\mathrm{d}\boldsymbol{H}_c/\mathrm{d}P_c$ appearing in eq. (67.15) is of the order of magnitude -10^{-8} Oersted dyn^{-1} cm^2.† The fractional volume change obtained from this is $(v^{(s)} - v^{(n)})/v^{(s)} \approx 10^{-7}$.

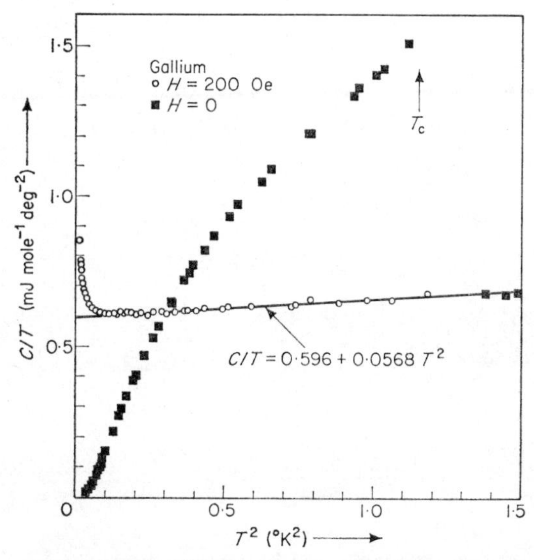

Fig. 58. The specific heat of gallium in the normal and the superconducting state

We finally comment on the nature of the transition at $\boldsymbol{H}_c = 0$. The above equations show clearly that all first derivatives of the Gibbs free energy are continuous. We have, moreover, shown explicitly [eq. (67.14)] that the specific heat shows a finite discontinuity at this point. Analogous results can be deduced easily for the other second derivatives. There can be little doubt, therefore, that we are dealing with a second-order transition of the Ehrenfest type (§ 48). When we consider the validity of the Ehrenfest equations we must, however, bear in mind which one of the variables

† H. M. Rosenberg, *Low Temperature Solid State Physics*, Cambridge, 1963.

is kept constant. For P_c = const. and for T_c = const. the second-order transition occurs only at a single point. We have, therefore, exactly the case which was excluded in the derivation of the Ehrenfest equations. Only in the representation on the P–T-plane does an equilibrium curve exist for the second-order transition at $\boldsymbol{H}_c = 0$. We can, therefore, apply the Ehrenfest equations (48.4) and (48.5). These equations can also be derived from the equations in this section. Since the experimental accuracy is very limited for the reasons given above, an experimental verification is very difficult. Within these limits, however, there is good agreement between theory and experiment.

CHAPTER XI

Electrochemical systems

§ 68 Definition and general properties of electrochemical systems

Experience shows that aqueous solutions of certain substances called *electrolytes* are relatively good conductors of electricity. The substances are those called acids, bases, and salts in chemistry. Electrolytic conduction differs from metallic conduction in the following characteristic ways:

(a) The passage of the current is accompanied by a macroscopic transport of matter.
(b) The passage of the current is coupled with a chemical reaction.
(c) A circuit which includes electrolytic conductors is necessarily a heterogeneous system consisting of several open phases.

Systems with these characteristics are called *electrochemical systems.* We include, for simplicity, homogeneous electrolyte solutions in this concept.†

These properties of electrochemical systems were explained first by Arrhenius in 1887 by the hypothesis (completely proved by experiment) that electrolytes dissociate into electrically charged ions in aqueous solution. Another definition of electrochemical systems is therefore given by the statement that the transport of charge constituting a current is due to the movement of ions whereas in metals and semiconductors it is due to the movement of electrons.

Solutions of electrolytes in certain non-aqueous solvents, salt melts, and solid salts are also electrochemical systems. The study of electrochemical systems comprises a large part of physical chemistry and goes far beyond the bounds of thermodynamics. According to the plan of this book we confine our discussion to the consideration of a few basic questions occurring in the application of thermodynamics to this class of systems; for other details the reader is referred to the numerous monographs in this field. Since the differences between the various types of electrolyte are without significance as far as our

† Some authors separate the two kinds.

discussion is concerned, we shall confine ourselves mainly to aqueous electrolyte solutions which, in any case, are the type which has been most thoroughly investigated. We immediately introduce a further simplification. We have so far not mentioned the degree of electrolytic dissociation. We differentiate conventionally between strong and weak electrolytes although the division between them is naturally not sharp. Quantitative statements require in every case a thorough investigation of the thermodynamic, electrical, and optical properties of the system. This problem is of minor importance for our discussion. We shall, therefore, ignore it and assume throughout that the electrolyte is completely dissociated and the solution thus contains only ions and neutral water molecules.† This assumption is justified to a sufficient accuracy particularly for the halides and perchlorates of alkali metals, alkaline earths, and a few transition metals. In any case, it is not difficult to take into consideration the presence of neutral electrolyte molecules.

We now look at a few general properties of electrochemical systems which are important for our discussion. In the absence of an electric potential difference the condition of electrical neutrality applies to electrolyte solutions. If, therefore, we denote the mole numbers of the ions by n_i and their (positive or negative) electrovalency by z_i, we have

$$\sum_i z_i n_i = 0 \tag{68.1}$$

where the sum is to be taken over every kind of ion present. If the solution contains only a binary electrolyte‡ which dissociates into ν_+ cations and ν_- anions, we have

$$z_+ n_+ + z_- n_- = 0 \tag{68.2}$$

and

$$z_+ \nu_+ + z_- \nu_- = 0. \tag{68.3}$$

We see from these expressions (which can readily be generalized for other types of electrolyte) together with the definitions of § 2 that a solution of a single electrolyte has only two independent components. Its composition is therefore fixed by a single concentration variable.§ It is, however, convenient and for certain problems necessary|| to

† The hydration of the ions can be ignored for the reasons given in §§ 2 and 33.

‡ A binary electrolyte is an electrolyte which dissociates into two kinds of ions.

§ This is so even if dissociation is incomplete.

|| The situation here is analogous to that of chemical reactions. Cf. § 36.

introduce the ions (and, if necessary, the undissociated part of the electrolyte) and the neutral solvent as components; the neutrality condition is then introduced subsequently.

If electric potential differences occur (this is possible only in heterogeneous systems) the neutrality condition is no longer strictly applicable. The deviations are, however, so small that, for any potential differences met in this connexion, they lie far below the limits of any experimental demonstration.† This fact justifies an approximation which constitutes the formal basis for the thermodynamic treatment of electrochemical systems. If we neglect the change in composition associated with the potential difference, the change in internal energy for a virtual displacement can be divided into two parts. The first part depends on the change in the variables of state as discussed earlier. The second part constitutes electrostatic work done by the transport of ions through the difference in potential. The electrostatic work done in electrochemical systems can, however, not be introduced into a thermodynamic discussion in the form of a work co-ordinate in the sense of the definition (14.1). This constitutes a fundamental difference between electrochemical systems and the dielectric systems discussed in Chapter IX. The reason for this difference is that the electrostatic work in electrochemical systems necessarily involves transport of ions between open phases and this is inconsistent with the definition of a work co-ordinate.

Finally, there is a further fact which is important for the thermodynamics of electrochemical systems. This fact was first pointed out by Gibbs and, more recently, by Guggenheim: electric potential differences in electric conductors are measurable only between media of the same chemical composition (e.g. between two points on a copper wire) whereas the potential difference between two conducting media of different chemical compositions (e.g. between an electrolyte solution and an electrode dipped into the solution) is not a measurable quantity. We shall see that a direct consequence of this is that the chemical potential of a single ionic species is also not a measurable quantity. The reasons why, in spite of this, the use of these quantities is justified within the framework of a phenomenological theory will appear from the following sections and will not be discussed here.

† Consider a homogeneously charged sphere of radius 1 cm *in vacuo*. If this sphere bears an excess charge of 1 mole of positive univalent ions, the potential at its surface is about 10^{17} volt. The potentials met in electrochemical systems will always be less than 10^{7} volt. The deviation from strict neutrality will thus be less than 10^{-10} mol cm^{-3}.

§ 69 General conditions for electrochemical equilibrium

A suitable generalization of the fundamental equation is necessary for a derivation of the general equilibrium conditions for a heterogeneous electrochemical system. According to the arguments of § 68 we may write for the variation in internal energy

$$\delta \tilde{U} = \delta U + \delta U^{(e)} \tag{69.1}$$

where $\tilde{U}$ is the total internal energy and $U^{(e)}$ is the electrical part. The quantity U depends on the variables of state in the way discussed previously and is therefore obtained from eqs. (20.4) and (20.5). In order to find an explicit expression for $\delta U^{(e)}$ we start from the well-known electrostatic relationship

$$U^{(e)} = \frac{1}{8\pi} \int \boldsymbol{E}^2 \, \mathrm{d}V. \tag{69.2}$$

A necessary condition for thermodynamic equilibrium is that the electric field inside a conductor vanishes since the irreversible process of conduction would otherwise take place. The integral (69.2) must, therefore, extend over the whole space outside the conductor and we shall suppose, for simplicity, that the conductor is *in vacuo*. Equation (69.2) then gives

$$\delta U^{(e)} = \frac{1}{4\pi} \int \boldsymbol{E} . \delta \boldsymbol{E} \, \mathrm{d}V = -\frac{1}{4\pi} \int \operatorname{grad} \phi . \delta \boldsymbol{E} \, \mathrm{d}V, \tag{69.3}$$

where ϕ is the electric potential. But we know that

$$\int \operatorname{grad} \phi . \delta \boldsymbol{E} \, \mathrm{d}V = \int \operatorname{div} (\phi \delta \boldsymbol{E}) \, \mathrm{d}V - \int \phi \operatorname{div} \delta \boldsymbol{E} \, \mathrm{d}V. \tag{69.4}$$

The second integral on the right-hand side vanishes since *in vacuo*

$$\operatorname{div} \boldsymbol{E} = 0. \tag{69.5}$$

With the aid of Gauss' theorem the first integral can be transformed into an integral over the surface of the conductor and over an infinitely removed surface. The latter integral vanishes since the field strength comes sufficiently close to zero at infinity. Since ϕ is constant at the surface of the conductor eq. (69.3) becomes

$$\delta U^{(e)} = \frac{1}{4\pi} \phi \int \delta \boldsymbol{E} . \mathrm{d}\boldsymbol{f}. \tag{69.6}$$

Use of eq. (57.2)† finally gives

$$\delta U^{(e)} = \phi \delta e, \tag{69.7}$$

where e is the total charge on the conductor. We now have

$$e = \sum_i z_i n_i \mathfrak{F}, \tag{69.8}$$

where

$$\mathfrak{F} = 96{,}487 \text{ Coulomb mole}^{-1} \tag{69.9}$$

is the Faraday, i.e. the charge on 1 mole of univalent ions (the charge on one equivalent) and the summation is to be taken over all electrically charged species.‡ We therefore have

$$\delta U^{(e)} = \mathfrak{F} \sum_i z_i \delta n_i. \tag{69.10}$$

We now obtain the required generalization of the fundamental equation from (69.1) and (69.10) in the form§

$$\mathrm{d}\tilde{U} = T\,\mathrm{d}S - P\,\mathrm{d}V + \sum_{i=1} (\mu_i + z_i \mathfrak{F}\phi)\,\mathrm{d}n_i. \tag{69.11}$$

The quantity

$$\tilde{\mu}_i = \mu_i + z_i \mathfrak{F}\phi \tag{69.12}$$

was called the *electrochemical potential* of component i by Guggenheim. It reduces to the chemical potential for electrically neutral components for which $z_i = 0$.

We now consider a closed system with fixed work co-ordinates which consists of m components and σ phases. We assume that at least one phase is an electrolyte solution and this phase is in direct contact with at least one further electrically conducting phase. Apart from this we make the same assumptions as in § 27. In particular, the phases are assumed to be completely open so that entropy, volume, and mole numbers of each phase can be varied, the last with the restriction that every component need not necessarily appear in every phase. Furthermore, we exclude dissociation processes of electrolytes (which can, according to our assumptions, only be

† D_n must be replaced by E_n since we have assumed a vacuum. The sign of the right-hand side must be reversed since the normal to the surface is taken in the opposite direction.

‡ If the phase under consideration is an electronic conductor, the electrons must be included in the summation in (69.8).

§ For solid substances, we assume hydrostatic pressure in all directions and the absence of uniaxial normal stresses and of tangential stresses. Cf. the remarks at the end of § 56.

regarded as heterogeneous reactions) and other chemical reactions. We shall consider dissociation processes in § 70, rigid phase boundaries in § 71, and chemical reactions in § 72.

We denote the phases by small Greek letters as in § 27 and also use α as the general superscript. The general equilibrium condition (17.2) for our case is then

$$\sum_{\alpha} \delta \tilde{U}^{(\alpha)} = 0 \tag{69.13}$$

with the secondary conditions

$$\sum_{\alpha} \delta S^{(\alpha)} = 0, \tag{69.14}$$

$$\sum_{\alpha} \delta V^{(\alpha)} = 0, \tag{69.15}$$

$$\sum_{\alpha} \delta n_i^{(\alpha)} = 0 \quad (\text{all } i). \tag{69.16}$$

The procedure described in § 27 applied to (69.13)–(69.16) gives the explicit equilibrium conditions

$$T^{(\alpha)} = T^{(\beta)} = \ldots = T^{(\sigma)}, \tag{69.17}$$

$$P^{(\alpha)} = P^{(\beta)} = \ldots = P^{(\sigma)}, \tag{69.18}$$

$$\tilde{\mu}_i^{(\alpha)} = \tilde{\mu}_i^{(\beta)} = \ldots = \tilde{\mu}_i^{(\sigma)}. \tag{69.19}$$

Equations (69.17) and (69.18) are identical with (27.6) and (27.7) while (69.19) reduces to (27.8) for neutral molecules. Conditions (69.17)–(69.18) thus differ from (27.6)–(27.8) in that, for ions, electrochemical potentials replace chemical potentials. The main difficulty is due to the fact that electrochemical potentials have no simple relationship with measurable quantities. Because of its complexity we shall develop this problem step by step by means of examples in the following sections. We shall, at the same time, formulate the equilibrium conditions for some other cases which we have ignored in the above derivation. These cases are, however, important for the discussion of our problem.

§ 70 Solutions of strong electrolytes

We shall first introduce some definitions which are generally used nowadays in the thermodynamics of electrochemical systems. These definitions largely constitute an adaptation of the definitions of § 34(b) to the particular problem of electrolyte solutions.

For the chemical potential of the solvent (denoted by the subscript 1) we write

$$\mu_1 = RT \ln x_1 f_1 + \mu_{10}(T, P) \tag{70.1}$$

with

$$\lim_{x_1 \to 1} f_1 = 1, \tag{70.2}$$

where f_1 is the activity coefficient of the solvent in the solution and $\mu_{10}(T, P)$ is the chemical potential of the pure solvent in the standard state.

Equation (69.12) for the dissolved ions is written explicitly as†

$$\tilde{\mu}_i = RT \ln x_i f_i + z_i \mathfrak{F}\phi + \mu_i^0(T, P) \tag{70.3}$$

with

$$\mu_i^0(T, P) = \lim_{x_1 \to 1} (\mu_i - RT \ln x_i) \tag{70.4}$$

and

$$\lim_{x_1 \to 1} f_i = 1. \tag{70.5}$$

The choice of standard state and the standardization of the activity coefficients thus correspond to case (β) in § 34(b).

We now consider two aqueous electrolyte solutions $'$ and $''$ in contact with each other. Both solutions contain the ionic species i. At equilibrium we find from (69.19) and (70.3) that

$$z_i \mathfrak{F}(\phi'' - \phi') = RT \ln \frac{x_i' f_i'}{x_i'' f_i''}. \tag{70.6}$$

The potential difference $\phi'' - \phi'$ is not measurable as mentioned in § 68. The activity coefficients are measurable neither singly nor in the combination $\ln f_i' - \ln f_i''$. Equation (70.6) cannot, therefore, contribute an experimentally verifiable relationship between measurable quantities. In principle, however, f_i can be calculated from statistical thermodynamics and we may therefore regard eqs. (70.6) as a physically meaningful definition of the potential difference $\phi'' - \phi'$. In practice, however, the situation is somewhat different since the exact calculation of the f_i has only been successful for the limiting case of infinite dilution. Useful approximate formulae exist for dilute electrolyte solutions permitting an approximate definition of $\phi'' - \phi'$. For concentrated electrolyte solution we must, for the present, be content

† The concentration variables used in experimental electrochemistry are mole l^{-1} (molar scale) and mole/kg solvent (molal scale). The activity coefficients are then defined in a formally analogous way.

with the statement that $\phi''-\phi'$ is, at least in principle, a physically definable quantity. Analogous arguments apply to the particularly important case of one phase being an electrolyte solution and the other a metallic conductor. The potential difference $\phi''-\phi'$ is then called a *single electrode potential*. We shall return to this problem in § 72.

In order to obtain measurable quantities we consider groups of various ionic species which obey the neutrality condition

$$\sum \nu_i z_i = 0, \tag{70.7}$$

where the ν_i are integers. The corresponding linear combination of electrochemical potentials is, according to (70.3),

$$\sum \nu_i \tilde{\mu}_i = RT \sum \nu_i \ln x_i + RT \sum \nu_i \ln f_i + \sum \nu_i \mu_i^0 (T, P). \tag{70.8}$$

The terms containing the potential ϕ cancel because of (70.7). We shall show later that combinations of electrochemical potentials of the type of the left-hand side of eq. (70.8) can be reduced to measurable quantities. We can, therefore, conclude that combinations of ion activity coefficients of the type

$$\prod f_i^{\nu_i} \tag{70.9}$$

[where the ν_i obey eq. (70.7)] can also be determined by experiment and are thus unequivocally defined. This is the basis for the introduction of the *mean activity coefficient of an electrolyte* $f_\pm$ by Lewis and Randall. For simplicity we shall consider only binary electrolytes which dissociate into ν_+ cations of valency z_+ and ν_- anions of valency z_-. The neutrality condition (70.7) then becomes

$$\nu_+ z_+ + \nu_- z_- = 0. \tag{70.10}$$

The mean activity coefficient is then defined by the equation

$$f_\pm^\nu \equiv f_\pm^{\nu_+ + \nu_-} = f_+^{\nu_+} f_-^{\nu_-}, \tag{70.11}$$

where f_+ and f_- are the activity coefficients of the cations and anions respectively and

$$\nu = \nu_+ + \nu_-. \tag{70.12}$$

The definition (70.11) is a special case of (70.9). The mean activity coefficients are, therefore, thermodynamically unequivocally defined. Equation (70.11) leads to the remarkable conclusion that the mean activity coefficients of electrolytes in solutions of several electrolytes having certain ionic species in common are not independent of each other. If, for example, we have a solution containing two kinds of univalent cations A^+ and B^+ and two kinds of univalent anions C^- and D^-, (70.11) gives

$$\frac{f_{A,C}}{f_{A,D}} = \frac{f_{B,C}}{f_{B,D}} \tag{70.13}$$

for the mean activity coefficients. In this context, a physical meaning can be ascribed to ion activity coefficients within the framework of thermodynamics.

Analogously to eq. (70.11) the *mean mole fraction of an electrolyte* is defined by the equation

$$x_{\pm}^{\nu} \equiv x_{\pm}^{\nu_+ + \nu_-} = x_+^{\nu_+} x_-^{\nu_-}. \tag{70.14}$$

The *mole fraction of an electrolyte* is also used. It is defined by the equation

$$x = \frac{n_2}{n_1 + n_+ + n_-}, \tag{70.15}$$

where n_2 is the mole number obtained from the chemical formula of the electrolyte and we have to bear in mind that $x_1 + x \neq 1$. But we have (at complete dissociation) the simple relationships

$$x_+ = \nu_+ x, \quad x_- = \nu_- x. \tag{70.16}$$

We now find from (70.14) and (70.16) that

$$x_{\pm} = \nu_{\pm} x \tag{70.17}$$

with

$$\nu_{\pm}^{\nu} = \nu_+^{\nu_+} \nu_-^{\nu_-}. \tag{70.18}$$

Finally, the *stoichiometric mole fraction of the electrolyte*

$$x_2^* = \frac{n_2}{n_1 + n_2} \tag{70.19}$$

is a third one which is used. If x_1^* denotes the stoichiometric mole fraction of the solvent, it is true here that $x_1^* + x_2^* = 1$ but the relationships (70.16) are not valid.

We find directly from eq. (33.16) that for a homogeneous dissociation equilibrium the chemical potential μ_2 of the undissociated electrolyte is related to the electrochemical potentials of the ions by the equation

$$\mu_2 = \nu_+ \tilde{\mu}_+ + \nu_- \tilde{\mu}_-. \tag{70.20}$$

The right-hand side of this equation is a linear combination of the type occurring in (70.8). We can, therefore, formally define this linear combination as the *chemical potential of the electrolyte* even in the case of complete dissociation. We shall show later that this definition is physically meaningful. We write accordingly that

$$\mu_2^0(T,P) = \nu_+\mu_+^0(T,P)+\nu_-\mu_-^0(T,P). \qquad (70.20a)$$

Equation (70.8) together with the definitions (70.11), (70.12), (70.14), (70.20), and (70.20a) gives the chemical potential of the electrolyte as

$$\mu_2 = \nu RT\ln x_\pm f_\pm+\mu_2^0(T,P). \qquad (70.21)$$

Using (70.17) and (70.18) we get the alternative form

$$\mu_2 = \nu RT\ln xf_\pm+\mu_2^{0*}(T,P), \qquad (70.22)$$

where we have put

$$\mu_2^0(T,P)+\nu RT\ln\nu_\pm = \mu_2^{0*}(T,P). \qquad (70.23)$$

The *osmotic coefficient* g† introduced by Bjerrum is often used instead of the activity coefficient f_1 in connexion with the chemical potential of the solvent. The osmotic coefficient is defined by the equation

$$\mu_1 = gRT\ln x_1+\mu_{10}(T,P) \qquad (70.24)$$

with the standardization

$$\lim_{x_1\to 1} g = 1. \qquad (70.25)$$

The relationship between the osmotic coefficient and the activity coefficient of the ions at constant temperature and pressure is given by the equation

$$-\Big(1-\sum_i x_i\Big)\,\mathrm{d}\Big[(1-g)\ln\Big(1-\sum_i x_i\Big)\Big]+\sum_i x_i\,\mathrm{d}\ln f_i = 0, \qquad (70.26)$$

due to Bjerrum. The summation is to be taken over all ionic species. Equation (70.26) is obtained from the Gibbs–Duhem equation (20.43) together with (70.3), (70.24), and the neutrality condition (70.7).

We now illustrate the above definitions by means of two simple examples involving heterogeneous equilibria. Our first example is the case of the distribution of a binary electrolyte between two immiscible solvents ′ and ″. We assume equal temperatures, equal pressures, complete dissociation, and electrical neutrality for the two phases. We therefore have to consider only the variation in the mole numbers

† The name 'osmotic coefficient' is explained by the relationship between $\mu_1-\mu_{10}\ (T,P)$ and the osmotic pressure mentioned at the end of § 28.

of the ions and the equilibrium condition (69.13) reduces to the form†

$$\sum_i (\mu_i' + z_i \mathfrak{F}\phi')\,\delta n_i' + \sum_i (\mu_i'' + z_i \mathfrak{F}\phi'')\,\delta n_i'' = 0 \qquad (70.27)$$

with the secondary conditions

$$\delta n_i' + \delta n_i'' = 0 \quad (\text{all } i), \qquad (70.28)$$

$$\sum_i z_i\,\delta n_i' = 0, \quad \sum_i z_i\,\delta n_i'' = 0. \qquad (70.29)$$

The expressions (70.29) show immediately that the terms involving the electric potential ϕ in eq. (70.27) cancel out. Furthermore, the second of the two conditions (70.29) is not independent and we can thus ignore it. We thus obtain for a binary electrolyte that

$$\mu_+'\,\delta n_+' + \mu_-'\,\delta n_-' + \mu_+''\,\delta n_+'' + \mu_-''\,\delta n_-'' = 0 \qquad (70.30)$$

with the secondary conditions

$$\left.\begin{aligned} \delta n_+' + \delta n_+'' = 0, \quad \delta n_-' + \delta n_-'' = 0, \\ z_+\,\delta n_+' + z_-\,\delta n_-' = 0. \end{aligned}\right\} \qquad (70.31)$$

We now use again the method of Lagrange multipliers.‡ If we multiply the first of the secondary conditions (70.31) by λ_1, the second by λ_2, the third by λ_3, and add the resulting equations to (70.30) we get

$$(\mu_+' + \lambda_1 + z_+\lambda_3)\,\delta n_+' + (\mu_-' + \lambda_2 + z_-\lambda_3)\,\delta n_-' + (\mu_+'' + \lambda_1)\,\delta n_+'' + (\mu_-'' + \lambda_2)\,\delta n_-'' = 0. \qquad (70.32)$$

We choose the undetermined multipliers $\lambda_1, \lambda_2, \lambda_3$ such that the expressions in the first three sets of parentheses vanish. The variation $\delta n_-''$ is then independent and (70.32) can be fulfilled only if the expression in the fourth set of parentheses is also equal to zero. We therefore have

$$\mu_+' + \lambda_1 + z_+\lambda_3 = 0, \qquad (70.33)$$

$$\mu_-' + \lambda_2 + z_-\lambda_3 = 0, \qquad (70.34)$$

$$\mu_+'' + \lambda_1 = 0, \qquad (70.35)$$

$$\mu_-'' + \lambda_2 = 0. \qquad (70.36)$$

† We thus really use the equilibrium condition (23.11) which is formulated using the Gibbs free energy. From now on the subscript i refers only to the ions.

‡ We again use two simple examples to illustrate the application of this method which is indispensable for more complicated cases, although the present problems could be solved much more easily by a simple elimination of the dependent variables.

We now multiply eqs. (70.33) and (70.35) by ν_+, and eqs. (70.34) and (70.36) by ν_-. Because of (70.10) the addition of (70.33) and (70.34) gives

$$\nu_+ \mu'_+ + \nu_- \mu'_- + \nu_+ \lambda_1 + \nu_- \lambda_2 = 0. \tag{70.37}$$

By means of (70.35) and (70.36) we finally arrive at

$$\nu_+ \mu'_+ + \nu_- \mu'_- = \nu_+ \mu''_+ + \nu_- \mu''_- \tag{70.38}$$

or, if we introduce the chemical potential of the electrolyte defined by eq. (70.20),

$$\mu'_2 = \mu''_2. \tag{70.39}$$

The distribution equilibrium is thus determined by the equality of the chemical potentials of the electrolyte. Even if we assume the phase $'$ to be a very dilute solution, for which the ion activity coefficients are theoretically known to sufficient accuracy, we could use the distribution equilibrium to determine experimentally merely the mean activity coefficient for the phase $''$ and this only with the exclusion of a factor depending on T and P. This last limitation is quite unimportant for our problem since only a difference in the standard value occurring in (70.21) for the two phases is involved. This is seen even more clearly if we use (70.11), (70.14), and (70.21) to write eq. (70.39) in the form

$$\frac{(x''_+)^{\nu_+} (x''_-)^{\nu_-} (f''_\pm)^{\nu_+ + \nu_-}}{(x'_+)^{\nu_+} (x'_-)^{\nu_-} (f'_\pm)^{\nu_+ + \nu_-}} = K(T, P) \tag{70.40}$$

with

$$RT \ln K(T, P) = \mu_2^{0\prime}(T, P) - \mu_2^{0\prime\prime}(T, P). \tag{70.41}$$

The quantity $K(T, P)$ can be found by extrapolating the measurements to infinite dilution. The case where $K(T, P) = 1$ and the limitation mentioned above thus disappears is considered in § 71.

Our second example is the vapour pressure equilibrium of an aqueous solution of a volatile binary electrolyte. We assume that the electrolyte in the solution phase denoted by $'$ is completely dissociated while the electrolyte in the vapour phase denoted by $''$ is present exclusively in the form of neutral molecules. These assumptions are justified to a sufficient degree of accuracy for aqueous solutions of HCl, HBr, and HI. As in the previous example we assume equal temperature and pressure for the two phases. However, electrolytic dissociation must now be taken into consideration as a heterogeneous reaction. The equilibrium condition (69.13) then becomes

$$(\mu'_+ + z_+ \mathfrak{F}\phi')\,\delta n'_+ + (\mu'_- + z_- \mathfrak{F}\phi')\,\delta n'_- + \mu'_1\,\delta n'_1 + \mu''_2\,\delta n''_2 + \mu''_1\,\delta n''_1 = 0 \quad (70.42)$$

with the secondary conditions

$$\delta n'_1 + \delta n''_1 = 0, \quad (70.43)$$

$$\delta n'_+ + \nu_+\,\delta n''_2 = 0, \quad (70.44)$$

$$\delta n'_- + \nu_-\,\delta n''_2 = 0. \quad (70.45)$$

The conditions (70.44) and (70.45) are completely analogous to the conditions (33.9) for chemical reactions. Using Lagrange multipliers we obtain

$$\mu'_1 + \lambda_1 = 0, \quad (70.46)$$

$$\mu''_1 + \lambda_1 = 0, \quad (70.47)$$

$$\mu'_+ + z_+ \mathfrak{F}\phi' + \lambda_2 = 0, \quad (70.48)$$

$$\mu'_- + z_- \mathfrak{F}\phi' + \lambda_3 = 0, \quad (70.49)$$

$$\mu''_2 + \nu_+ \lambda_2 + \nu_- \lambda_3 = 0. \quad (70.50)$$

Equations (70.46) and (70.47) immediately give

$$\mu'_1 = \mu''_1. \quad (70.51)$$

If (70.48) is multiplied by ν_+, (70.49) by ν_- and the two resulting equations are added together, terms involving the electric potential disappear because of (70.10). By means of (70.50) we then get

$$\nu_+ \mu'_+ + \nu_- \mu'_- = \mu''_2. \quad (70.52)$$

The equilibrium is completely determined by (70.51) and (70.52) since the solution contains only two independent components. The linear combination on the left-hand side of (70.52) must, therefore, be interpreted as the chemical potential of the electrolyte even in the case of complete dissociation. This interpretation has already been used earlier.

If the vapour phase can be regarded with sufficient accuracy as an ideal gas, the partial pressure of the solvent is found from (70.51), (34.3), and (70.1) to be

$$p_1 = p_{10}\, x_1 f_1, \quad (70.53)$$

where p_{10} is the vapour pressure of the pure solvent at the same temperature. The activity coefficient of the solvent in the electrolyte solution can thus be determined experimentally by measuring the partial pressure p_1.† The derivation of the expression for the partial pressure of the electrolyte is (because of the problem of standardization) rather involved and cannot be considered here. We merely remark that the relationship between the partial pressure p_2 and the mean activity coefficient of the electrolyte in the solution contains a temperature-dependent factor which can be determined (similarly to the previous example) by extrapolation of the measurements to infinite dilution.

§ 71 Membrane equilibria of electrolyte solutions

We now consider equilibria between two aqueous electrolyte solutions separated by a rigid semipermeable membrane. We again denote the two phases by $'$ and $''$. For simplicity we assume that phase $'$ contains only one binary electrolyte whose cations and anions can pass through the membrane. Phase $''$ is assumed to contain in addition a polyelectrolyte‡ whose cations are identical with those of the electrolyte in phase $'$. The membrane is assumed to be impassable to the macromolecular anions. From the general theory of membrane equilibria (§ 28) and the theory of electrochemical equilibrium (§ 69) we obtain the equilibrium conditions

$$T' = T'', \tag{71.1}$$

$$\tilde{\mu}_i' = \tilde{\mu}_i'' \quad (\text{all } i) \tag{71.2}$$

where the subscript i now applies to all the components capable of passing through the membrane. The condition (69.18) does not appear here, i.e. equality of pressure of the two phases is not necessary for equilibrium. Whether a pressure difference is possible or perhaps necessary for equilibrium depends only on whether eq. (71.2) can be fulfilled under the appropriate experimental conditions.

(a) *Non-osmotic equilibrium.* We now suppose that the membrane is impassable to the solvent and assume equality of pressure. The conditions (71.2) are then explicitly

$$RT \ln x_+' f_+' + z_+ \mathfrak{F}\phi' = RT \ln x_+'' f_+'' + z_+ \mathfrak{F}\phi'', \tag{71.3}$$

$$RT \ln x_-' f_-' + z_- \mathfrak{F}\phi' = RT \ln x_-'' f_-'' + z_- \mathfrak{F}\phi''. \tag{71.4}$$

† This method has hardly any practical significance.

‡ Polyelectrolytes are macromolecules with a large number of groups capable of electrolytic dissociation in aqueous solution. Many physiologically important substances (e.g. proteins, nucleic acids) and certain synthetic polymers (e.g. polyacrylic acid) are polyelectrolytes.

The terms for the standard state are the same for both phases and cancel. If we now use the neutrality condition (70.10) to eliminate terms containing electric potentials and apply (70.11) we get

$$\frac{(x''_+)^{\nu_+}\,(x''_-)^{\nu_-}\,(f''_\pm)^{\nu_++\nu_-}}{(x'_+)^{\nu_+}\,(x'_-)^{\nu_-}\,(f'_\pm)^{\nu_++\nu_-}} = 1. \tag{71.5}$$

Equation (71.5) shows that a membrane equilibrium is possible with the two phases at the same pressure for a system which conforms to our original assumptions. Such an equilibrium is called a *non-osmotic equilibrium*. Equation (71.5) shows that such an equilibrium is completely determined by the chemical potentials of the permeating electrolyte. If the mean activity coefficient for the phase $'$ is known, that for the phase $''$ can be found by means of (70.5).

The quantity $\phi''-\phi'$ defined by the equation

$$\phi''-\phi' = \frac{RT}{z_+\mathfrak{F}}\ln\frac{x'_+f'_+}{x''_+f''_+} = \frac{RT}{z_-\mathfrak{F}}\ln\frac{x'_-f'_-}{x''_-f''_-} \tag{71.6}$$

is called the *membrane potential*. Equation (71.6) is obviously a special case of (70.6) and our earlier considerations are largely applicable to the present case. Two qualifying remarks need to be made, however. Firstly, although the membrane potential is not measurable directly, it can be determined at least approximately by means of a suitable experimental arrangement. We shall return to this problem in § 72. Secondly, no suitable way to a reliable calculation (not even an approximate one) of the activity coefficients for the phase $''$ containing the macromolecular anion is known.

(b) *Osmotic equilibrium* (*Donnan equilibrium*). We consider the same system we considered under (a) except that we now assume the membrane to be permeable not only to the ions of the binary electrolyte but also to the solvent. In this case it is generally necessary for the two phases to be at different pressures in order that equilibrium may be attained.† Simple examples of such equilibria were originally discussed by Donnan. The general theory was developed by Donnan and Guggenheim.

† This becomes obvious if we consider the extreme case of the polyelectrolyte being present at very high concentration and the permeating electrolyte at very low concentration. The amount of permeating electrolyte is then not enough to reduce the chemical potential of the solvent in phase $'$ [cf. eq. (28.10)] to the same value as that in phase $''$.

From §§ 28 and 69 we immediately obtain the equilibrium conditions

$$\mu_1'(T, P', x_1') = \mu_1''(T, P'', x_1''), \tag{71.7}$$

$$\tilde{\mu}_+'(T, P', x_+', x_-') = \tilde{\mu}_+''(T, P'', x_+'', x_-''), \tag{71.8}$$

$$\tilde{\mu}_-'(T, P', x_+', x_-') = \tilde{\mu}_-''(T, P'', x_+'', x_-''). \tag{71.9}$$

We assume the phase $'$ to be at normal pressure P_0 and put $P'' = P_0 + \Pi$. According to eq. (26.27) we then have

$$\mu_{10}(T, P_0 + \Pi) = \mu_{10}(T, P_0) + \int_{P_0}^{P_0+\Pi} v_{10}\, \mathrm{d}P$$

$$\equiv \mu_{10}(T, P_0) + \bar{v}_{10}\, \Pi, \tag{71.10}$$

where $\bar{v}_{10}$ is the mean of the partial molar volume v_{10} over the pressure difference Π. Using (70.1), eq. (71.7) can thus be written

$$\Pi = -\frac{RT}{\bar{v}_{10}} \ln \frac{x_1'' f_1''}{x_1' f_1'}. \tag{71.11}$$

For the permeating ions we have

$$\mu_i^0(T, P_0 + \Pi) = \mu_i^0(T, P_0) + \int_{P_0}^{P_0+\Pi} v_i^0\, \mathrm{d}P$$

$$\equiv \mu_i^0(T, P_0) + \bar{v}_i^0\, \Pi, \tag{71.12}$$

where v_i^0 is the partial molar volume of the ionic species i in the standard state defined by eq. (70.4). Using (70.3), eqs. (71.8) and (71.9) can then be written

$$\phi'' - \phi' = \frac{RT}{z_+ \mathfrak{F}} \left(\ln \frac{x_+' f_+'}{x_+'' f_+''} + \frac{\bar{v}_+^0\, \Pi}{RT} \right), \tag{71.13}$$

$$\phi'' - \phi' = \frac{RT}{z_- \mathfrak{F}} \left(\ln \frac{x_-' f_-'}{x_-'' f_-''} + \frac{\bar{v}_-^0\, \Pi}{RT} \right). \tag{71.14}$$

If we define the mean partial molar volume of the electrolyte in the standard state by the equation

$$\bar{v}_\pm^0 = \nu_+ \bar{v}_+^0 + \nu_- \bar{v}_-^0 \tag{71.15}$$

we find from (71.13) and (71.14) with (70.10) and (70.11) that

$$\Pi = -\frac{RT}{\bar{v}_\pm^0} \ln \frac{(x_+'')^{\nu_+} (x_-'')^{\nu_-} (f_\pm'')^{\nu_+ + \nu_-}}{(x_+')^{\nu_+} (x_-')^{\nu_-} (f_\pm')^{\nu_+ + \nu_-}}. \tag{71.16}$$

Elimination of Π from eqs. (71.11) and (71.16) finally gives

$$\frac{(x'_+)^{\nu_+}(x'_-)^{\nu_-}(f'_\pm)^{\nu_++\nu_-}}{(x'_1 f'_1)^r} = \frac{(x''_+)^{\nu_+}(x''_-)^{\nu_-}(f''_\pm)^{\nu_++\nu_-}}{(x''_1 f''_1)^r} \tag{71.17}$$

with

$$r = \bar{v}^0_\pm/\bar{v}_{10}. \tag{71.18}$$

All activity coefficients of the phase $''$ are defined at the pressure $P_0+\Pi$ in these equations. We shall now use the solvent as an example for the method of converting to activity coefficients at normal pressure. Equations (26.27), (70.1), and (71.10) show that

$$\frac{\partial \mu_1}{\partial P} = v_1 = v_{10} + RT\frac{\partial \ln x_1 f_1}{\partial P}. \tag{71.19}$$

Integration between P_0 and $P_0+\Pi$ gives

$$RT \ln f''_1(P_0+\Pi) = RT \ln f''_1(P_0) + (\bar{v}_1 - \bar{v}_{10})\,\Pi, \tag{71.20}$$

where $f''_1(P_0)$ is the activity coefficient defined for the temperature and composition of the phase $''$ and the pressure P_0. The dependence of the activity coefficient on pressure can often be ignored since, in practice, only very small pressure differences need to be considered and the second term on the right-hand side of (71.20) can therefore be neglected.

The pressure difference Π in eq. (71.11) or eq. (71.16) is called the *osmotic pressure* (cf. §28). In contrast with case (a), the Donnan equilibrium is often described as an *osmotic membrane equilibrium.* The potential difference defined by eq. (17.13) or eq. (71.14) is known as the *membrane potential* as in the case discussed under (a). The remarks made there about the physical meaning of this concept are naturally also valid for an osmotic equilibrium.

§72 Galvanic cells

We still have to discuss how the immeasurable potential difference $\phi''-\phi'$ between neighbouring phases is related to measurable quantities. In this connexion we consider a closed system with fixed work co-ordinates. The system consists of σ phases and is subject to the following specifications:

(a) All phases are electrical conductors.

(b) The phases to which we attach definite numbers are arranged in such a way that each of the phases 2 to $\sigma-1$ is in contact with exactly two other phases.† Phase 1 is in contact only with phase 2, and phase σ only with phase $\sigma - 1$.

(c) The end phases 1 and σ are chemically identical metallic conductors.

(d) There is at least one ionic conductor among the phases 2 to $\sigma - 1$.

(e) Each pair of touching phases has at least one kind of electrically charged particle (ions or electrons) in common. The phase boundary is permeable to the common species.

Such a system is called a *galvanic cell.* The whole of it is a conductor which can be a part of an electric circuit. An electric current will then flow from phase 1 via phases 2 to $\sigma - 1$ to phase σ. The conduction of current necessarily entails chemical reactions since the current is carried by electrons in some phases and by ions in others.

The potential difference $\phi^{(\sigma)} - \phi^{(1)}$ is a measurable quantity since the two end phases 1 and σ are chemically identical. Thermodynamics is concerned only with the equilibrium value of this quantity. It must therefore always be determined in a state in which no current is flowing since the passage of current is an irreversible process. We also exclude any other irreversible process in the cell. This means that *the chemical reactions occurring in the cell must be coupled in a definite way with the conduction of current* such that all chemical reactions take place in the opposite direction when the current is reversed. Furthermore, all stationary irreversible processes (e.g. diffusion, heat conduction, thermal diffusion) are excluded.‡ The last requirement is not fulfilled for all galvanic cells and we therefore distinguish between

(a′) *reversible galvanic cells* in which *every* process occurring in the cell is reversed when the current is reversed,

(b′) *irreversible galvanic cells* which involve to a considerable extent stationary irreversible processes whose direction is independent of the direction of the current.

The exact theory of irreversible galvanic cells can be developed only on the basis of the thermodynamics of irreversible process. We therefore confine our attention to reversible galvanic cells.

We must first define more exactly what we understand by thermodynamic equilibrium in a galvanic cell. The important consideration here is that the overall chemical reaction occurring in the cell is

† We ignore contacts with non-conducting phases (e.g. the vapour phase). These contacts may exist but are of no interest for the present problem.

‡ Non-stationary irreversible processes do not lead to measurable quantities independent of time and are thus excluded *ab initio*.

coupled with the conduction of the current in a definite way. We therefore distinguish between the following three cases:

(a″) The cell is open, i.e. there is no direct conducting contact between the end phases 1 and σ. No electric current can, therefore, flow and no chemical reaction can take place. The system may be in thermodynamic equilibrium. The chemical reaction is, however, only inhibited and thus does not appear (since the corresponding variations are not possible) in the equilibrium conditions (cf. § 17). The potential difference $\phi^{(\sigma)} - \phi^{(1)}$ is thus not measurable under these conditions.

(b″) The end phases 1 and σ are shorted via a conductor of constant chemical composition. The current now flows and the chemical reaction proceeds irreversibly until either chemical equilibrium is reached or one of the reactants disappears. Thermodynamic equilibrium is therefore impossible in the initial state.

(c″) An external potential difference Φ' is applied to the end phases 1 and σ by means of a potentiometer circuit. If

$$|\Phi'| > |\phi^{(\sigma)} - \phi^{(1)}|$$

the electric current will flow in a certain direction and the chemical reaction will take place. If $|\Phi'| < |\phi^{(\sigma)} - \phi^{(1)}|$ the current flows in the opposite direction and the reverse chemical reaction takes place. The case $|\Phi'| = |\phi^{(\sigma)} - \phi^{(1)}|$ therefore corresponds to a state in which no current flows. This state constitutes a thermodynamic equilibrium since no irreversible processes can occur in the system. The chemical reaction is now not inhibited since it can proceed in either direction as a quasi-static or reversible process by changing Φ' infinitesimally.

A meaningful definition of thermodynamic equilibrium for a galvanic cell must obviously include the chemical reaction and the potential difference $\phi^{(\sigma)} - \phi^{(1)}$ in the equilibrium conditions and must be consistent with the general definitions of §§ 16 and 17. Cases (a″) and (b″) are therefore excluded and we define the equilibrium described under (c″) as the *thermodynamic equilibrium for a galvanic cell*. The quantity

$$|\Phi'| = |\phi^{(\sigma)} - \phi^{(1)}| \equiv \Phi \tag{72.1}$$

is called the *electromotive force* (e.m.f.) of the galvanic cell.

We wish to make two observations relevant to this definition. Firstly, we can see that the above considerations show remarkable analogy with the thermodynamic treatment of a gas confined in a cylinder by a frictionless piston. If the piston is arrested, expansion is an inhibited process and the pressure of the gas is not measurable [case (a″)]. If the piston is freely moveable and its movement is not opposed by an external pressure, expansion of the gas occurs as an irreversible process [case (b″)]. If an external pressure acts on the piston and the external pressure is equal to the gas pressure in the cylinder, we have thermodynamic equilibrium and the equilibrium pressure is a measurable quantity. The volume of the gas can be changed in either direction by infinitesimal changes in the external pressure and the volume changes then occur as quasi-static processes [case (c″)]. The main difference between the gas and the galvanic cell lies in the fact that the volume is a work co-ordinate in the sense of § 14 in the case of the gas whereas electrical work cannot be described by a work co-ordinate here since our galvanic cell is defined as a heterogeneous system consisting of open phases.

The second observation concerns the nature of the thermodynamic equilibrium for a galvanic cell. We are here dealing with an electrochemical equilibrium which must be distinguished from the chemical equilibrium for the reaction involved (if such a chemical equilibrium exists). The reversible course of the reaction described under case (c″) accordingly differs from the course of the reaction at maintained chemical equilibrium as described in § 14.

We shall carry out the derivation of the equilibrium conditions by two methods. The first method introduces only observable quantities and ignores the detailed mechanism. We saw in §§ 70 and 71 that all material equilibria between neighbouring phases are determined by the chemical potentials of the undissociated electrolytes and those of the neutral molecules. We therefore regard these as the components of the system and accordingly write the total reaction (cf. § 33) as

$$\sum \nu_j \mathrm{X}_j = 0, \tag{72.2}$$

where the X_j are electrically neutral chemical compounds and the ν_j are the stoichiometric reaction numbers. The total reaction described by (72.2) is not, however, necessarily a chemical reaction in the usual sense.† Since one, and only one, measurable potential difference $\phi^{(\sigma)}-\phi^{(1)}$ is involved, an additional electrically charged component must be introduced. The choice of this component is fixed by our definition of the end phases as electronic conductors.

† The reaction (72.2) can, for example, consist of the transfer of a component i from a phase ′ to a phase ″ of different composition. We shall give an example of such a cell at the end of this section.

We denote the mole number of electrons in phase α by $n_{\mathrm{e}}^{(\alpha)}$. The electrochemical potential $\tilde{\mu}_{\mathrm{e}}$ of the electrons now appears in the fundamental equation together with the chemical potentials of the neutral components. We now have to formulate the coupling of the chemical reaction with the passage of the current. We therefore write the change in mole numbers for the r components which take part in the reaction, as we did in § 36, in the form

$$\delta n_j = \nu_j \, \delta\xi, \tag{72.3}$$

where ξ is the progress variable. When an infinitesimal current passes, δn_{e} mole of electrons disappear from phase σ and appear in phase 1 (or vice versa). The coupling is then expressed by the equation

$$\delta n_{\mathrm{e}} = \nu^* \, \delta\xi, \tag{72.4}$$

where the proportionality factor ν^* depends on the reaction used in the construction of the cell.

We now assume that the entropy and the volume of every phase can be varied.† We make no special assumptions about the permeability of the phase boundaries apart from those made for the definition of a galvanic cell [point (e)]. We shall, however, show later that not all the phase boundaries can be permeable to all the components. The summations over the mole number variations must therefore be understood to be subject to these limitations and we shall denote this by using the symbol $\sum'$. We finally assume that all the components participating in the reaction (72.2) are present in the system. We then have the equilibrium condition (cf. § 33)

$$\sum_{\alpha} T^{(\alpha)} \, \delta S^{(\alpha)} - \sum_{\alpha} P^{(\alpha)} \, \delta V^{(\alpha)} + \sum_{\alpha} \left[\sum_{i=1}^{m} \mu_i^{(\alpha)} \, \delta n_i^{(\alpha)} + \tilde{\mu}_{\mathrm{e}}^{(\alpha)} \, \delta n_{\mathrm{e}}^{(\alpha)} \right] = 0 \tag{72.5}$$

with the secondary conditions

$$\sum_{\alpha} \delta S^{(\alpha)} = 0, \tag{72.6}$$

$$\sum_{\alpha} \delta V^{(\alpha)} = 0, \tag{72.7}$$

$\sum_{\alpha}' \delta n_i^{(\alpha)} = 0$ for components not participating in reaction (72.2), (72.8)

$\sum_{\alpha} \delta n_j^{(\alpha)} = \nu_j \, \delta\xi$ for the r components which participate in reaction (72.2), (72.9)

$$\delta n_{\mathrm{e}}^{(\alpha)} = 0 \quad (\alpha = 2, 3, \ldots, \sigma - 1), \tag{72.10}$$

$$\delta n_{\mathrm{e}}^{(1)} = -\delta n_{\mathrm{e}}^{(\sigma)} = \nu^* \, \delta\xi. \tag{72.11}$$

† The assumption about the volume is unnecessary and is made only for the sake of simplicity. The important condition for the following argument is merely that the two end phases 1 and σ are at the same pressure.

Equations (72.5)–(72.8) immediately give

$$T^{(\alpha)} = T^{(\beta)} = \ldots T^{(\sigma)}, \tag{72.12}$$

$$P^{(\alpha)} = P^{(\beta)} = \ldots P^{(\sigma)}, \tag{72.13}$$

$$\mu_i' = \mu_i'', \tag{72.14}$$

where ′ and ″ indicate each pair of neighbouring phases whose common boundary is permeable to component i. Using (72.9)–(72.11), eq. (72.5) reduces to the condition

$$\left[\sum_{\alpha}\sum_{j=1}^{r} \nu_j \mu_j^{(\alpha)} + \nu^*(\tilde{\mu}_{\mathrm{e}}^{(1)} - \tilde{\mu}_{\mathrm{e}}^{(\sigma)})\right] \delta\xi = 0. \tag{72.15}$$

Since ξ is assumed to be variable at will, (72.15) gives

$$\sum_{\alpha}\sum_{j=1}^{r} \nu_j \mu_j^{(\alpha)} + \nu^*(\tilde{\mu}_{\mathrm{e}}^{(1)} - \tilde{\mu}_{\mathrm{e}}^{(\sigma)}) = 0. \tag{72.16}$$

The chemical potentials of the two end phases 1 and σ must be equal since we have assumed (or shown) that they are chemically identical, and are at the same temperature and pressure. We therefore find, according to eq. (69.12), that

$$\tilde{\mu}_{\mathrm{e}}^{(1)} - \tilde{\mu}_{\mathrm{e}}^{(\sigma)} = -\mathfrak{F}(\phi^{(1)} - \phi^{(\sigma)}). \tag{72.17}$$

From (72.16) and (72.17) we get

$$\nu^* \mathfrak{F}(\phi^{(1)} - \phi^{(\sigma)}) = \sum_{\alpha}\sum_{j=1}^{r} \nu_j \mu_j^{(\alpha)}. \tag{72.18}$$

Equation (72.18) is the general condition for electrochemical equilibrium in a galvanic cell in which the heterogeneous overall reaction (72.2) occurs with a reversible passage of current. If, as is often the case, each participant in reaction (72.2) occurs in only one phase, eq. (72.18) can be written

$$\nu^* \mathfrak{F}(\phi^{(1)} - \phi^{(\sigma)}) = \sum_{j=1}^{r} \nu_j \mu_j. \tag{72.19}$$

If we introduce the free energy change ΔG† defined by eq. (36.10)

† We here consider ΔG with respect to a reaction corresponding to one stoichiometric formula, in accordance with the earlier definition. In the literature ΔG is often related to a reaction corresponding to one equivalent. This corresponds to the quantity $\Delta G/\nu^*$ in our notation.

and the e.m.f. of the cell defined by eq. (72.1), eq. (72.18) becomes

$$-\nu^* \mathfrak{F}\Phi = \Delta G. \qquad (72.20)$$

The e.m.f. obtained when all the participants in the reaction are in their standard states is called the *standard e.m.f.* for the cell. If we denote this quantity by Φ^0 we find from (34.36), (34.41), (34.43), and (72.18) that

$$\Phi = \Phi^0 - \frac{RT}{\nu^* \mathfrak{F}} \ln \left[\prod_{\alpha} \prod_{j=1}^{r} (a_j^{(\alpha)})^{\nu_j} \right] \qquad (72.21)$$

with

$$\Phi^0 = -\Delta G^0 = -\sum_{\alpha} \sum_{j=1}^{r} \nu_j \mu_j^{0(\alpha)}. \qquad (72.22)$$

If all the participants in the reaction are present as pure phases, we simply have $\Phi = \Phi^0$.

The above method is strict and general. It suffers from the disadvantages, however, that the connexion with the previous section is not apparent and that there is no relation between galvanic cells which have some partial reactions in common. Moreover, it cannot be used in the theory of irreversible cells.

We now outline a second method for the derivation of the equilibrium conditions that avoids these disadvantages. The principle of the method is that the overall reaction (72.2) is divided into localized material equilibria between the phases and similarly localized chemical equilibria within the phases. Such a division does not, however, introduce measurable quantities and is thus not always free from arbitrariness. We use a simple example to illustrate the method since the general formulation is clumsy and involved. We therefore consider the cell

$$\overset{(1)}{\mathrm{Pt}}\,|\,\overset{(2)}{\mathrm{M(s)}}\,|\,\overset{(3)}{\mathrm{MX_2(s)}}\,|\,\overset{(4)}{\mathrm{KX(aq.)}}\,|\,\overset{(5)}{\mathrm{AgX(s)}}\,|\,\overset{(6)}{\mathrm{Ag(s)}}\,|\,\overset{(7)}{\mathrm{Pt}}, \qquad (72.23)$$

where X is a halogen which forms a salt sparingly soluble in water with both silver and the bivalent metal M.† KX (aq.) denotes an aqueous solution of the appropriate potassium halide. We assume that the phase boundaries 1–2 and 6–7 can be passed only by electrons, 3–4 and 4–5 only by X^- ions, 2–3 only by M^{2+} ions, and 5–6 only by Ag^+ ions. The material equilibria then, according to

† The assumption of sparing solubility is necessary to make diffusion and irreversible chemical reactions negligible.

(69.12) and (69.19) obey

$$\left.\begin{aligned} \mathfrak{F}(\phi^{(1)}-\phi^{(2)}) &= \mu_{\mathrm{e}}^{(1)}-\mu_{\mathrm{e}}^{(2)}, \\ 2\mathfrak{F}(\phi^{(2)}-\phi^{(3)}) &= \mu_{\mathrm{M}^{++}}^{(3)}-\mu_{\mathrm{M}^{++}}^{(2)}, \\ \mathfrak{F}(\phi^{(3)}-\phi^{(4)}) &= \mu_{\mathrm{X}^-}^{(3)}-\mu_{\mathrm{X}^-}^{(4)}, \\ \mathfrak{F}(\phi^{(4)}-\phi^{(5)}) &= \mu_{\mathrm{X}^-}^{(4)}-\mu_{\mathrm{X}^-}^{(5)}, \\ \mathfrak{F}(\phi^{(5)}-\phi^{(6)}) &= \mu_{\mathrm{Ag}^+}^{(6)}-\mu_{\mathrm{Ag}^+}^{(5)}, \\ \mathfrak{F}(\phi^{(6)}-\phi^{(7)}) &= \mu_{\mathrm{e}}^{(6)}-\mu_{\mathrm{e}}^{(7)}. \end{aligned}\right\} \tag{72.24}$$

According to (33.16) we have for the homogeneous equilibria within the phases that

$$\left.\begin{aligned} \mu_{\mathrm{M}^{++}}^{(2)}+2\mu_{\mathrm{e}}^{(2)} &= \mu_{\mathrm{M}}^{(2)}, \\ \mu_{\mathrm{M}^{++}}^{(3)}+2\mu_{\mathrm{X}^-}^{(3)} &= \mu_{\mathrm{MX}_2}^{(3)}, \\ \mu_{\mathrm{Ag}^+}^{(5)}+\mu_{\mathrm{X}^-}^{(5)} &= \mu_{\mathrm{AgX}}^{(5)}, \\ \mu_{\mathrm{Ag}^+}^{(6)}+\mu_{\mathrm{e}}^{(6)} &= \mu_{\mathrm{Ag}}^{(6)}. \end{aligned}\right\} \tag{72.25}$$

If we add the eqs. (72.24) together and use (72.25) to introduce the chemical potentials of the neutral components, we obtain

$$2\mathfrak{F}(\phi^{(1)}-\phi^{(7)}) = \mu_{\mathrm{MX}_2}+2\mu_{\mathrm{Ag}}-\mu_{\mathrm{M}}-2\mu_{\mathrm{AgX}}. \tag{72.26}$$

The overall reaction which occurs when current passes through the cell is

$$\mathrm{M}+2\mathrm{AgX} \to \mathrm{MX}_2+2\mathrm{Ag}. \tag{72.27}$$

If, therefore, we use (36.10) and (72.1) we again obtain eq. (72.20) with $\nu^* = 2$.

This second derivation illustrates clearly that the potential differences between neighbouring phases mentioned a number of times in previous sections are not measurable singly. Only their combination into a suitable galvanic cell becomes measurable. This is true in particular for the single electrode potentials mentioned in § 70. The single electrode potentials occurring in the example (72.23) are $\phi^{(3)}-\phi^{(4)}$ and $\phi^{(4)}-\phi^{(5)}$. In spite of this it is convenient to be able to compare different electrode potentials. The comparison becomes possible when a *half-cell* corresponding to the electrode potential is combined with the half-cell of a *standard electrode* to form a galvanic cell whose e.m.f. can be measured. The standard electrode used is platinum saturated with hydrogen at 1 atm pressure† and dipping into a solution containing H^+ ions at unit activity of the dissolved electrolyte. The values measured in this way are called *single electrode potentials* or, more correctly, *standard e.m.f. values for the appropriate half-cell.* They have been tabulated for all the more important half-cells. We illustrate the use of these values by means of the cell

$$\mathrm{Pt}\,|\,\mathrm{Pb(s)}\,|\,\mathrm{PbCl_2(s)}\,|\,\mathrm{KCl(aq.)}\,|\,\mathrm{AgCl(s)}\,|\,\mathrm{Ag(s)}\,|\,\mathrm{Pt} \qquad \text{(I)}$$

which is a particular case of the cell (72.23). The standard values of the e.m.f. of the two half-cells are defined by the cells

$$\mathrm{Pt}\,\left|\,\mathrm{Pt{-}H_2}\,\right|\,\begin{matrix}\text{solution}\\ \text{containing}\\ \mathrm{H^+}\end{matrix}\,\left|\,\begin{matrix}\text{solution}\\ \text{containing}\\ \mathrm{Cl^-}\end{matrix}\,\right|\,\mathrm{AgCl(s)}\,|\,\mathrm{Ag(s)}\,|\,\mathrm{Pt}, \qquad \text{(II)}$$

$$\mathrm{Pt}\,\left|\,\mathrm{Pt{-}H_2}\,\right|\,\begin{matrix}\text{solution}\\ \text{containing}\\ \mathrm{H^+}\end{matrix}\,\left|\,\begin{matrix}\text{solution}\\ \text{containing}\\ \mathrm{Cl^-}\end{matrix}\,\right|\,\mathrm{PbCl_2(s)}\,|\,\mathrm{Pb(s)}\,|\,\mathrm{Pt}. \qquad \text{(III)}$$

The two neighbouring solutions in cells (II) and (III) are assumed to have practically the same composition.‡ According to the last described method we obtain§

$$2\mathfrak{F}\Phi^0_{\mathrm{II}} = \mu^0_{\mathrm{H_2}} - 2\mu^0_{\mathrm{H^+}} + 2\mu_{\mathrm{AgCl}} - 2\mu_{\mathrm{Ag}} - 2\mu^0_{\mathrm{Cl^-}}, \qquad (72.28)$$

$$\mathfrak{F}\Phi^0_{\mathrm{III}} = \mu^0_{\mathrm{H_2}} - 2\mu^0_{\mathrm{H^+}} + \mu_{\mathrm{PbCl_2}} - \mu_{\mathrm{Pb}} - 2\mu^0_{\mathrm{Cl^-}}. \qquad (72.29)$$

From (72.26), (72.28), and (72.29) we immediately get

$$2\Phi_{\mathrm{I}} = 2\Phi^0_{\mathrm{II}} - \Phi^0_{\mathrm{III}} \qquad (72.30)$$

since, according to (72.26), the e.m.f. of the cell (I) is determined only by pure phases. We can, therefore, calculate the e.m.f. of any galvanic cell which is made up of half-cells whose standard values are known.

The potential difference between neighbouring phases can be determined at least approximately by experiment in certain cases when the potential difference of interest is large compared with all other potential differences occurring in the cell. This allows the approximate determination of membrane potentials (§ 71). We cannot go into details here.

† Hydrogen at unit fugacity would be preferable for the definition of the standard electrode from a theoretical point of view.

‡ Cells (II) and (III) are generally written with only one electrolyte solution. The present notation is due to Guggenheim. It is theoretically clearer.

§ Phase superscripts are superfluous since we are dealing with ions.

The derivation of eq. (72.26) shows that although there is always electrochemical equilibrium between two neighbouring phases, this is not true for the two end phases since $\tilde{\mu}_e^{(1)} \neq \tilde{\mu}_e^{(\sigma)}$. This is explained for an open cell by the fact that one or more phase boundaries do not allow electrons to pass through. Electrochemical equilibrium does, however, exist between the phases 1 and σ if the potential difference $\phi^{(1)} - \phi^{(\sigma)}$ is compensated by an external potential difference Φ'; this external potential difference has not been explicitly introduced here.

The local equilibria described here play an important role in the theory of irreversible galvanic cells. Details can be found in textbooks of the thermodynamics of irreversible processes.

We now consider briefly a few further thermodynamic aspects of reversible galvanic cells. If we combine the basic equation (72.20) with the Gibbs–Helmholtz equation (24.8) we obtain the expression

$$\nu^* \mathfrak{F}\left[\Phi - T\left(\frac{\partial \Phi}{\partial T}\right)_P\right] = -\Delta \mathrm{H} \tag{72.31}$$

for the temperature dependence of the e.m.f. This equation was first derived by Helmholtz. According to (72.20) and (21.37) we have

$$\nu^* \mathfrak{F}\left(\frac{\partial \Phi}{\partial T}\right)_P = \Delta S, \tag{72.32}$$

where ΔH and ΔS are enthalpy and entropy changes corresponding to a reaction of one stoichiometric formula as in (72.2). The quantities ΔG, ΔH, and ΔS for the reaction (72.2) can thus be found by measuring the e.m.f. and its temperature dependence. Since the three quantities are functions of state their values do not depend (at constant temperature and pressure) on an irreversible [case (b″)] or reversible [case (c″)] course of the reaction. In contrast, the heat absorbed by the system depends on the path taken in phase space and is ΔH for a reversible process and $T\,\Delta S$ for an irreversible process. In the latter case ΔH is, according to (72.31), equal to the sum of the heat absorbed and the electrical work done by the potentiometer on the system. The thermodynamic investigation of heterogeneous reactions by means of reversible galvanic cells plays an important part in the experimental study of the Nernst heat theorem (§ 38).

E.m.f. measurements on galvanic cells are also often used in the determination of mean activity coefficients of electrolytes. We illustrate this by a simple example, the cell

$$\mathrm{Pt}\,|\,\mathrm{M(s)}\,|\,\mathrm{MCl(aq.)}\,|\,\mathrm{AgCl}\,|\,\mathrm{Ag}\,|\,\mathrm{Pt}, \tag{72.33}$$

where M is a univalent metal whose chloride is soluble in water.† The overall reaction for the cell is

$$M(s)+AgCl(s) \rightarrow MCl(aq.)+Ag(s). \qquad (72.34)$$

$\nu^* = 1$ for this reaction. Since each participant in the reaction occurs in one phase only, the e.m.f. of the cell is, according to (72.21), given by

$$\Phi = \Phi^0 + \frac{RT}{\mathfrak{F}} \ln \frac{a_{M(s)}\, a_{AgCl(s)}}{a_{MCl(aq.)}\, a_{Ag(s)}}. \qquad (72.35)$$

The activities of the pure solids are unity by definition (§ 34b). Using (70.22) (with $\nu = 2, \nu_{\pm} = 1$), (72.35) can therefore be written

$$\Phi = \Phi^0 - \frac{RT}{\mathfrak{F}} \ln a_{MCl(aq.)} = \Phi^0 - \frac{2RT}{\mathfrak{F}} \ln x f_{\pm}, \qquad (72.36)$$

where x is the mole fraction defined by (70.15) and $f_{\pm}$ is the mean activity coefficient of the electrolyte MCl defined by (70.11). If we now give the definition

$$\Phi^* = \Phi + \frac{2RT}{\mathfrak{F}} \ln x \qquad (72.37)$$

and use (72.36), we find that

$$\Phi^* = \Phi^0 - \frac{2RT}{\mathfrak{F}} \ln f_{\pm}. \qquad (72.38)$$

The right-hand side of (72.37) involves only measurable quantities. We can, therefore, plot Φ^* as a function of $x^{\frac{1}{2}}$ from experimental data. Since $\lim_{x \to 0} f_{\pm} = 1$, the intercept of the extrapolated curve with the ordinate gives the quantity Φ^0 according to (72.38). This equation also shows that $f_{\pm}$ is directly obtained as a function of $x^{\frac{1}{2}}$ from the curve. The reliability of the extrapolation can be much improved by using the theoretical formulae for $f_{\pm}$ mentioned in § 70.

E.m.f. measurement on galvanic cells is one of the few methods capable of yielding activities and activity coefficients for solid solutions, including binary alloys.‡ We use as our example the cell

$$Pt\,|\,Al(s)\,|\,AlCl_3.NaCl(m)\,|\,Al\text{–}Zn(s)\,|\,Pt \qquad (72.39)$$

† The irreversible reaction $Ag^+ + M \rightarrow Ag + M^+$ must in practice be assumed to be absent because of the low solubility of AgCl.

‡ The two other possible methods are the investigation of chemical equilibria discussed in § 35 and the measurement of partial pressures. The latter method is, however, applicable only if at least one component is sufficiently volatile in the appropriate temperature range. This is so at higher temperatures for the systems Ag–Zn and Cu–Zn for example.

where $AlCl_3.NaCl(m)$ denotes the salt melt acting as the electrolyte and Al–Zn(s) is the solid binary alloy. The overall reaction in the cell is simply the transfer of 1 mole Al from the pure phase into the binary alloy. This can be represented by

$$\mathrm{Al}_{\mathrm{Al}} \longrightarrow \mathrm{Al}_{\mathrm{Al\text{-}Zn}}. \tag{72.40}$$

$\nu^* = 3$ and (72.19) therefore gives the e.m.f. of the cell as

$$3\mathfrak{F}\Phi = -(\mu_{\mathrm{Al}} - \mu_{\mathrm{Al}0}) = -\Delta\mu_{\mathrm{Al}} = -RT\ln a_{\mathrm{Al}} \tag{72.41}$$
$$= -RT\ln x_{\mathrm{Al}} f_{\mathrm{Al}}.$$

The e.m.f. measurement therefore gives directly the difference between the chemical potential of Al in the alloy and that in the pure metal or, amounting to the same thing, the activity or activity coefficient of Al in the alloy.

Quite apart from their scientific interest galvanic cells are of the highest technical importance. They are used as sources of current (e.g. accumulators) and also for the performance of chemical reactions which are difficult or impossible to realize by other methods. Well-known examples are the electrolysis of alkali metal chlorides, the electrolytic production of aluminium, and the electrolytic deposition of surface layers of metals (galvanization). All these are large-scale processes commonly used in industry.

CHAPTER XII

Gravitational field. Centrifugal field. The determination of molecular weights

§ 73 Systems in a gravitational field

Nowhere in the formulation of our theories have we taken account of the fact that practically all the measurements we have mentioned have to be carried out in the earth's gravitational field. We must therefore try to find out how the presence of the gravitational field may affect the equilibrium conditions. The following properties of the gravitational field† are important in this connexion:

(a) Both strength and direction of gravity are constant with respect to time and to position.

(b) The presence of the system does not modify the gravitational field.

(c) The action of gravity is on the mass of the system and is independent of the chemical composition and the thermodynamic state of the system.

The gravitational field is, therefore, represented by the potential

$$\psi(z) = gz, \tag{73.1}$$

where g is the acceleration due to gravity and z is the height above an arbitrarily chosen datum level (e.g. the laboratory bench). The equipotential surfaces of the gravitational field are therefore parallel to the datum level.

Equation (73.1) shows that the contribution of the gravity field to the total energy changes continuously with the height above the datum level for a system with finite extension in the vertical direction. If, therefore, we denote by dm that part of the mass of the

† Any deviations from (a) and (b) are far below the limits of accuracy of thermodynamic measurement.

system lying between two neighbouring equipotential surfaces infinitesimally far apart from each other, the total energy contribution of the gravitational field is given by

$$U^{(g)} = \int \psi \, \mathrm{d}m. \tag{73.2}$$

The total mass m of a system consisting of m components can be represented by

$$m = \sum_{i=1}^{\mathrm{m}} M_i n_i, \tag{73.3}$$

where M_i is the molecular weight† of component i. We then have

$$\mathrm{d}m = \sum_{i=1}^{\mathrm{m}} M_i \, \mathrm{d}n_i, \tag{73.4}$$

where $\mathrm{d}n_i$ is the number of moles of component i found between two neighbouring equipotential surfaces infinitesimally far apart.

The contribution to the total energy of the system due to the gravitational field is a function of the positional co-ordinate. We must, therefore, initially regard all quantities of state as functions of position. This leads to the following consequences:

(a′) The definitions of homogeneous substances and of phases given in § 2 are not realizable in the strict sense.

(b′) The spatial distribution of mass must be included in any consideration of energy variation.

Re(a′) From now on we consider a *homogeneous substance* to be one whose inhomogeneity is due solely to a gravitational field. A *phase* is then defined in the same way as in § 2. This terminology is justified because the inhomogeneities, or at least those which cannot be completely neglected (e.g. for solids), are continuous functions of the positional co-ordinates and are generally easily distinguished from the discontinuous inhomogeneities which occur at phase boundaries.‡ The present terminology is also convenient since we can retain the nomenclature used in earlier chapters without modification.

† More details concerning molecular weights and their determination can be found in § 75.

‡ Difficulties do arise near the critical point of liquid–vapour equilibria. We cannot discuss these here.

Re(b′) The new definition of a homogeneous substance no longer assumes the spatial constancy of quantities of state. The equilibrium conditions for homogeneous substances in the absence of chemical reactions must, therefore, also be derived explicitly. The generalization for heterogeneous systems is not difficult but is of little interest as far as thermodynamics is concerned and will not be discussed here.

We consider a closed fluid system with fixed work co-ordinates. The system consists of m components and is homogeneous as defined above. Other external fields are excluded, as are chemical reactions. Let us write

$$U^* = U + U^{(g)}, \tag{73.5}$$

where U is the internal energy in the absence of gravity and $U^{(g)}$ is defined by (73.2). The equilibrium condition (17.2) then becomes (for the equality sign only)

$$(\delta U^*)_s = 0 \tag{73.6}$$

or

$$\delta \int \mathrm{d}U + \delta \int \psi \, \mathrm{d}m = 0 \tag{73.7}$$

with the secondary conditions

$$\delta \int \mathrm{d}S = 0, \tag{73.8}$$

$$\delta \int \mathrm{d}n_i = 0 \quad (i = 1, 2, \ldots, \mathrm{m}). \tag{73.9}$$

The only virtual displacements which are possible with the assumptions we have made are variations in entropy and mole number in each layer bounded by two infinitesimally neighbouring equipotential surfaces. By means of (20.5) and (73.4) we can, therefore, write the equilibrium condition in the form

$$\int T(z)\,\delta\left(\frac{\mathrm{d}S}{\mathrm{d}z}\right)\mathrm{d}z + \sum_{i=1}^{\mathrm{m}} \int \mu_i(z)\,\delta\left(\frac{\mathrm{d}n_i}{\mathrm{d}z}\right)\mathrm{d}z + \sum_{i=1}^{\mathrm{m}} \int \psi(z)\,M_i\,\delta\left(\frac{\mathrm{d}n_i}{\mathrm{d}z}\right)\mathrm{d}z = 0. \tag{73.10}$$

The corresponding secondary conditions are

$$\int \delta\left(\frac{\mathrm{d}S}{\mathrm{d}z}\right)\mathrm{d}z = 0, \tag{73.11}$$

$$\int \delta\left(\frac{\mathrm{d}n_i}{\mathrm{d}z}\right)\mathrm{d}z = 0 \quad (i = 1, 2, \ldots, \mathrm{m}). \tag{73.12}$$

We multiply (73.11) by the undetermined multiplier λ_0, (73.12) by λ_i and then subtract these equations from (73.10). This gives

$$\int [T(z)-\lambda_0]\,\delta\left(\frac{\mathrm{d}S}{\mathrm{d}z}\right)\mathrm{d}z+\sum_{i=1}^{\mathrm{m}}\int [\mu_i(z)+M_i\,\psi(z)-\lambda_i]\,\delta\left(\frac{\mathrm{d}n_i}{\mathrm{d}z}\right)\mathrm{d}z = 0. \tag{73.13}$$

In this equation the variations are no longer subject to secondary conditions. The equation can, therefore, be obeyed generally only if the expressions in square brackets are identically equal to zero. We therefore find that, for all z,

$$T(z) = \lambda_0 \equiv T, \tag{73.14}$$

$$\mu_i(z)+M_i\,\psi(z) = \lambda_i \equiv \mu_i^* \quad (i = 1, 2, \ldots, \mathrm{m}) \tag{73.15}$$

or

$$\frac{\mathrm{d}T}{\mathrm{d}z} = 0, \tag{73.16}$$

$$\frac{\mathrm{d}\mu_i^*}{\mathrm{d}z} = 0 \quad (i = 1, 2, \ldots, \mathrm{m}). \tag{73.17}$$

Equations (73.14) and (73.15), or (73.16) and (73.17), represent the explicit equilibrium conditions for a homogeneous fluid system without chemical reaction in a gravitational field. It is immediately apparent that the influence of the gravitational field is formally expressed by substituting the quantities μ_i^* defined by (73.15) for the chemical potentials. We can, therefore, immediately answer the question whether a *chemical equilibrium* is influenced by a gravitational field. According to the method described in § 33 we now obtain the condition for chemical equilibrium in the form

$$\sum_i \nu_i\,\mu_i^* = 0, \tag{73.18}$$

where the summation is to be taken, as in § 33, over all the participants in the reaction. However, mass is conserved in chemical reactions and we therefore have

$$\sum_i \nu_i\,M_i = 0. \tag{73.19}$$

Because of (73.15), eq. (73.18) then reduces to the condition

$$\sum_i \nu_i\,\mu_i(z) = 0. \tag{73.20}$$

According to (34.14) and (34.23), eq. (73.20) for homogeneous gas reactions can be written

$$\Pi[p_i^*(z)]^{\nu_i} = K_p(T). \tag{73.21}$$

Since the right-hand side of this equation depends only on the temperature, eq. (73.16) shows that the equilibrium constant K_p is not influenced by the gravitational field. The same is true for the equilibrium constant K_c defined by eq. (34.13). However, the equilibrium constant K_x and the equilibrium constant $K(T, P)$ for homogeneous reactions in solution defined by (34.42) depend on the pressure. We shall see later that the pressure is not constant in the z-direction. The last two equilibrium constants thus also are not constant. Under ordinary conditions, however, the influence of gravity is negligible in both cases.

In order to obtain an expression for the pressure change in the z-direction we start with the Gibbs–Duhem equation (20.43) which, because of (73.16), can be written

$$V \,\mathrm{d}P = \sum_{i=1}^{\mathrm{m}} n_i \,\mathrm{d}\mu_i. \tag{73.22}$$

With (73.15) this becomes

$$V \frac{\mathrm{d}P}{\mathrm{d}z} = -\sum_{i=1}^{\mathrm{m}} n_i M_i \frac{\mathrm{d}\psi}{\mathrm{d}z} \tag{73.23}$$

or, using (73.1) and (73.3),

$$\frac{\mathrm{d}P}{\mathrm{d}z} = -\frac{m}{V}\frac{\mathrm{d}\psi}{\mathrm{d}z} = -\rho \frac{\mathrm{d}\psi}{\mathrm{d}z} = -\rho g, \tag{73.24}$$

where ρ is the mean density of the system. Equation (73.24) is the general *condition for hydrostatic equilibrium.*

In order to investigate the change in mole fractions in the z-direction, we start with the relationship†

$$\mathrm{d}\mu_i = -s_i \,\mathrm{d}T + v_i \,\mathrm{d}P + \sum_{j=2}^{\mathrm{m}} \left(\frac{\partial \mu_i}{\partial x_j}\right)_{T,P,x_{k \neq j}} \mathrm{d}x_j. \tag{73.25}$$

If we abbreviate by writing

$$G_{ij} = \left(\frac{\partial \mu_i}{\partial x_j}\right)_{T,P,x_{k \neq j}} \tag{73.26}$$

† The summation in eq. (73.25) is taken from $j = 2$ to m with a view to later application.

we find from (73.16), (73.17), and (73.25) that

$$v_i \frac{\mathrm{d}P}{\mathrm{d}z} + M_i \frac{\mathrm{d}\psi}{\mathrm{d}z} + \sum_{j=2}^{\mathrm{m}} G_{ij} \frac{\mathrm{d}x_j}{\mathrm{d}z} = 0 \quad (i = 1, 2, \ldots, \mathrm{m}). \tag{73.27}$$

The pressure can be eliminated from these equations by means of (73.24). We must, however, remember that (73.24) and (73.27) are not independent from each other. Thus, if we multiply eq. (73.27) by n_i, sum over all i and introduce the relationships

$$\sum_{i=1}^{\mathrm{m}} n_i v_i = V, \tag{73.28}$$

$$\sum_{i=1}^{\mathrm{m}} n_i G_{ij} = 0 \tag{73.29}$$

which are obtained from (26.10) and (26.16) we get eq. (73.24). If, therefore, we use eq. (73.24), only $\mathrm{m}-1$ of the eqs. (73.27) are still independent from each other. Elimination of the pressure then gives

$$(M_i - \rho v_i) \frac{\mathrm{d}\psi}{\mathrm{d}z} + \sum_{j=2}^{\mathrm{m}} G_{ij} \frac{\mathrm{d}x_j}{\mathrm{d}z} = 0 \quad (i = 2, 3, \ldots, \mathrm{m}). \tag{73.30}$$

This with eq. (73.24) again gives us a set of m equations called the *equations for the sedimentation equilibrium.*

We now briefly consider the application of the general theory to one-component systems where the equilibrium is completely determined by eq. (73.24). If we denote the molecular weight by M and the molar volume by v, we have

$$\rho = M/v. \tag{73.31}$$

For an ideal gas we have

$$Pv = RT. \tag{73.32}$$

Introduction of (73.31) and (73.32) into (73.24) gives

$$\frac{\mathrm{d}P}{P} = -\frac{M}{RT}\mathrm{d}\psi = -\frac{M}{RT} g\,\mathrm{d}z. \tag{73.33}$$

Integration between the limits 0 and h gives the *barometric height equation*

$$RT \ln (P^{(h)}/P^{(0)}) = -M(\psi^{(h)} - \psi^{(0)}) = -Mgh. \tag{73.34}$$

For an incompressible liquid eq. (73.24) directly gives

$$P^{(h)} - P^{(0)} = -\rho(\psi^{(h)} - \psi^{(0)}) = -\rho g h. \tag{73.35}$$

We can conclude from eq. (73.34) that the pressure difference due to gravity in gases and, therefore, according to (73.32) also the density differences, are significant only when great differences in height are involved or when the compressibility of the gas is considerably greater than that of an ideal gas. The latter is the case, particularly, near the critical point of condensation (cf. §§47 and 51). The compressibility and, therefore, the density change can be completely neglected in liquid systems for all likely pressure differences. The hydrostatic pressure difference given by eq. (73.35), however, plays an important part, e.g. in the usual arrangements for osmotic pressure measurements (cf. §28).

For an ideal gas mixture we can easily derive a relationship analogous to eq. (73.34) for the partial pressure of each component. The upper layers of the atmosphere are therefore richer in the lighter gases. This effect is not observable in the laboratory since the gravitational field is too weak.

§74 Systems in a centrifugal field

We consider a fluid system of m components in a cylindrical vessel which rotates round the axis of the cylinder at a constant angular velocity ω. Unlike gravitational force, the strength of the centrifugal force† acting on unit mass varies with position and is proportional to the distance r from the axis of rotation. Apart from this, however, a centrifugal field has the properties (a), (b), and (c) of a gravitational field as given in §73. If, therefore, we represent the centrifugal field by the potential

$$\psi(r) = -\tfrac{1}{2}\omega^2 r^2, \tag{74.1}$$

the general derivations of §73 can be adopted for the centrifugal field. In particular, eqs. (73.24) and (73.30) also apply to the *sedimentation equilibrium in a centrifugal field* if ψ is replaced by the expression (74.1). These equations determine the spatial distribution of the components as a function of the distance r from the axis of rotation.

We illustrate these conditions in greater detail by using a binary system as our example. We can accept that the solvent is unambiguously defined since the theory is applied in practice only to dilute solutions. We denote the solvent, as usual, by the subscript 1. The

† Coriolis forces vanish at thermodynamic equilibrium.

equations for the sedimentation equilibrium are then

$$\mathrm{d}P = \rho\omega^2 r\,\mathrm{d}r, \tag{74.2}$$

$$(M_2 - \rho v_2)\,\omega^2 r = \left(\frac{\partial \mu_2}{\partial x_2}\right)_{T,P} \frac{\mathrm{d}x_2}{\mathrm{d}r}. \tag{74.3}$$

The partial derivative on the right-hand side is necessarily positive according to (41.35). The sign of the concentration gradient therefore depends only on the expression in parentheses on the left-hand side. This expression becomes easily comprehensible if we introduce the partial specific volumes

$$v_1/M_1; \quad v_2/M_2 \tag{74.4}$$

and the densities ρ_1 and ρ_2 of the components. Using Euler's theorem (§ 19b) we then get

$$(M_2 - \rho v_2) = M_2\left[1 - (\rho_1 + \rho_2)\frac{v_2}{M_2}\right] = M_2\,\rho_1\left(\frac{v_1}{M_1} - \frac{v_2}{M_2}\right). \tag{74.5}$$

The expression in parentheses is therefore positive if the partial specific volume of the solvent is greater than that of the solute which is usually the case. Under these conditions the concentration of the solute increases with the distance from the axis of rotation at equilibrium. The gradient also increases with increasing distance from the axis of rotation and is proportional to the square of the angular velocity.

By using (34.36), we can write (74.3) as

$$\frac{\mathrm{d}\ln x_2}{\mathrm{d}r} = (M_2 - \rho v_2)\left(1 + \frac{\partial \ln f_2}{\partial \ln x_2}\right)^{-1} \frac{\omega^2 r}{RT}, \tag{74.6}$$

where the activity coefficient is standardized according to § 34(b β). Since we have, therefore, that

$$\lim_{x_1 \to 1} f_2 = 1 \tag{74.7}$$

eq. (74.6) can be immediately integrated for the limiting case of infinite dilution. We thus obtain the expression

$$RT \ln (x_2'/x_2'') = \tfrac{1}{2}(M_2 - \rho v_2)\,\omega^2(r'^2 - r''^2) \tag{74.8}$$

which was originally derived by Svedberg. The sedimentation equilibrium of the solute in a centrifugal field is, therefore, under the given conditions, similar to the sedimentation equilibrium of an ideal gas in a gravitational field as described by eq. (73.34).

Centrifugal fields are extraordinarily important in physics and physical chemistry since the *ultracentrifuge* introduced by Svedberg in 1924 can generate accelerations of up to about $10^6 g$. Although the sedimentation equilibrium is quite unimportant in the gravitational field, the equilibrium in a centrifugal field can thus be used to separate the components of a system (preparative ultracentrifuge) or to determine the molecular weight according to eq. (74.8) (analytical ultracentrifuge). For experimental reasons, however, measurements of sedimentation velocity are used almost exclusively nowadays for the determination of molecular weights. The theory of this method is based on the thermodynamics of irreversible processes. We do not, therefore, give details here but refer the reader to the appropriate textbooks.

§75 The determination of molecular weights

As is usual nowadays in thermodynamics we have used the mole as our mass unit throughout this book. When a system is made up or analysed, however, the mass of the components is usually determined by weighing, i.e. in gram units. The mass in grams m_i of a component is related to the mass in moles of a component by the expression

$$m_i = M_i n_i. \tag{75.1}$$

The expressions used so far therefore imply that the molecular weight of all the components is known.

The use of mole numbers has great advantages particularly in the application of thermodynamics to chemical and electrochemical problems and in interrelating phenomenological and statistical thermodynamics. This quantity suffers, however, from the following disadvantages:

(a) The molecular weight is not definable within the framework of thermodynamics.

(b) The molecular weight is defined as a measurable quantity only for two limiting states, i.e. for an infinitely dilute one-component gas and (as the molecular weight of the solute) for an infinitely dilute binary solution. We can, of course, generally assume that the concept of molecular weight retains its physical meaning for reasonably neighbouring states, e.g. a slightly compressed real gas. For far distant states, however, particularly if these are separated from the defined state by one or more phase transitions, the molecular weight becomes merely a conventional calculated quantity. For example, NaCl has a definite molecular weight in the sufficiently dilute gas phase but no molecular weight can be defined for NaCl in the crystalline state.

(c) Many substances (e.g. graphite, diamond, Al_2O_3, silicates, three-dimensional cross-linked polymers) can never be brought into a state which allows us to measure their molecular weight.

All this shows that the masses m_i of the components are the natural or primary variables of state in phenomenological thermodynamics from both a logical and a practical point of view. The question how the dimension mass is divided into the two factors on the right-hand side of eq. (75.1) is therefore of lesser significance for thermodynamics. The new international agreements formulated in the recommendations of the International Union of Pure and Applied Physics (IUPAP)† and the International Union of Pure and Applied Chemistry (IUPAC)‡ introduce the *amount of substance* as a new basic quantity. The corresponding basic unit is the *mole* which is defined as the amount of substance in a system consisting of a number of molecules (or ions, or atoms, or electrons, or other relevant particles) equal to the number of atoms in exactly 12 grams of the pure carbon nuclide ^{12}C. This definition together with eq. (75.1) shows that the molecular weight must be defined as mass per unit amount of substance.

We shall now consider how molecular weight can be introduced into thermodynamics and then discuss the most important methods for the experimental determination of molecular weights as far as they are based on measurements involving thermodynamic equilibria. We shall exclude electrolytes and chemical reactions (particularly association and dissociation reactions) since we shall discuss only basic concepts.

The definition of molecular weight is based on certain equations of statistical thermodynamics which apply exactly if some very general assumptions are made.§ These equations are:

The *thermal equation of state of a one-component gas*

$$P = \frac{RT}{M}\rho\left[1 + B\frac{\rho}{M} + C\left(\frac{\rho}{M}\right)^2 + D\left(\frac{\rho}{M}\right)^3 + \dots\right], \qquad (75.2)$$

where ρ is the density in g cm^{-3}.

† *IUPAP Symbols, Units and Nomenclature in Physics*, 1965.

‡ *IUPAC Information Bulletin No.* 24.

§ A direct introduction as a fact of experience is less satisfactory since it does not have the wide empirical basis of the basic equations of statistical thermodynamics.

The equation for the *osmotic pressure of a binary solution*

$$\Pi = \frac{RT}{M_2} c_g \left[1 + B^* \frac{c_g}{M_2} + C^* \left(\frac{c_g}{M_2}\right)^2 + D^* \left(\frac{c_g}{M_2}\right)^3 + \ldots\right], \qquad (75.3)$$

when c_g is the concentration in g cm^{-3} and the subscript 2 refers to the solute.†

The thermodynamic definition of molecular weight in the gas phase is, therefore,

$$M = RT \lim_{\rho \to 0} \frac{\rho}{P} \qquad (75.4)$$

and the definition for the *solute in a binary solution* is

$$M_2 = RT \lim_{c_g \to 0} \frac{c_g}{\Pi}. \qquad (75.5)$$

The existence of the limiting value is proved by statistical thermodynamics. The definitions (75.4) and (75.5) show that an experimental determination of molecular weight basically involves an extrapolation of experimental data to vanishing density or infinite dilution. Measurements at several densities or concentrations are therefore required.‡ The main methods used are as follows:

(α) *Measurements of the pressure and the density of a one-component gas.* This method for which there are various detailed experimental arrangements represents the direct application of the definition (75.4). The method is useful only if the substance concerned vaporizes without decomposition and the saturated vapour pressure is not too low in the experimentally accessible temperature range.

(β) *The measurement of the osmotic pressure of a binary solution.* This method constitutes the direct application of the definition (75.5). It is practically useless for substances of low molecular weight since a suitable semipermeable membrane is available only for exceptional cases. For macromolecular substances the difference in size between the solute molecules and the molecules of the solvent (of low molecular weight) are so great that this difficulty hardly ever exists. Osmotic

† Equations (75.2) and (75.3) are called virial equations and the coefficients are called virial coefficients.

‡ If measurements are feasible at a density or a concentration which does not depart considerably from the limiting laws (75.4) or (75.5) and no great accuracy is required, these measurements can be made at a single density or concentration.

pressure measurements have, therefore, been for a long time the most important method used in this field for molecular weight determinations.† The method has, however, lost importance in recent times since other, more accurate methods which are experimentally simpler have become available. In practice the most important method of measuring molecular weights in solution are based on the fact that, according to eq. (28.15), the osmotic pressure is mainly a measure of the free energy of dilution. In principle, therefore, any effect which depends on the free energy of dilution or on its derivative with respect to concentration can be used for molecular weight determination on the basis of the definition (75.5).

(γ) *The lowering of the vapour pressure of the solvent.* We assume, for simplicity, that the ideal gas laws apply. p_1 is the partial pressure of the solvent over the solution and p_{10} is the partial of the pure solvent. We then have for the vapour pressure equilibrium, from (27.8) together with (26.30) and (34.3), that

$$\Delta\mu_1 = RT \ln \frac{p_1}{p_{10}}. \tag{75.6}$$

From eqs. (28.15) and (75.6) we obtain

$$\int_{p_0}^{p_0+\Pi} v_1 \, \mathrm{d}p = \bar{v}_1 \Pi = RT \ln \frac{p_{10}}{p_1}. \tag{75.7}$$

By means of (75.3) we therefore find that

$$\frac{p_1}{p_{10}} = \exp\left[-\frac{\bar{v}_1}{M_2} c_g \left(1 + B^* \frac{c_g}{M_2} + \ldots\right)\right]. \tag{75.8}$$

Expansion of the exponential function and passage to the limit $c_g \to 0$ finally give

$$\lim_{c_g \to 0} \frac{p_{10} - p_1}{p_{10} \bar{v}_1 c_g} = \frac{1}{M_2}. \tag{75.9}$$

Since, at the limit of infinite dilution, $\bar{v}_1 \to V/n_1$ and $x_2 \to n_2/n_1$, eq. (75.9) can also be written in the more familiar form

$$\lim_{x_2 \to 0} \frac{p_{10} - p_1}{p_{10} x_2} = 1. \tag{75.10}$$

This equation is known as *Raoult's law*. The molecular weight of the solute can thus be determined by measuring the lowering of the

† The Donnan equilibrium must be taken into account if the measurements involve polyelectrolytes.

vapour pressure of the solvent. The measurement of the elevation of the boiling point is, however, usually simpler and more accurate. We shall return to this in the following subsection.

Unlike eq. (75.5), eq. (75.9) involves the partial molar volume and thus implicitly the molecular weight of the solvent. We have seen earlier that this is not a measurable quantity. The derivation of eq. (75.9) shows, however, that by using the equilibrium condition we can ascribe to the solvent the same molecular weight in the liquid and the gaseous phase. The latter quantity is, however, experimentally accessible according to eq. (75.4). This ensures the unambiguousness of molecular weights determined according to eq. (79.9).

(δ) *The elevation of the boiling point of the solvent.* We suppose that only the solvent has a measurable vapour pressure and assume that the external pressure is kept constant (say at 1 atm). We now investigate the dependence of the boiling point of the solvent on the concentration of the solution. If we denote the chemical potential of the solvent in the vapour phase† by $\mu_1^{(v)}$, the equilibrium condition becomes

$$\mu_1^{(v)} = \mu_1. \tag{75.11}$$

If we change the temperature and the composition at constant pressure it must be true that

$$\mathrm{d}\mu_1^{(v)} = \mathrm{d}\mu_1 \tag{75.12}$$

or, with (26.27), that

$$-s_1^{(v)}\,\mathrm{d}T = -s_1\,\mathrm{d}T + \frac{\partial \mu_1}{\partial x_2}\,\mathrm{d}x_2. \tag{75.13}$$

The system has only one thermodynamic degree of freedom according to the phase rule (29.3). A change in temperature therefore determines the change in concentration unambiguously. By using (26.28) and (75.11) we can transform (75.13) into

$$\frac{h_1 - h_1^{(v)}}{T}\,\mathrm{d}T = \frac{\partial \mu_1}{\partial x_2}\,\mathrm{d}x_2 \tag{75.14}$$

or, by introducing the molar heat of vaporization,

$$L_v = h_1^{(v)} - h_2, \tag{75.15}$$

into

$$-\frac{L_v}{T}\,\mathrm{d}T = \frac{\partial \mu_1}{\partial x_2}\,\mathrm{d}x_2. \tag{75.16}$$

† For simplicity we write the quantities relating to the liquid phase from now on without superscript.

This relationship is merely the differential equation for the co-existence curve in the T–x_2-plane. The relationship is exact but unsuitable for practical purposes since the quantity L_v depends on the concentration. We therefore have to modify the expression by writing the heat of vaporization in the form

$$L_v = L_v^0 + h_{10} - h_1 \tag{75.17}$$

where L_v^0 is the molar heat of vaporization of the pure solvent. Equation (75.16) then becomes

$$-\frac{h_{10}-h_1}{T}\mathrm{d}T - \frac{L_v^0}{T}\mathrm{d}T = \frac{\partial \mu_1}{\partial x_2}\mathrm{d}x_2. \tag{75.18}$$

At constant pressure we have the general expressions

$$(\mathrm{d}\mu_1)_P = -s_1\,\mathrm{d}T + \frac{\partial \mu_1}{\partial x_2}\mathrm{d}x_2 \tag{75.19}$$

and

$$(\partial \mu_{10})_P = -s_{10}\,\mathrm{d}T. \tag{75.20}$$

From (75.18)–(75.20) with (26.28) we obtain

$$-\frac{L_v^0}{T^2}\mathrm{d}T = \mathrm{d}\left(\frac{\Delta\mu_1}{T}\right)_P. \tag{75.21}$$

The required elevation of the boiling point is obtained by integrating this equation between the boiling point T_0 of the pure solvent and the boiling point T of the solution.

The temperature dependence of the heat of vaporization of the pure solvent is given by *Kirchhoff's equation*

$$\frac{\mathrm{d}L_v^0}{\mathrm{d}T} = C_P^{(v)} - C_P \tag{75.22}$$

which is obtained immediately from the definitions (21.26) and (31.7). The temperature dependence of the right-hand side of (75.22) can be neglected since only very small temperature ranges are involved. If, therefore, we denote by L_{0v}^0 the heat of vaporization of the pure solvent at its boiling point T_0, integration of eq. (75.22) gives

$$L_v^0 = L_{0v}^0 + (C_P^{(v)} - C_P)(T - T_0). \tag{75.23}$$

If we introduce this expression into eqn. (75.21) and integrate between T_0 and the boiling point T_1 of the solution, we obtain

$$\frac{\Delta\mu_1}{T_1} = L_{0v}^0\left(\frac{1}{T_1} - \frac{1}{T_0}\right) + (C_P^{(v)} - C_P)\left(\ln\frac{T_0}{T_1} + 1 - \frac{T_0}{T_1}\right). \tag{75.24}$$

If we introduce the boiling point elevation

$$\Delta T = T_1 - T_0 \tag{75.25}$$

and expand the right-hand side of eq. (75.24) in powers of $\Delta T/T_0$ we get

$$-\frac{\Delta\mu_1}{RT_1} = \frac{\Pi\bar{v}_1}{RT_1} = -\ln x_1 f_1$$

$$= \frac{L^0_{0v}}{RT_0^2}\Delta T + \left(-\frac{L^0_{0v}}{RT_0} + \frac{1}{2}\frac{C_P^{(v)} - C_P}{R}\right)\left(\frac{\Delta T}{T_0}\right)^2 \tag{75.26}$$

if we terminate the expansion with the square term. Equation (75.26) relates the boiling point elevation to the free energy of dilution, the osmotic pressure, and the activity coefficient of the solvent. The last of these can, therefore, be determined from measurements of the boiling point elevation. The second term on the right-hand side of (75.26) can be neglected anyway since we must extrapolate to infinite dilution for the calculation of the molecular weight of the solute. We can therefore write

$$\Delta T = \frac{RT_0^2}{L^0_{0v}}\frac{\Pi\bar{v}_1}{RT_1}. \tag{75.27}$$

If we now introduce the definition (75.5) we get

$$\lim_{c_g\to 0}\frac{\Delta T}{c_g} = \frac{RT_0^2}{L^0_{0v}}\frac{1}{RT_1}\lim_{c_g\to 0}\frac{\Pi\bar{v}_1}{c_g} = \frac{RT_0^2}{L^0_{0v}}\frac{v_{10}}{M_2}. \tag{75.28}$$

It can be seen that the molecular weight of the solute cancels out on the right-hand side. The heat of vaporization per unit mass (specific heat of vaporization)

$$L^*_{0v} = L^0_{0v}/M_1 \tag{75.29}$$

is therefore usually introduced and the quantity

$$m_2^* = 10^3\frac{m_2}{m_1} \tag{75.30}$$

is used as the concentration variable. For very high dilutions we have

$$\bar{v}_1 \approx v_{10} = \frac{M_1 V}{m_1} \tag{75.31}$$

and we thus obtain

$$\lim_{m_2^*\to 0}\frac{\Delta T}{m_2^*} = \frac{RT_2^0}{10^3 L^*_{0v}}\lim_{c_g\to 0}\frac{\Pi}{c_g} = \frac{RT_0^2}{10^3 L^*_{0v}}\frac{1}{M_2}. \tag{75.32}$$

The quantity

$$\Theta_v = \frac{RT_0^2}{10^3 L_{0v}^*} \tag{75.33}$$

is called the *ebullioscopic constant* and has been tabulated for more important solvents.

(ε) *The lowering of the freezing point of the solvent.* If the liquid binary solution is in equilibrium with the solid phase of the pure solvent an argument completely analogous to that of the previous subsection can be used. We shall merely quote the result. If T_0 is the melting point of the pure solvent, T_1 that of the solution and

$$\Delta T = T_0 - T_1 \tag{75.34}$$

is the lowering of the freezing point we have

$$\lim_{m_2^* \to 0} \frac{\Delta T}{m_2^*} = \frac{RT_0^2}{10^3 L_{0f}^*} \frac{1}{M_2} \tag{75.35}$$

where L_{0f}^* is the specific heat of fusion of the pure solvent. The quantity

$$\Theta_f = \frac{RT_0^2}{10^3 L_{0f}^*} \tag{75.36}$$

is called the *cryoscopic constant*. It, too, has been tabulated for the more important solvents.

We finally mention two other methods for the determination of molecular weights. These methods also depend on the definition (75.5) but are used in practice only for macromolecules (like the direct measurement of osmotic pressure). The first of these methods is the sedimentation equilibrium in the ultracentrifuge discussed in § 74. This method is of very little significance nowadays. The second method uses the scattering of light in solutions. The general basis of the theory can be found in advanced textbooks of statistical thermodynamics. Applications to solutions of macromolecules are given in specialized monographs.

Problems

(Problems marked with an asterisk are more difficult and use of the hints or, possibly, the solution may be advisable.)

CHAPTER I

§3

1. The thermal equation of state for an ideal gas is

$$Pv = RT \tag{P1}$$

where R is the gas constant. The state of the gas is changed quasi-statically from P_1, T_1 to P_2, T_2 along the paths I and II shown in Fig. P1. $dT = 0$ along AB and $dP = 0$ along BC.

(a) Calculate for both paths the volume change Δv and the work W done on the system.

(b) Which properties of the quantities v and W explain the results?

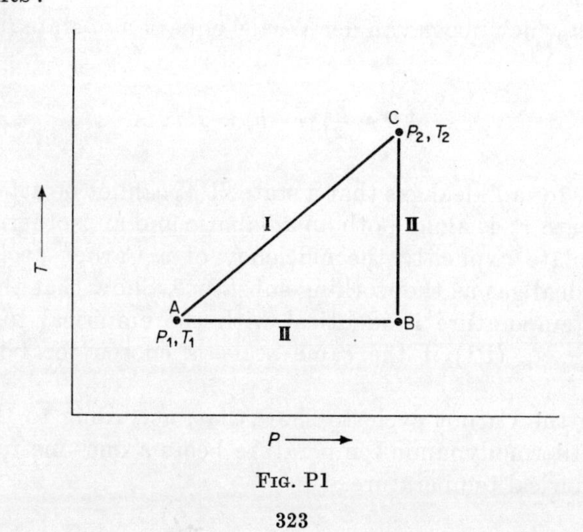

FIG. P1

*2. Let

$$dQ = X\,dx + Y\,dy. \tag{P2}$$

Use this expression to derive Clausius' relation

$$\frac{\partial X}{\partial y} - \frac{\partial Y}{\partial x} = \frac{\partial(\partial W/\partial y)_x}{\partial x} - \frac{\partial(\partial W/\partial x)_y}{\partial y}. \tag{P3}$$

3. What form does eqn. (P3) assume for quasi-static processes?

§4

4. Derive Carnot's theorem from Clausius' principle.

*5. Show that a perpetual motion machine of the second kind could be constructed if the Rayleigh radiation equation

$$\varepsilon \equiv \frac{U}{V} = \beta T \quad (\beta = \text{const.}) \tag{P4}$$

were true.

6. Derive the relationship

$$\left(\frac{\partial U}{\partial V}\right)_T + P = T\left(\frac{\partial P}{\partial T}\right)_V. \tag{P5}$$

7. Assume that C_V is independent of temperature and use (P1) to derive the adiabatic for

(a) an ideal gas,

(b) a gas which obeys van der Waals' equation of state (referred to 1 mole)

$$\left(P + \frac{a}{v^2}\right)(v - b) = RT. \tag{P6}$$

8. Show for an ideal gas that a state P_2, V_2 cannot be attained from another state P_1, V_1 along both an adiabatic and an isotherm.

9. Calculate explicitly the efficiency of a Carnot cyclic process using an ideal gas as the working substance. Show that the thermodynamic temperature is identical with the empirical temperature defined by eq. (P1) if the same scale is chosen for both. Prove eq. (4.14).

10. Use the Carnot cycle to show that, according to the Second Law, the thermodynamic temperature bears a one–one relationship to the empirical temperature.

*11. Show, by a reduction to the sum of infinitesimal Carnot cycles as represented in Fig. 2 (p. 13), that eq. (4.15) is valid for any cyclic process.

*12. Use Stokes' theorem of vector analysis to prove eq. (4.15) without division into Carnot cycles.

13. The diesel engine is based on a reversible cyclic process consisting of two adiabatic parts, one isobaric (P = const.) part, and one isochoric (V = const.) part (Fig. P2). The parts occur in the order $1 \rightarrow 2 \rightarrow 3 \rightarrow 4$. Calculate the dependence of the efficiency on (a) the compression ratio V_1/V_2 and (b) the isobaric expansion ratio V_3/V_2. Assume that the working substance is an ideal gas whose molar heat capacities are independent of temperature.

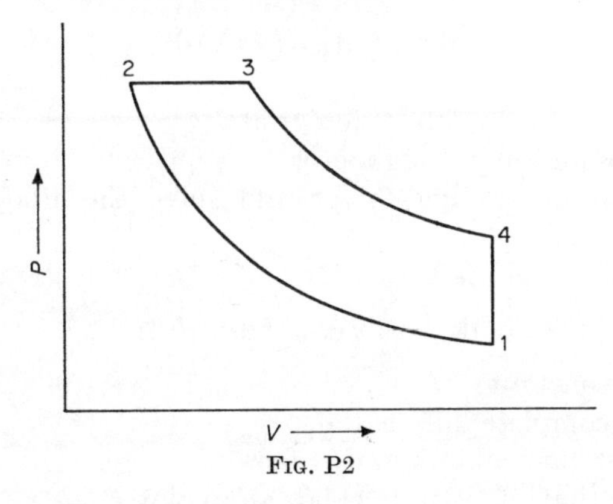

FIG. P2

*14. Two adiabatically insulated substances have constant heat capacities C_1 and C_2. The temperatures of the two substances are T_1 and T_2. They are brought into thermal contact, the adiabatic insulation towards the surroundings being maintained. The volume is kept constant. Their common final temperature is T_3. Calculate T_3, the entropy change ΔS and prove that $\Delta S > 0$.

15. Prove (4.27) for the case of Joule heat being generated by an electric current.

§8

16. The molar heat capacity C of a homogeneous substance is generally defined by the equation

$$C = \frac{dQ}{dT}. \tag{P7}$$

(a) Use this to derive the definition of the molar heat capacity at constant volume C_V given at the end of § 8.

(b) Use the function

$$U + PV \equiv H = H(T, P) \tag{P8}$$

to derive an analogous definition of the molar heat capacity at constant pressure C_P.

17. If C_x is the molar heat capacity measured at a constant variable of state x, derive equations relating C_x to

(a) the molar heat capacity at constant volume C_V,

(b) the molar heat capacity at constant pressure C_P.

18. Show that $\mathrm{d}Q$ is not a complete differential.

19. Let

$$\mathrm{d}Q = y\,\mathrm{d}x - (y + x)\,\mathrm{d}y. \tag{P9}$$

Show that

(a) (P9) is an incomplete differential,

(b) (P9) has an integrating factor.

(c) Find an integrating factor and solve the Pfaff differential equation $\mathrm{d}Q = 0$.

20. Show that the Pfaff equation

$$\mathrm{d}Q = -y\,\mathrm{d}x + x\,\mathrm{d}y + k\,\mathrm{d}z \tag{P10}$$

(where k is a constant)

(a) is an incomplete differential,

(b) does not have an integrating factor.

*21. Show that, if (P10) is valid, any point $P_1 = (x_1, y_1, z_1)$ can be reached from any other point $P_0 = (x_0, y_0, z_0)$ along a path where $\mathrm{d}Q = 0$ throughout.

§ 10

22. Show without using the Second Law that, for an ideal gas, the entropy defined by eq. (10.24) exists as a function of state.

§ 11

*23. Show that the quantities S, U, and T calculated from empirical adiabatics and isotherms are independent of the choice of the empirical scales of s and t.

§ 13

*24. Use the Gay–Lussac experiment (streaming of an ideal gas into a vacuum) as the empirical basis for the derivation of (13.4).

*25.
(a) Show that the theory developed in § 10 does not exclude the possibility that a non-static adiabatic process involving the parts of the system leaves the entropy of the total system unchanged.
(b) Show that this possibility is excluded by the extension of the theory in § 13.

CHAPTER III

§ 20

26. The thermal equation of state for a monatomic ideal gas is

$$PV = nRT \tag{P11}$$

and the caloric equation of state is

$$U = \tfrac{3}{2}nRT. \tag{P12}$$

Derive the fundamental equation per mole in the entropy representation
(a) by integration of eq. (20.42),
(b) by using the Gibbs–Duhem equation.

27. For an ideal gas, what is the relationship between
(a) temperature and volume, or pressure, along an adiabatic,
(b) entropy and volume, or pressure, along an isotherm if the molar heat capacities C_P and C_V are assumed to be independent of temperature?

28. The expressions

$$\varepsilon \equiv \frac{U}{V} = f(T), \quad P = \tfrac{1}{3}\varepsilon \tag{P13}$$

apply to black body radiation (radiation from an enclosure). Derive
(a) the Stefan–Boltzmann law

$$\varepsilon = aT^4, \tag{P14}$$

(b) the function $S(T, V)$.

29. Can the fundamental equation have the form

$$U = \frac{aS^2}{cn(V-b)} \tag{P15}$$

(where a, b, and c are constants)?

30. From the thermal equation of state

$$P = \left(\frac{an}{27cV}\right)^{\frac{1}{2}} T^{\frac{3}{2}} \tag{P16}$$

and the caloric equation of state

$$U = \left(\frac{anV}{27c}\right)^{\frac{1}{2}} T^{\frac{3}{2}} \tag{P17}$$

(where a and c are constants), derive the fundamental equation in the energy representation

(a) by integration of eq. (20.42),

(b) by using the Gibbs–Duhem equation.

§ 21

31. Derive the equation

$$\left(\frac{\partial H}{\partial P}\right)_T - V = -T\left(\frac{\partial V}{\partial T}\right)_P. \tag{P18}$$

32. Derive expressions

(a) for the dependence of the molar heat capacity C_V on the volume,

(b) for the dependence of the molar heat capacity C_P on the pressure.

(c) Apply the expressions derived under (a) and (b) to an ideal gas; apply the expression derived under (a) to a gas obeying the van der Waals' equation (P6).

§ 22

33.

(a) Use eqs. (P11) and (P12) to derive the explicit expression for the molar Helmholtz free energy of a monatomic ideal gas.

(b) Verify that the answer to (a) is the Legendre transform of the fundamental equation.

34. For a van der Waals gas, derive the explicit (except for a pure temperature function) expression for the molar Helmholtz free energy.

35. Derive the molar Helmholtz free energy from eqs. (P16) and (P17).

36. What conclusions about the quantities $(\partial U/\partial V)_T$ and $(\partial C_P/\partial v)_T$ can be drawn from the thermal equation of state

(a) for an ideal gas,

(b) for a van der Waals gas.

§ 24

*37. Use eq. (24.10c) to calculate the molar Gibbs free energy
(a) for an ideal gas,
(b) for a van der Waals gas.

38. Let the molar internal energy of a substance be

$$u = \mathrm{A}(v) + \mathrm{B}(T) \tag{P19}$$

where A and B are known functions. Calculate $P(T, V)$, $s(T, V)$, $f(T, V)$, $h(T, V)$, and C_P and C_V. Can a characteristic function be derived from these expressions?

39. Derive eq. (P5) by means of a Massieu–Planck function.

40. Derive the expression

$$\left(\frac{\partial S}{\partial V}\right)_{T,\mu} = \left(\frac{\partial P}{\partial T}\right)_{V,\mu}. \tag{P20}$$

§ 25

41. Reduce the following to standard derivatives:
(a) $(\partial V/\partial U)_F$,
(b) $(\partial P/\partial U)_G$,
(c) C_P/C_V.

42. Calculate the coefficient of thermal expansion of a van der Waals gas.

43.
(a) Apply the solution of 17(a) to C_P and that of 17(b) to C_V. Show that the equations obtained are identical.
(b) Use the solution of (a) to derive eq. (25.30).

44. Derive an expression for the enthalpy change associated with an isothermal (differential) volume change $\mathrm{d}V$.

45. What is the thermal equation of state for a gas which shows no cooling effect
(a) in the Gay–Lussac experiment,
(b) in the Joule–Thomson experiment,
(c) in both experiments.

46. State the thermodynamic equation of the inversion curve.

47. Calculate the Joule–Thomson coefficient δ for a van der Waals gas limited to the second virial coefficient B' defined by the equation

$$Pv = RT + B'P + \ldots. \tag{P21}$$

What is the relationship here between the inversion temperature and the Boyle temperature T_{B} defined by $B' = 0$?

48. If van der Waals' equation of state per mole is written in the form

$$Pv = RT + Pb - \frac{a}{v} + \frac{ab}{v^2} \tag{P22}$$

and we put, on the right-hand side,

$$\frac{1}{v} = \frac{P}{RT} \tag{P23}$$

limited to moderate pressures, we obtain the approximation

$$Pv = RT + \left(b - \frac{a}{RT}\right) P + \frac{ab}{(RT)^2} P^2. \tag{P24}$$

(This approximation is not self-consistent from the point of view of the virial expansion since it does not take the third virial coefficient completely into account. The use of the approximation is justified, however, because it is simpler than the correct approximation and the van der Waals equation gives only a qualitative picture in any case.)

Use eq. (P24) to calculate

(a) the Joule–Thomson coefficient δ,

(b) the equation of the inversion curve.

49. At the critical point we have (cf. § 46)

$$\frac{\partial P}{\partial v} = 0, \quad \frac{\partial^2 P}{\partial v^2} = 0. \tag{P25}$$

(a) Express the constants a and b of van der Waals' equation of state in terms of the critical quantities of state T_c, P_c.

(b) State the equation of the inversion curve [problem 48(b)] in terms of the reduced quantities of state $P^* \equiv P/P_c$ and $T^* = T/T_c$.

(c) Draw the inversion curve from (b) on a P^*–T^*-diagram. (This curve gives a qualitatively correct picture. The experimental curve shows a steeper gradient and a higher maximum.)

*50. Derive a relationship between the thermodynamic temperature T, the corresponding empirical temperature t of the real gas thermometer, and the course of the Joule–Thomson coefficient δ between the ice point t_0 and the temperature t.

*51. Assume equal initial conditions and equal final volumes; calculate the differential cooling effect in comparable forms for

(a) adiabatic streaming into a vacuum (Gay–Lussac experiment),

(b) the Joule–Thomson effect,
(c) isentropic expansion.
Compare (a) and (c) with the Joule–Thomson effect.

52. Calculate the maximum value of the Joule–Thomson coefficient δ at given pressure.

53. For the integral (total) Joule–Thomson effect corresponding to a finite pressure difference $P_{\mathrm{I}} - P_{\mathrm{II}}$ with $P_{\mathrm{II}} = 0$, derive an expression containing the Joule–Thomson coefficient δ. Assume that the limiting value of the molar heat capacity C_P for $P \to 0$ has the temperature independent value C_{P0} over the relevant temperature range.

54. Show that the total cooling effect becomes an extremum for that pressure at which the differential Joule–Thomson effect vanishes for the same initial temperature.

55. Derive the thermodynamic condition for the inversion of the total cooling effect and compare it with the corresponding condition for the differential Joule–Thomson effect.

56. Use the approximation (P24) to derive the condition for the inversion of the total cooling effect for a van der Waals gas and compare it with the corresponding condition for the differential Joule–Thomson effect (problem 48).

§ 26

*57. For a multi-component system let

$$\frac{V}{RT}\left(\frac{\partial \mu_i}{\partial n_j}\right)_{T,V,n_k} = \frac{|B|_{ij}}{|B|} \tag{P26}$$

where $|B|_{ij}$ is the cofactor of the element B_{ij} of the determinant $|B|$. Calculate the isothermal compressibility and the partial molar volumes.

CHAPTER IV

§ 28

58. The volume fractions

$$\phi_i = c_i v_i, \quad (c_i = n_i/V)$$

are often used as concentration variables in the theory of liquid mixtures. Derive a relationship, for a binary solution, between the derivative of the osmotic pressure with respect to the volume fraction of the solute and the derivative of the chemical potential of the solute with respect to the mole fraction of the solute.

59. Use (P26) to derive an expression for the partial derivatives $(\partial\Pi/\partial c_i)_{T,\mu_1,c_j}$ (subscript 1 = solvent) for a system of m components.

§ 31

60. Derive the differential equation of the co-existence curve in the μ–T-plane.

*61. The Einstein condensation (for details see textbooks on statistical thermodynamics) which occurs in the ideal Bose–Einstein gas is determined by the condition $\mu/T = 0$ valid at all transition temperatures. Figure P3 shows an isotherm diagrammatically. The horizontal part of the curve corresponds to the co-existence of vapour and liquid in the case of ordinary condensation; it must be interpreted here as the co-existence of an infinite number of infinitesimally different 'neighbouring' phases.

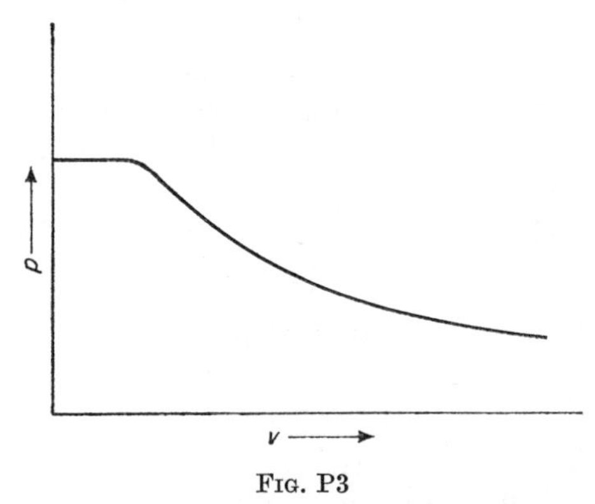

FIG. P3

(a) Show that all neighbouring phases have the same energy density and the same entropy density at a given temperature.

(b) Show that the assumption of the co-existence of an infinite number of neighbouring phases does not contradict the phase rule. Derive the equation for the temperature dependence of the 'saturation pressure' which here takes the place of the Clausius–Clapeyron equation.

62. State the differential equation which determines the temperature dependence of a function of state $Z(T, P)$ along the co-existence curve for a two-phase equilibrium in a one-component system.

63. Derive a differential equation for the temperature dependence of the molar volume difference ΔV^* along the co-existence curve for a two-phase equilibrium in a one-component system.

64. Apply the solution of problem 63 to the vapour–liquid equilibrium; assume that the molar volume, the coefficient of expansion, and the compressibility of the condensed phase are negligible compared with the corresponding quantities of the vapour, and that the vapour can be treated as an ideal gas.

65. Derive a differential equation for the temperature dependence of the heat of transition L for a two-phase equilibrium in a one-component system. (This equation is due to M. Planck.)

66. Derive differential equations for the temperature dependence of the molar entropies along the co-existence curve for a two-phase equilibrium in a one-component system.

67. Use (P7) to define the molar heat capacities at the saturation vapour pressure for the vapour–liquid equilibrium in a one-component system. (The molar heat capacity of liquid helium is usually measured under these conditions for experimental reasons.)

68. Use the assumptions of problem 63 to derive explicit expressions for the molar heat capacities C_s measured at the saturated vapour pressure. What conclusion can you draw from this about the C_s of the condensed phase (example: liquid helium)? (The expression for the molar heat capacity of the saturated vapour is due to Clausius.)

69. Use the solution of problem 17(b) to solve problem 68.

*70. Use eq. (P3) to solve problem 65.

*71. Suppose that the assumptions of problem 64 apply. Deduce a vapour pressure equation by integrating the Clausius–Clapeyron equation between a standard temperature T_0 and a temperature T. Neglect the temperature dependence of the difference $\Delta C_P \equiv C'_P - C''_P$ between T_0 and T, and the pressure dependence of the entropy of the condensed phase at the temperature T_0. If T and T_0 do not differ by much, i.e. $(T-T_0)/T_0 \ll 1$, what is the form of the vapour pressure equation?

§32

*72. For a binary two-phase system derive differential equations giving the relationships between the variables T, P, and x'_1, and between the variables T, P, and x''_1.

73. From the solution to problem 72 derive the differential equations for an isothermal vapour pressure diagram. Use eq. (28.10) to show that these differential equations lead to Konowalow's rules.

*74. Make the same assumptions as in problem 64† for binary vapour–liquid equilibria. Prove the following statement: 'If the azeotropic point corresponds to a minimum (maximum) on the isobaric boiling point curve, an increase in pressure causes the azeotropic mixture to be enriched with the component having the greater (smaller) partial heat of vaporization.' (Wrewsky's theorem.)

75. Derive the differential equations of the isobaric $T(x_1')$- and $T(x_1'')$-curves from the solution to problem 72.

76. Apply the solution of problem 75 to a binary melt in equilibrium

(a) with pure solid component 1,

(b) with pure solid component 2.

Assume (usually correctly) that the partial heats of fusion are

$$L_{f1} = H_1^{*''} - H_1^{*'} > 0, \quad L_{f2} = H_2^{*''} - H_2^{*'} > 0 \qquad \text{(P27)}$$

and use (28.10). Use your results to draw a schematic isobaric melting point diagram. What does the phase rule state about the point of intersection of the two co-existence curves?

CHAPTER V

§ 33

77. Use (28.10) to show that a dystectic point (equilibrium of a solid chemical compound with a melt of variable composition) is necessarily represented by a maximum on the isobaric melting point curve if the heat of fusion of the compound is positive (this is so in all known cases). A dystectic point is also called the melting point of a congruently melting compound since the solid compound co-exists with a melt of the same stoichiometric composition.

78. The typical diagram for a dystectic point shown in Fig. 22 (p. 128) sometimes changes into the form shown in Fig. P4. The length E_1D' of the curve represents the co-existence of the melt with the solid compound of composition a, length $D'M_B$ represents the co-existence of the melt with pure solid component B. Use problems 76 and 77 to show that

(a) $dT/dx < 0$ along the length $M_A E_1$ of the curve,

(b) $dT/dx > 0$ along the lengths $E_1 D'$ and $D'M_B$,

(c) the isobaric melting point curve generally has a kink at the point D'.

† The assumptions in problem 64 about coefficients of expansion and compressibility are irrelevant here.

The point D′ is called the melting point of an incongruently melting compound. Explain this name. What statement does the phase rule make about D′? (Example: the system Na_2SO_4 (solid), $Na_2SO_4.10H_2O$ (solid), aqueous solution of Na_2SO_4. The point D′ in this system is frequently used as a thermometric fixed point.)

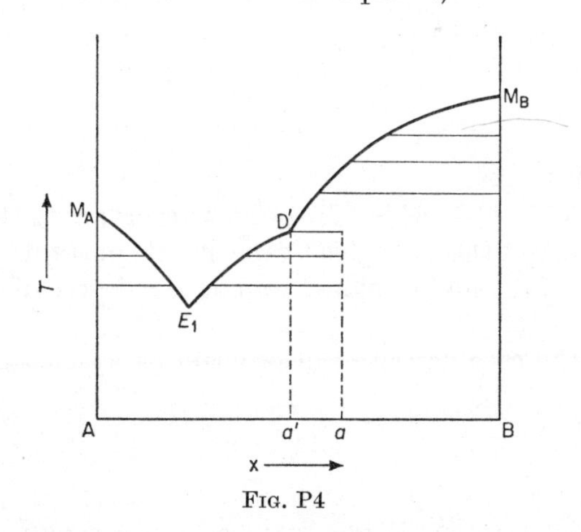

FIG. P4

79. Assume that CO_2 can be treated as an ideal gas. Derive the differential equation for the decomposition pressure of $CaCO_3$

(a) from the generalized Clausius–Clapeyron equation (31.23),

(b) from the condition for chemical equilibrium.

80. The heat of reaction at constant volume ΔU and the heat of reaction at constant pressure ΔH are both equal to the appropriate sum of the heats of reaction of the (arbitrarily chosen) reaction steps (Hess' law).

(a) Prove Hess' law.

(b) Use Hess' law to show that the heat of formation of benzene from graphite and H_2 can be found from the experimentally easily accessible heats of combustion.

*81. Derive differential equations for the temperature dependence of the heat of reaction Q for conditions of

(a) constant volume,

(b) constant pressure,

(c) concurrently changing volume and pressure.

(The equations for cases (a) and (b) are known as Kirchhoff's equations.)

*82. Derive the solutions to problem 81 as special cases from eq. (P3).

§ 34

83. What is the form of eqs. (34.5) and (34.8) for an ideal gas which dissociates according to

$$A_2 \rightleftharpoons 2A \tag{P28}$$

when the degree of dissociation

$$\alpha = \frac{\frac{1}{2}n_A}{n_{A_2}+\frac{1}{2}n_A} \tag{P29}$$

is introduced?

84. For an ideal gas which dissociates according to (P28) calculate the dependence of the molar heat capacity at constant volume (based on one mole of the undissociated gas) on the degree of dissociation α defined by (P29).

*85. For the case described in problem 84, calculate as functions of α

(a) $C_P - C_V$,
(b) C_P/C_V.

86.

(a) Whatan alytic properties of the functions $[\alpha(T)]_P$ and $(\partial\alpha/\partial T)_P$ are obtained from the solution to problem 85 if the temperature independent value for ΔH is that given under (c) here and T is of the order of magnitude 5×10^2 K?

(b) Use the results of (a) to draw a schematic representation of the function $\alpha(T)$. Use the values $\alpha = 0{\cdot}2$ at $T = 300$ K, $\alpha = 0{\cdot}5$ at $T = 340$ K, and $\alpha = 0{\cdot}8$ at $T = 380$ K.

(c) Use the curve drawn for (b), and problems 84 and 85 to draw the curves $C_P(T)$ as well as C_{PA_2} and $C_{PA_2}+\alpha(2C_{PA}-C_{PA_2})$ with the values

$$C_{PA_2} = 4R, \quad C_{PA} = 4R, \quad \Delta H = 12900 \text{ cal.}$$

(The values given correspond approximately to the system $N_2O_4 \rightleftharpoons 2NO_2$.)

87. If a chemical compound is prepared using a homogeneous gaseous equilibrium, it is often convenient to introduce the degree of formation y into the mass action law expression. For a reaction

$$A_2 + B_2 \rightleftharpoons 2AB \tag{P30}$$

where A_2 and B_2 are the reactants, the definitions

$$n_{AB} = 2y, \quad n_{A_2} = a - y, \quad n_{B_2} = b - y \tag{P31}$$

apply; a and b are the original numbers of moles of A_2 and B_2 respectively. Investigate whether, and for what values of a and b, y becomes a maximum.

88. For the ammonia synthesis

$$3H_2 + N_2 \rightleftharpoons 2NH_3 \tag{P32}$$

starting with a reaction mixture of stoichiometric composition, the definitions

$$n_{N_2} = n_0(1-y), \quad n_{H_2} = 3n_0(1-y), \quad n_{NH_3} = 2n_0 y \tag{P33}$$

apply; n_0 is the number of moles of N_2 in the initial reaction mixture.

(a) How does the equilibrium constant K_p depend on the temperature between 375 °C and 500 °C where, within a good approximation, $0 > \Delta H = \text{const}$.

(b) In the same temperature range, how does the degree of formation y depend on temperature and pressure if the assumption $y \ll 1$ is made.

§ 38

89. Derive an expression for the vapour pressure of a crystalline solid by integrating the Clausius–Clapeyron equation (31.6) between 0 and T with the following assumptions:

(1) The vapour can be treated as an ideal gas.

(2) The molar volume of the solid can be neglected compared with the molar volume of the vapour.

(3) The temperature dependence of the molar heat capacity of the vapour can be represented in the form†

$$\begin{aligned} C''_P(T) &= C''_{P0} + C''_{Ps}(T) \quad \text{for} \quad T > T_0 \\ C''_P(T) &= C''_{P0} \quad \text{for} \quad T \leqslant T_0. \end{aligned} \tag{P34}$$

(4) The molar heat capacity of the solid decreases sufficiently steeply with temperature to ensure the existence of the integral at the lower limit. (Cf. § 38 which also discusses the interpretation of the integral to $T \to 0$.)

The constant of integration usually denoted by i is called the vapour pressure constant.

† Many misleading statements about assumption 3 are found in the literature. We only point out here that this assumption remains strictly valid in quantum statistics if the ideal gas state is defined by $n/V \to 0$. Details can be found in textbooks of statistical thermodynamics.

90.
(a) Derive an expression for the equilibrium constant K_p of the mass action law for an ideal gas by integrating the van't Hoff equation (34.29) from 0 to T. Use assumption 3 of problem 89.
(b) Show that the constant of integration I has the form

$$I = \sum_i \nu_i j_i \equiv \Delta j \quad \text{(P35)}$$

where the j_i are constants of the material, called chemical constants.

91.
(a) Make the assumptions of problems 89 and 90 for a reaction between pure crystalline phases. Derive an expression for the quantity ΔG from the condition that the partial pressures of the participants in the reaction must obey both the law of mass action and the vapour pressure equation.
(b) Show that the Nernst heat theorem may be formulated as a statement concerning vapour pressure constants and chemical constants.

92. Show that

$$\frac{\partial \kappa}{\partial T} \to 0 \quad \text{for} \quad T \to 0 \quad \text{(P36)}$$

can be deduced from the Nernst heat theorem.

CHAPTER VI

§41

93.
(a) Explain why the stability conditions make statements about the sign of the molar heat capacity C_P and the compressibility κ but not about the coefficient of thermal expansion α.
(b) Derive the inequality

$$\alpha \left(\frac{\partial S}{\partial P}\right)_T < 0. \quad \text{(P37)}$$

94. Show that heat must be introduced into a system during a quasi-static isothermal expansion of a gas. Which property of gases is important in this connexion and why must the property be taken into consideration?

95. Show that the path of the adiabatics and isotherms in the P–V-plane shown in Fig. 1 (p. 12) is derived with complete generality from the stability conditions.

§ 43

96. Derive the solution to problem 95 from the Le Châtelier–Braun principle. Investigate which stability conditions play an important rôle in this connexion.

97. Suppose that a gas is in thermal equilibrium with a heat reservoir *via* a diathermic wall. Let the pressure of the gas be controlled by a manostat in experiment 1. A temperature change dT now causes an entropy change $d_1 S$. Let the gas in experiment 2 be surrounded by a rigid wall. The same temperature change now causes an entropy change $d_2 S$.

(a) Apply the Le Châtelier–Braun principle to both experiments. What immediate conclusion can be drawn from the stability conditions?

(b) Which stability conditions are used in the application of the Le Châtelier–Braun principle?

(c) Can the Le Châtelier–Braun principle in the form (43.2) be used as the empirical basis for the derivation of the stability conditions? What, therefore, is the relationship between the Le Châtelier–Braun principle and the stability conditions?

§ 44

98. Show that a chemical equilibrium in an ideal gaseous mixture necessarily obeys the stability condition (44.2).

CHAPTER VII

§ 46

99. Show that at the critical point of vaporization of a one-component system

(a) the condition of thermal stability $C_V > 0$ is necessarily fulfilled,

(b) in general† it is also true that $C_V < +\infty$.

*100. Prove the statements of problem 99 for multi-component systems.

† Here, and in problem 102, 'in general' means that exceptions are of an accidental nature and have no thermodynamic connexion with the stability limit.

§ 47

101. What assumption has to be made for the mechanical stability condition $(\partial^2 F^*/\partial V^{*2})_{T,x} > 0$ not to be fulfilled for the critical phase of the vapour–liquid equilibrium in a binary system?

102. Show that the mechanical stability condition is, in general, fulfilled for the critical solution point of a binary system.

§ 48

103. Derive the Ehrenfest equations for a third order transition.

CHAPTER VIII

§ 55

104. The thermal strain coefficients α_i are defined by the equation

$$\alpha_i = \left(\frac{\partial S_i}{\partial T}\right)_{\{T_j\}}, \quad (i,j = 1, 2, \ldots, 6) \tag{P38}$$

and the thermal stress coefficients β_i by

$$\beta_i = \left(\frac{\partial T_i}{\partial T}\right)_{\{S_j\}}, \quad (i,j = 1, 2, \ldots, 6). \tag{P39}$$

Express α_i by means of the β_j and β_i by means of the α_j.

105. Show that

$$\alpha_i \to 0, \quad \beta_i \to 0 \quad \text{as} \quad T \to 0, \quad (i = 1, 2, \ldots, 6) \tag{P40}$$

can be deduced from the Nernst heat theorem.

106. The molar heat capacity at constant strain is defined by

$$C_{\{S_j\}} = T\left(\frac{\partial S}{\partial T}\right)_{\{S_j\}} \tag{P41}$$

and the molar heat capacity at constant stress by

$$C_{\{T_j\}} = T\left(\frac{\partial S}{\partial T}\right)_{\{T_j\}}. \tag{P42}$$

Use problem 104 to calculate the difference $C_{\{T_j\}} - C_{\{S_j\}}$.

107. Show that

$$\frac{\partial c_{ij}}{\partial T} \to 0 \quad \text{as} \quad T \to 0 \tag{P43}$$

follows from the Nernst heat theorem.

108. The isentropic elastic stiffness coefficients are

$$c_{ij}^{(S)} = \left(\frac{\partial T_i}{\partial S_j}\right)_{S,S_{k\neq j}}. \tag{P44}$$

Use the methods described in § 25 to derive a relationship between isentropic and isothermal elastic stiffness coefficients.

*109. Express the ratio $C_{\{T_j\}}/C_{\{S_j\}}$ in terms of isentropic and isothermal elastic stiffness coefficients [cf. eq. (41.25)].

§ 56

*110. Derive the matrix of the elastic stiffness coefficients for a tetragonal crystal.

*111. Derive the matrix of the elastic stiffness coefficients for a cubic crystal.

*112. Derive the matrix of the elastic compliance coefficients for a cubic crystal. Express the elastic compliance coefficients in terms of the elastic stiffness coefficients.

113. Cubic crystals whose building elements interact only by central forces obey the Cauchy relation

$$c_{12} = c_{44} \tag{P45}$$

which is borne out quite well by experiments on alkali metal halides. Express the Cauchy relation and the isotropy condition in terms of the elastic compliance coefficients. How does an isotropic substance which obeys Cauchy's relation behave?

114. Express the isothermal compressibility
(a) in terms of the elastic compliance coefficients for a triclinic crystal,
(b) in terms of the elastic stiffness coefficients for a cubic crystal,
(c) in terms of Lamé coefficients, Young's modulus, and Poisson's ratio for an isotropic substance.

115. What conclusions can be drawn from the stability conditions
(a) for the elastic stiffness coefficients of a triclinic crystal,
(b) for the elastic stiffness coefficients of a tetragonal crystal,
(c) for the elastic stiffness coefficients of a cubic crystal,
(d) for the Lamé coefficients?

116. What conclusions can be drawn from the stability conditions
(a) for the elastic compliance coefficients of a tetragonal crystal,
(b) for the elastic compliance coefficients of a cubic crystal,
(c) for Young's modulus and Poisson's ratio.

117. Heat is evolved when rubber is stretched isothermally, whereas rubber contracts on heating (Gough–Joule effect).

Make the simplifying assumption that all strain components except the change in length can be ignored and explain this effect.

CHAPTER IX

§ 58

*118. What can be deduced from the stability conditions
(a) for the (scalar) electric susceptibility,
(b) for the dielectric constant.

§ 59

119. Calculate to the order $\boldsymbol{E}^2$ the dependence of
(a) the partial molar entropy,
(b) the partial molar enthalpy,
(c) the partial molar volume on the electric field strength.
(d) Use the solution of (c) to derive eq. (60.3).

120.
(a) Derive the differential equation of the co-existence curve in the T–$\boldsymbol{E}$-plane ($P = \text{const.}$) for a two-phase equilibrium in a one-component system. Assume that the phase boundary has the same direction as $\boldsymbol{E}$.
(b) Integrate the equation between 0 and $\boldsymbol{E}$. Assume that $(T - T_0)/T_0 \ll 1$ for the fractional displacement of the equilibrium temperature by the electric field and that, therefore, the temperature dependence of the molar heat capacities and their dependence on the electric field strength can be neglected.
(c) Use eq. (59.9) to derive the same result.

121.
(a) For a binary vapour–liquid equilibrium derive differential equations which describe the dependence of the change in composition of the two phases on the field strength at $T = \text{const.}$, $P = \text{const.}$
(b) What conclusions can be drawn from the application of the equations derived in (a) to a mixture which is azeotropic at $\boldsymbol{E} = 0$.

122. Find the dependence of the molar heat capacity $C_{P,E}$ on the electric field strength. Assume that the Debye equation

$$v\chi \equiv \bar{\chi} = \bar{\chi}_0 + \frac{[\mu]^2}{3RT} \tag{P46}$$

where $\bar{\chi}_0$ is independent of temperature and $[\mu]$ is the molar dipole moment, can be applied to the electric susceptibility.

123. Investigate the effect of an electric field on a chemical equilibrium in an ideal gaseous mixture.

124. Find the dependence of the molar heat capacity $C_{P,H}$ on $\boldsymbol{H}$. Assume the validity of the Curie–Weiss law (66.6).

125. Use the solution to problem 124 to derive an explicit expression for cooling by adiabatic demagnetization. Assume that $\theta \approx 0$ and that

$$C_{P,H}(T, 0) = \frac{A}{T^2}. \tag{P47}$$

126. Use the equations of § 67 to derive the Ehrenfest equations (48.4) and (48.5) for the setting in of superconduction at vanishing magnetic field strength.

Hints for solving the problems

*2. Use eq. (3.2) in the differential form. Divide the work done into two steps, one at constant x, the other at constant y. Expressions for X and Y are thus obtained. When these are differentiated and then subtracted, eq. (P3) is obtained.

*5. Use the equation

$$P = \tfrac{1}{3}\varepsilon, \tag{H1}$$

obtained from electrodynamics, for the radiation pressure. Then use the expression (§ 8)

$$\mathrm{d}Q = \left(\frac{\partial U}{\partial T}\right)_V \mathrm{d}T + \left[\left(\frac{\partial U}{\partial V}\right)_T + P\right] \mathrm{d}V \tag{H2}$$

to derive an explicit equation for $\mathrm{d}Q$. Find the integrating factor $\mathrm{N}(T)$ for $\mathrm{d}Q$ and calculate the efficiency of the Carnot cycle.

6. Start from eq. (4.17) and represent S as a function of T and V. Since $\mathrm{d}S$ is a complete differential, the mixed second derivatives must be the same. Equation (P5) is then obtained directly.

10. If T is not a reversible one–one, i.e. strictly monotonic, function of t, there must be at least one pair of values t_1, t_2 with $t_1 > t_2$ such that $T(t_1) = T(t_2)$. Now use the Carnot process to show that eq. (4.14) is not obeyed under these circumstances and that the entropy is therefore not a function of state, which contradicts the Second Law.

*11. Let the given process correspond to n Carnot processes. Formulate the statement to be proved and show that it is equivalent to the statement: the sum of the areas of the curved triangles by which the two processes differ vanishes for $n \to \infty$. For the proof, use the argument that the curvatures can be neglected for sufficiently large n and that neighbouring adiabatics may, therefore, be regarded as being parallel.

*12. Stokes' theorem for a vector $\boldsymbol{A}$ is

$$\oint \boldsymbol{A} . \mathrm{d}\boldsymbol{s} = \int \operatorname{rot} \boldsymbol{A} . \mathrm{d}\boldsymbol{f} \tag{H3}$$

where d$\boldsymbol{s}$ is the line element and d$\boldsymbol{f}$ the vectorial area element. The integration on the left-hand side must be carried out over a closed curve, the integral on the right over the area bounded by this curve. Use (3.6) to transform the integral $\oint(\mathrm{d}Q/T)$ stretching over an arbitrary reversible cyclic process in the P–V-plane into the form of the left-hand side of eq. (H3). Now apply Stokes' theorem and use the continuity properties of thermodynamic functions (homogeneous system!) and eq. (4.14) to show that the integrand vanishes everywhere in the P–V-plane. Equation (4.15) follows immediately.

13. Define efficiency as

$$\eta_D = -\frac{W}{Q_{23}}. \tag{H4}$$

Calculate W from the part processes and remember that the heat Q_{23} is introduced at constant pressure. Eliminate the temperatures with the aid of the equation of state (P1).

14. Calculate the entropies of the substances in the initial and final state by means of eq. (4.17). Calculate the total entropy change using (4.20) and (4.21). In order to show that $\Delta S > 0$, first prove the auxiliary theorem

$$\alpha A + \beta B > A^{\alpha} B^{\beta} \quad \text{for} \quad \begin{cases} \alpha+\beta = 1,\ A \neq B \\ 0 < \alpha,\ \beta < 1,\ A, B > 0. \end{cases} \tag{H5}$$

Use (H5) in the expressions obtained for T_3 and for ΔS.

15. Return the system reversibly to the original entropy surface and then use the First Law.

*21. Divide the path into three parts, such that y = const. along the first part, z = const. along the second, and z = const. also along the third.

*23. Consider the change from empirical variables of state s, t measured on an arbitrary scale to the values s^*, t^* measured on a different scale. Let the two scales be related by the expressions

$$s = s(s^*), \quad t = t(t^*). \tag{H6}$$

Write eqs. (11.5) and (11.8) for the asterisked variables and express the Jacobi determinant of the right-hand side of (11.8) in terms of the Jacobi determinant of the original variables.

*24. Refer to theorem 4 of § 13 and show that the half volumes generated by an adiabatic may be distinguished physically by means of the empirical entropy for the special case of an ideal gas even though they are generally indistinguishable. The sign of Φ, and therefore that of τ, is thus clearly fixed for the present case as can be seen from (10.23), (11.10), and (11.20).

*25.

(a) Use eq. (10.3) and the definition preceding eq. (13.1),

(b) Use eqs. (10.3)–(10.5) together with the theorem of experience formulated at the end of § 13.

28.

(a) Use the fact that the entropy is a complete differential.

30. First derive the equations of state (20.22) and (20.23) in the entropy representation. Use these to eliminate the intensive parameters of the entropy representation

(a) from eq. (20.42).

(b) from eq. (20.46).

32.

(a) Use (P5).

(b) Use (P18).

(c) Use (P1) and (P6).

*37.

(b) Eliminate the volume by means of the approximate expression

$$Pv = RT + BP \equiv RT + \left(b - \frac{a}{RT}\right) P. \tag{H7}$$

*50. Use

$$\Delta T = \frac{dT}{dt}\Delta t, \quad \alpha' = \alpha\frac{dT}{dt}, \quad C'_P = C_P\frac{dT}{dt} \tag{H8}$$

and integrate eq. (25.36) between the ice point T_0, or t_0, and the unknown thermodynamic temperature T. Since the difference between the ice point and the steam point is 100°, T_0 can be eliminated by means of the integral to this steam point.

*51.

(a) In order to obtain comparable forms, introduce the pressure change dP corresponding to the volume change dv in the Joule–Thomson experiment.

(b) Derive the expression for the Joule–Thomson coefficient from eqs. (21.15) and (20.20).

(c) Use (25.9), (24.24), and (25.36).

Remember for purposes of comparison that, in general, $[\partial(Pv)/\partial P]_h > 0$.

*57. Use (25.3) and (26.16).

*61.

(a) Derive the analogues to the Clausius–Clapeyron equation for the pairs of variables μ/T, $1/T$ and μ, T. Remember that the finite difference quotients on the right-hand sides become differential quotients for neighbouring phases.

(b) Apply the condition $\mu/T = 0$ and the result of (a) to the infinite number of Gibbs–Duhem equations.

*70. Put

$$y = T, \quad X = L, \quad x = n$$

$$Y = n(C'_s - C''_s) \tag{H9}$$

in the form of eq. (P3) valid for quasi-static processes. Remember that the saturation pressure is independent of n. Use (A117) and the Clausius–Clapeyron equation.

*71. Introduce the standard pressure P^+ and eliminate L_0/T_0 by means of (31.5), (31.7), and (A50).

*72. Start with the differential form of the equilibrium conditions (27.8) and represent the chemical potentials as functions of T, P, and the x_i. The derivatives of μ_2 with respect to the mole fractions are eliminated by using (26.16).

*74. Derive first a differential equation for the change in composition of the azeotropic mixture with temperature. Start with the equations mentioned in the hints for problem 72 and use the condition

$$x'_1 = x''_1 \tag{H10}$$

for the azeotropic point. Find the second derivative of the isobaric boiling point curve at the point (H10) and thus deduce a criterion which shows whether the boiling point curve has a maximum or a minimum at the azeotropic point. Simplify the differential equation derived first by introducing the partial heats of vaporization

$$L_1 = h'_1 - h''_1 > 0, \quad L_2 = h'_2 - h''_2 > 0 \tag{H11}$$

and by using the assumptions mentioned in the problem.

*81. Consider on the one hand a reaction at the temperature T followed by heating of the reaction products to the temperature $T+\mathrm{d}T$ and, on the other, heating of the reactants from T to $T+\mathrm{d}T$ followed by reaction at $T+\mathrm{d}T$. Take the work done into consideration for the cases (b) and (c). In case (c), replace the pressure change by a constant mean value. Use the progress variable ξ defined by eq. (14.3) as the variable for the chemical reaction.

*82. In the form of eq. (P3) valid for quasi-static processes, put

$$y = T, \quad X = Q, \quad x = \xi$$

$$Y = \sum_i \nu_i \xi_i C_i \tag{H12}$$

where the summation is to be taken over the participants in the reaction with the sign convention of eq. (33.15), and the C_i denote the molar heat capacities for the conditions given in problem 81.

*85.

(a) Start with eq. (A31) with $x \equiv P$. Remember that $(\partial u/\partial v)_T \neq 0$ for the dissociating ideal gas. Calculate $(\partial \alpha/\partial v)_T$ from (A143) and $(\partial v/\partial T)_P$ from the thermal equation of state of the dissociating gas, (A142), and (34.29). Take (34.30) and (34.31) into consideration.

(b) Use (A146) and (A150).

98. First carry out the calculation for $V = \text{const.}$ and use the general theory to show that the result is also valid for $P = \text{const.}$

*99. Represent $\delta^2 u$ as a sum of squares by the method of completion of squares.

*100. Apply the method of problem 99 to the general case and calculate λ_1 assuming that the sequence of independent variables starts with s.

*109. Use the generalization of (25.19) for more than two variables.

*110. A tetragonal crystal has a four-fold axis of symmetry. If this is chosen as the z-axis, the transformation of coordinates corresponding to this symmetry property is

$$x \to -y, \quad y \to x, \quad z \to z.$$

The derivation is analogous to that of § 56 for a monoclinic crystal.

*111. A cubic crystal has three mutually perpendicular four-fold axes of symmetry.

*112. Use eq. (55.11) and the expression

$$\boldsymbol{A}^{-1} = \frac{\hat{\boldsymbol{A}}}{|A|} \tag{H17}$$

where the matrix $\hat{\boldsymbol{A}}$ is formed by substituting for each element A_{ij} of the matrix $\boldsymbol{A}$ the cofactor of A_{ij} from the determinant $|A|$ and then transposing the matrix so obtained. Remember that $|c|$ is a step determinant.

113. Express the elastic stiffness coefficients in terms of the elastic compliance coefficients (i.e. reverse problem 112).

114. Use (52.25), (55.12), and (56.26).

115. Use eq. (55.2).

116. Use eq. (55.5).

Put the variables in a sequence such that the temperature comes last.

117. Use eq. (55.2).

*118.

(b) The energy of the electric field in the dielectric must be included in the fundamental equation in order to obtain any statement about the dielectric constant. The second term of (57.19) must therefore be introduced into the fundamental equation and the stability conditions must then be applied. Use a thermodynamic potential analogous to (59.4).

121.

(a) Express the chemical potential of component 1 as a function of T, P, $\mathbf{E}$, and x_1. Use the differential form of the equilibrium condition (27.8).

(b) Find the limiting value of

$$\frac{\mathrm{d}x_1''}{\mathrm{d}\mathbf{E}}\Big/\frac{\mathrm{d}x_1'}{\mathrm{d}\mathbf{E}}$$

for

$$x_1' - x_1'' \to 0.$$

126. Use eqs. (67.14) and (67.15).

Solutions to the problems

1.

(a)
$$\Delta v_{\text{I}} = \Delta v_{\text{II}} = \frac{R(T_2 P_1 - T_1 P_2)}{P_1 P_2}, \tag{A1}$$

$$W_{\text{I}} = R\left(\frac{T_2 - T_1}{P_2 - P_1} P_1 - T_1\right) \ln \frac{P_1}{P_2}, \tag{A2}$$

$$W_{\text{II}} = -R\left(T_2 - T_1 + T_1 \ln \frac{P_1}{P_2}\right). \tag{A3}$$

(b) $\mathrm{d}v$ is an exact differential and v is thus a function of state (§ 2). $\mathrm{d}W$ is not an exact differential; W is thus not a function of state.

3.

$$\frac{\partial X}{\partial y} - \frac{\partial Y}{\partial x} = \left(\frac{\partial V}{\partial x}\right)_y \left(\frac{\partial P}{\partial y}\right)_x - \left(\frac{\partial V}{\partial y}\right)_x \left(\frac{\partial P}{\partial x}\right)_y. \tag{A4}$$

*5. The efficiency is obtained as

$$\eta = 1 - \left(\frac{T_2}{T_1}\right)^{\frac{1}{4}}. \tag{A5}$$

Use of Carnot's theorem then shows the possibility of a perpetual motion machine of the second kind.

7.

(a)
$$PV^{(R/C_V)+1} = \text{const.} \tag{A6}$$

(b)
$$\left(P + \frac{a}{v^2}\right)(v - b)^{(R/C_V)+1} = \text{const.} \tag{A7}$$

The assumption made in problem 7(a) and (b) that C_V is independent of temperature can be valid only for certain temperature ranges because of the excitation of internal degrees of freedom. Details can be found in textbooks of statistical thermodynamics.

8. A state P_2, V_2 can be reached from a state P_1, V_1 along an adiabatic and along an isotherm if the adiabatic and the isotherm through P_1, V_1 intersect in at least one other point. Equations (P1) and (A6) have, however, exactly one solution for P and V.

9.
$$\eta = \frac{T_1 - T_2}{T_1}. \tag{A8}$$

*11. For the sum of the Carnot processes the contributions of the internal curve parts vanish since they are travelled each time twice but in opposite directions. The statement to be proved is, therefore,

$$\lim_{n\to\infty}\left\{\oint_C \frac{dQ}{T} - \oint_{\hat{C}} \frac{dQ}{T}\right\} = 0 \tag{A9}$$

where C is the curve of the given cyclic process and $\hat{C}$ is the zig-zag curve which encloses the Carnot processes. (A9) is equivalent to the statement

$$\lim_{n\to\infty} \sum_1^{2n} \oint_\triangle \frac{dQ}{T} = 0 \tag{A10}$$

where $\oint_\triangle$ denotes the integral extending over the bounds of one of the curved triangles.

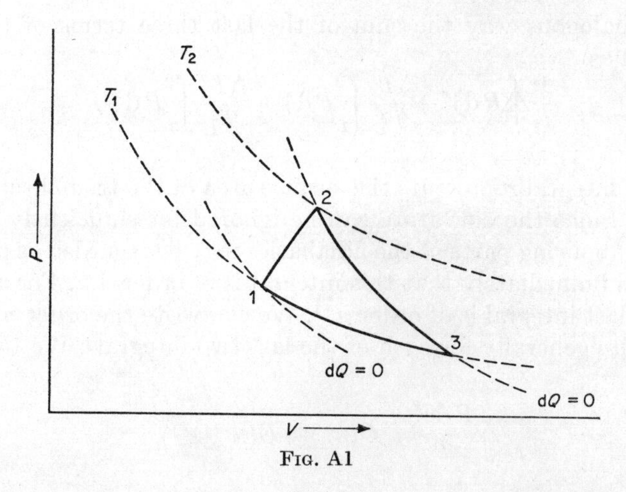

FIG. A1

Use of eq. (3.6) (the reversible character of the cyclic process is now being used!) gives

$$\oint_\triangle \frac{dQ}{T} = \int_1^2 \frac{dU}{T} + \int_2^3 \frac{dU}{T} + \int_3^1 \frac{dU}{T} + \int_1^2 \frac{P\,dV}{T} + \int_2^3 \frac{P\,dV}{T} + \int_3^1 \frac{P\,dV}{T}. \tag{A11}$$

The sum of the first three integrals (this does not vanish since $\mathrm{d}U/T$ is not an exact differential) may be written

$$\frac{1}{T'}(U_2-U_1)-\frac{1}{T_1}(U_3-U_1)-\frac{1}{T''}(U_2-U_3) \tag{A12}$$

or, for sufficiently large n,

$$\frac{1}{T_1}(U_2-U_1)-\frac{\Delta T}{T_1^2}(U_2-U_1)-\frac{1}{T_1}(U_3-U_1)-\frac{1}{T_1}(U_2-U_3)$$
$$+\frac{\Delta T''}{T_1^2}(U_2-U_3) = -\frac{\Delta T'}{T_1^2}(U_2-U_1)+\frac{\Delta T''}{T_1^2}(U_2-U_3). \tag{A13}$$

$\Delta T'$ and $\Delta T''$ are of order n^{-1}. The differences between the internal energies are also of order n^{-1} for an ideal gas. It is sufficient for the general case to state that they must be of order $n^{-\varepsilon}$ with $\varepsilon > 0$. We thus find that

$$\oint\frac{\mathrm{d}U}{T} = O(n^{-(1+\varepsilon)}). \tag{A14}$$

In an analogous way the sum of the last three terms of (A11) is obtained as

$$\frac{1}{T_1}\oint P\,\mathrm{d}V-\frac{\Delta T'}{T_1^2}\int_1^2 P\,\mathrm{d}V+\frac{\Delta T''}{T_1^2}\int_3^2 P\,\mathrm{d}V. \tag{A15}$$

The first integral represents the surface area of the triangle shown in Fig. A1. Since the curvature can be ignored for sufficiently large n and neighbouring parts of the adiabatics may be regarded as parallel, it follows immediately that this integral is of order n^{-2}. For an ideal gas, the last integral is of order n^{-1}. We can write the order $n^{-\varepsilon'}$ with $\varepsilon' > 0$ quite generally for each of the last two integrals. We therefore get

$$\oint\frac{P\,\mathrm{d}V}{T} = O(n^{-2})+O(n^{-(1+\varepsilon')}). \tag{A16}$$

The statement (A10) then follows from (A11), (A14), and (A16).

*12. We have

$$\oint\frac{\mathrm{d}Q}{T} = \oint\frac{1}{T}\left\{\left[\left(\frac{\partial U}{\partial V}\right)_P+P\right]\mathrm{d}V+\left(\frac{\partial U}{\partial P}\right)_V\mathrm{d}P\right\} \equiv \oint[A\,\mathrm{d}V+B\,\mathrm{d}P]. \tag{A17}$$

Stokes' theorem gives

$$\oint [A\,\mathrm{d}V + B\,\mathrm{d}P] = \iint \left[\frac{\partial A}{\partial P} - \frac{\partial B}{\partial V}\right] \mathrm{d}V\,\mathrm{d}P, \tag{A18}$$

and the equation

$$\frac{\partial A}{\partial P} - \frac{\partial B}{\partial V} = 0 \tag{A19}$$

must apply over the whole of the P–V-plane. Suppose that this were not so at, for example, a point ξ. Because of the continuity of thermodynamic functions, there would have to be a region round ξ in which $\partial A/\partial P - \partial B/\partial V \neq 0$ and this inequality would have to have the same sign throughout this region. We could, therefore, construct a Carnot cycle in this region. For this Carnot cycle, eqs. (A17) and (A18) would give

$$\oint \frac{\mathrm{d}Q}{T} \neq 0 \tag{A20}$$

which contradicts eq. (4.14). Equation (A19) must, therefore, be correct. Its correctness is easily verified explicitly by applying the equation to an ideal gas. We can now use eqs. (A19), (A17), and (A18) to obtain (4.15) directly.

13.

$$\eta_D = 1 - \frac{1}{\kappa (V_1/V_2)^{\kappa-1}} \frac{(V_3/V_2)^{\kappa} - 1}{(V_3/V_2) - 1} \tag{A21}$$

with

$$\kappa = \frac{C_P}{C_V}. \tag{A22}$$

*14.

$$T_3 = \frac{C_1 T_1 + C_2 T_2}{C_1 + C_2}. \tag{A23}$$

$$\Delta S = \ln \frac{T_3^{C_1+C_2}}{T_1^{C_1} T_2^{C_2}} = \ln \left[\left(\frac{C_1}{C_1+C_2} + \frac{C_2}{C_1+C_2}\frac{T_2}{T_1}\right)^{C_1} \left(\frac{C_2}{C_1+C_2} + \frac{C_1}{C_1+C_2}\frac{T_1}{T_2}\right)^{C_2}\right]. \tag{A24}$$

To prove the auxiliary theorem, prove that the function

$$F(\alpha) = (a-1)\alpha + 1 - a^{\alpha} \tag{A25}$$

is nowhere zero for $0 < \alpha < 1$ and $a > 0$, i.e. that in this region there is no point of intersection between the straight line $(a-1)\alpha$ and the curve $a^{\alpha} - 1$ (Figure!). The desired equation then follows from a consideration of the second derivative.

Application to (A23) gives

$$T_3 = \frac{C_1}{C_1+C_2}T_1 + \frac{C_2}{C_1+C_2}T_2 > T_1^{C_1/(C_1+C_2)}\,T_2^{C_2/(C_1+C_2)} \tag{A26}$$

or

$$\frac{T_3^{C_1+C_2}}{T_1^{C_1}T_2^{C_2}} > 1 \tag{A27}$$

and, therefore, with (A24), $\Delta S > 0$.

15. If an amount $\mathrm{d}Q'$ of Joule heat is generated in the system under conditions of adiabatic insulation the representative point of the system after internal equilibrium is attained is situated on an entropy surface which differs by an amount $\mathrm{d}S_i$ from the original entropy surface. In order to return the point to the original entropy surface, the adiabatic insulation must be removed and an amount of heat

$$\mathrm{d}Q = T\,\mathrm{d}S_i \tag{A28}$$

must be removed reversibly. Since no work is done, the First Law states that

$$|\mathrm{d}Q| = \mathrm{d}Q'. \tag{A29}$$

(4.27) then follows from (A28) and (A29).

16.

(b)
$$C_P = \left(\frac{\partial h}{\partial T}\right)_P. \tag{A30}$$

17.

(a)
$$C_x = \left(\frac{\mathrm{d}Q}{\mathrm{d}T}\right)_x = C_V + \left[\left(\frac{\mathrm{d}u}{\mathrm{d}v}\right)_T + P\right]\left(\frac{\mathrm{d}v}{\mathrm{d}T}\right)_x. \tag{A31}$$

(b)
$$C_x = \left(\frac{\mathrm{d}Q}{\mathrm{d}T}\right)_x = C_P + \left[\left(\frac{\mathrm{d}h}{\mathrm{d}P}\right)_T - v\right]\left(\frac{\mathrm{d}P}{\mathrm{d}T}\right)_x. \tag{A32}$$

18.

$$\frac{\partial}{\partial t}\left[\left(\frac{\partial U}{\partial V}\right)_T + P\right] \neq \frac{\partial^2 U}{\partial V\,\partial t}. \tag{A33}$$

19.

(a)
$$\left(\frac{\partial y}{\partial y}\right)_x \neq -\left(\frac{\partial(y+x)}{\partial x}\right)_y. \tag{A34}$$

(b) If (9.18) is reduced to two variables, e.g. by putting $k = j$, the equation is always obeyed.

(c)
$$\tau = y^2, \tag{A35}$$

$$\sigma(x, y) = \frac{x}{y} - \ln y. \tag{A36}$$

20.

(a)
$$\frac{\partial X}{\partial y} - \frac{\partial Y}{\partial x} = -\frac{\partial y}{\partial y} - \frac{\partial x}{\partial x} \neq 0$$

$$\frac{\partial Y}{\partial z} - \frac{\partial Z}{\partial y} = \frac{\partial x}{\partial z} - \frac{\partial k}{\partial y} = 0 \tag{A37}$$

$$\frac{\partial Z}{\partial x} - \frac{\partial X}{\partial z} = \frac{\partial k}{\partial x} + \frac{\partial y}{\partial z} = 0.$$

(b)
$$k\left(\frac{\partial x}{\partial x} + \frac{\partial y}{\partial y}\right) + x\left(-\frac{\partial y}{\partial z} - \frac{\partial k}{\partial x}\right) - y\left(\frac{\partial k}{\partial y} - \frac{\partial x}{\partial z}\right) \neq 0. \tag{A38}$$

21. Part 1:

$$y = y_0, \quad \frac{\mathrm{d}z}{\mathrm{d}x} = \frac{y_0}{k}, \tag{A39}$$

$$z - z_0 = \frac{y_0}{k}(x - x_0).$$

This part leads to the point x^*, y_0, z_1.
Part 2:

$$z = z_1, \quad \frac{\mathrm{d}y}{\mathrm{d}x} = \frac{y}{x}, \tag{A40}$$

$$y = \frac{y_0}{x^*} x.$$

The point $0, 0, z_1$ has now been reached.
Part 3:

$$z = z_1, \quad \frac{\mathrm{d}y}{\mathrm{d}x} = \frac{y}{x} \tag{A41}$$

$$y = \frac{y_1}{x_1} x_1.$$

This part ends at the point x_1, y_1, z_1. $\mathrm{d}Q = 0$ along every part.

22.

$$\frac{dQ}{T} = \frac{1}{T}\left(\frac{\partial U}{\partial T}\right)_V dT + \frac{1}{T}\left[\left(\frac{\partial U}{\partial V}\right)_T + P\right] dV = \frac{1}{T}\frac{\partial U}{\partial T} dT + \frac{P}{T} dV \tag{A42}$$

$$\frac{1}{T}\frac{\partial^2 U}{\partial V\,\partial T} = -\frac{P}{T^2} + \frac{1}{T}\left(\frac{\partial P}{\partial T}\right)_V = 0.$$

*23.

$$\frac{\partial(V,P)}{\partial(s^*,t^*)} = \frac{\partial(V,P)}{\partial(s,t)}\frac{ds}{ds^*}\frac{dt}{dt^*} \tag{A43}$$

which leads to

$$\frac{dT^*}{dt^*} = \frac{dT}{dt}\frac{dt}{dt^*}, \quad \frac{dS^*}{ds^*} = \frac{dS}{ds}\frac{ds}{ds^*} \tag{A44}$$

and, therefore,

$$T^* = T, \quad S^* = S$$

$$U^*(s^*, t^*) = U(s, t). \tag{A45}$$

24. The empirical entropy of an ideal gas is defined by (11.14). The empirical temperature defined by (11.13) remains constant during the Gay–Lussac experiment. The empirical entropy s therefore increases. (11.21) then shows that the metrical entropy also increases. The general validity of the result follows from theorem 4.

25.

(a) No general statements about the sign of τ can be derived from Carathéodory's principle. $\tau'/\tau = -\tau''/\tau$ could, therefore, be true; for non-static adiabatic processes of the part systems (according to the definition $d\sigma' > 0$, $d\sigma'' > 0$) $d\sigma = 0$ could apply.

(b) The theorem of experience shows that $\tau > 0, \tau' > 0, \tau'' > 0$. According to (10.4) and (10.5), σ is a monotonic function of σ' and σ'' with the same sign. $d\sigma' > 0$, $d\sigma'' > 0$ therefore implies $d\sigma > 0$ according to (10.3).

26.

(a), (b)

$$s = R(\tfrac{3}{2}\ln u + \ln v) + s_0. \tag{A46}$$

27.

(a)

$$Tv^{\kappa-1} = \text{const.}, \tag{A47}$$

$$\frac{P^{(\kappa-1)/\kappa}}{T} = \text{const.} \tag{A48}$$

(b)

$$s = R\ln v + s_0, \tag{A49}$$

$$s = -R\ln P + s_0'. \tag{A50}$$

28.

(b)
$$S = \tfrac{4}{3}aT^3 V. \tag{A51}$$

29. The form (P15) is impossible since it is not a homogeneous function of the first degree of the extensive parameters.

30.

(a)
(b)
$$U = \frac{cS^3}{anV}. \tag{A52}$$

32.

(a)
$$\left(\frac{\partial C_V}{\partial v}\right)_T = T\left(\frac{\partial^2 P}{\partial T^2}\right)_v. \tag{A53}$$

(b)
$$\left(\frac{\partial C_P}{\partial P}\right)_T = -T\left(\frac{\partial^2 v}{\partial T^2}\right)_P. \tag{A54}$$

(c) Ideal gas:
$$\left(\frac{\partial C_V}{\partial v}\right)_T = \left(\frac{\partial C_P}{\partial P}\right)_T = 0. \tag{A55}$$

Van der Waals gas:
$$\left(\frac{\partial C_V}{\partial v}\right)_T = 0. \tag{A56}$$

33.

(a)
$$f = -RT(\ln v + \tfrac{3}{2}\ln T) + T\,.\,\text{const.} \tag{A57}$$

34.
$$f = -RT\left[\ln(v-b) + \frac{a}{RTv}\right] + \phi(T). \tag{A58}$$

35.
$$f = -2\left(\frac{a}{27c}\right)^{\frac{1}{4}} T^{\frac{3}{4}} v^{\frac{1}{4}} + T\,.\,\text{const.} \tag{A59}$$

36.

(a) Ideal gas:
$$\left(\frac{\partial U}{\partial V}\right)_T = 0, \quad \left(\frac{\partial C_V}{\partial v}\right)_T = 0. \tag{A60}$$

(b) Van der Waals gas:
$$\left(\frac{\partial U}{\partial V}\right)_T = -\frac{a}{v^2}, \quad \left(\frac{\partial C_V}{\partial v}\right)_T = 0. \tag{A61}$$

Cf. problem 32.

*37.

(a) Ideal gas: $g = RT(\ln P - \frac{5}{2}\ln T) + T\,.\,\text{const.}$ (A62)

(b) Van der Waals gas:

$$g = RT\left[\ln P - \ln T - \ln\left(1 - \frac{aP}{(RT)^2}\right)\right]$$

$$+\left[\left(b - \frac{a}{RT}\right) - \frac{a}{RT + [b - (a/RT)]\,P}\right] P + \phi(T). \quad \text{(A63)}$$

38. $$P(v,T) = \gamma(v)\,T - \frac{\partial A}{\partial v} \equiv \gamma(v)\,T - A'(v) \quad \text{(A64)}$$

where $\gamma(v)$ is an unknown function.

$$s(T,v) = \int \frac{B'(T)}{T}\,\mathrm{d}T + \int \gamma(v)\,\mathrm{d}v + s_0, \quad \text{(A65)}$$

$$h(T,v) = A(v) + B(T) + Tv\gamma(v) - A'(v)\,v, \quad \text{(A66)}$$

$$f(T,v) = A(v) - T\int \gamma(v)\,\mathrm{d}v - T\int \frac{B(T)}{T^2}\,\mathrm{d}T + Ts_0, \quad \text{(A67)}$$

$$C_V = B'(T), \quad \text{(A68)}$$

$$C_P = B'(T) - \frac{T\gamma^2(v)}{T\gamma'(v) - A''(v)}. \quad \text{(A69)}$$

Since (P19) is not a characteristic function, no characteristic functions can be derived from it. Although $f(T,v)$ is formally a characteristic function, it contains the unknown function $\gamma(v)$.

41.

(a) $$\left(\frac{\partial V}{\partial U}\right)_F = \frac{\kappa S}{S(T\alpha - P\kappa) - P(nC_P\kappa - \alpha^2 VT)}. \quad \text{(A70)}$$

(b) $$\left(\frac{\partial P}{\partial U}\right)_G = \frac{S}{V[nC_V + (V\alpha - S\kappa)\,(T(\alpha/\kappa) - P)]}. \quad \text{(A71)}$$

(c) $$\frac{C_P}{C_V} = \frac{\kappa C_P}{\kappa C_P - Tv\alpha^2}. \quad \text{(A72)}$$

42. $$\alpha = \frac{1}{v}\,\frac{Rv^3(v-b)}{RTv^3 - 2a(v-b)^2}. \quad \text{(A73)}$$

44.
$$\left(\frac{\partial H}{\partial V}\right)_T = \frac{\delta C_P}{v\kappa}. \qquad (A74)$$

45.

(a)
$$P = Tf(V). \qquad (A75)$$

(b)
$$V = Tg(P). \qquad (A76)$$

(c)
$$PV = \text{const}.T = nRT. \qquad (A77)$$

46.
$$\left(\frac{\partial v}{\partial T}\right)_{P_{\text{inv}}} = \frac{v_{\text{inv}}}{T_{\text{inv}}}. \qquad (A78)$$

47.
$$\delta = \frac{(2a/RT)-b}{C_{P0}+(2a/RT^2)P} \qquad (A79)$$

where C_{P0} is the limiting value of C_P for $P \to 0$.

$$T_{\text{inv}} = \frac{2a}{Rb} = 2T_B. \qquad (A80)$$

48.

(a)
$$\delta = \frac{(2a/RT)-b-(3ab/(RT)^2)P}{C_{P0}+(2a/RT^2)P-(3ab/R^2T^3)P^2} \qquad (A81)$$

where C_{P0} is the limiting value of C_P for $P \to 0$.

(b)
$$P_{\text{inv}} = \tfrac{1}{3}RT_{\text{inv}}\left(\frac{2}{b}-\frac{RT_{\text{inv}}}{a}\right). \qquad (A82)$$

49.

(a)
$$a = \frac{27}{64}\frac{R^2T_c^2}{P_c}, \quad b = \frac{1}{8}\frac{RT_c}{P_c}. \qquad (A83)$$

(b)
$$P^*_{\text{inv}} = \tfrac{16}{3}(T^*_{\text{inv}} - \tfrac{4}{27}T^{*2}_{\text{inv}}). \qquad (A84)$$

*50.
$$T = \frac{100.e^J}{e^{J^*}-1}, \qquad (A85)$$

$$J = \int_{t_0}^{t} \frac{v\alpha'}{v+C_P\delta'}\,dt \qquad (A86)$$

where primed quantities are measured on the empirical scale and J^* is the integral (A86) with the normal boiling point of water as the upper limit.

*51.

(a)
$$\left(\frac{\partial T}{\partial v}\right)_u = -\frac{1}{C_V}\left(\frac{\partial u}{\partial v}\right)_T, \tag{A87}$$

$$\mathrm{d}P = \left(\frac{\partial P}{\partial v}\right)_h \mathrm{d}v, \tag{A88}$$

$$\left(\frac{\partial T}{\partial v}\right)_u \frac{\mathrm{d}v}{\mathrm{d}P} = -\frac{1}{C_V}\left(\frac{\partial u}{\partial v}\right)_T \left(\frac{\partial v}{\partial P}\right)_h. \tag{A89}$$

(b)
$$\left(\frac{\partial T}{\partial P}\right)_h \equiv \delta = -\frac{1}{C_V}\left(\frac{\partial u}{\partial v}\right)_T \left(\frac{\partial v}{\partial P}\right)_h - \frac{1}{C_V}\left[\frac{\partial (Pv)}{\partial P}\right]_h. \tag{A90}$$

(c)
$$\left(\frac{\partial T}{\partial P}\right)_s = \frac{T}{C_P}\left(\frac{\partial v}{\partial T}\right)_P = \left(\frac{\partial T}{\partial P}\right)_h + \frac{v}{C_P}. \tag{A91}$$

Cooling during the Gay–Lussac experiment depends only on intermolecular forces (and therefore disappears for an ideal gas). In the Joule–Thomson experiment this effect is modified by the work done and is generally reduced. For gases it is always true that $(\partial T/\partial P)_s > 0$. In general, therefore, the cooling effect in the Gay–Lussac experiment and in a reversible adiabatic decompression is greater than that in the Joule–Thomson experiment.

52.
$$\delta_{\max} = \frac{T(\partial^2 v/\partial T^2)_P}{(\partial C_P/\partial T)_P}. \tag{A92}$$

53.
$$\Delta T = \int_0^P \frac{C_P}{C_{P0}} \delta \, \mathrm{d}P \tag{A93}$$

where C_{P0} is the limiting value of C_P when $P \to 0$.

54. We have

$$\Delta T = -\frac{1}{C_{P0}} \int_0^P \left(\frac{\partial h}{\partial P}\right)_T \mathrm{d}P = -\frac{1}{C_{P0}} (h - h_0)_{T=\mathrm{const}}. \tag{A94}$$

The condition for the vanishing of the differential Joule–Thomson effect (problem 46) is

$$\left(\frac{\partial h}{\partial P}\right)_T = 0. \tag{A95}$$

The required statement then follows. (The extremum is a maximum.)

55.
$$(h - h_0)_{T=\mathrm{const}} = 0. \tag{A96}$$

The total cooling effect vanishes if the enthalpy at the initial temperature has the same value for the initial pressure and for $P \to 0$. The differential Joule–Thomson effect vanishes if, at the initial temperature, the enthalpy as a function of pressure passes through an extremum.

56. If P'_{inv} and T'_{inv} are the values of P and T for the inversion of the total cooling effect, we have

$$P'_{\text{inv}} = \frac{2RT'_{\text{inv}}}{3}\left(\frac{2}{b} - \frac{RT'_{\text{inv}}}{a}\right) \tag{A97}$$

and, therefore (cf. problem 48),

$$P'_{\text{inv}} = 2P_{\text{inv}}. \tag{A98}$$

*57.

$$\kappa = \frac{|B|}{RT \sum_{i=1}^{m} \sum_{j=1}^{m} c_i c_j |B|_{ij}}, \tag{A99}$$

$$v_i = \frac{\sum_{j=1}^{m} c_j |B|_{ij}}{\sum_{j=1}^{m} \sum_{k=1}^{m} c_j c_k |B|_{jk}} \tag{A100}$$

with $c_i \equiv n_i/V$.

58.

$$\phi_1^2 \phi_2 \left(\frac{\partial \Pi}{\partial \phi_2}\right)_{T,\mu_1} = c_1 x_2^2 \left(\frac{\partial \mu_2}{\partial x_2}\right)_{T,P}. \tag{A101}$$

59.

$$\left(\frac{\partial \Pi}{\partial c_i}\right)_{T,\mu_1,c_{k\neq i}} = RT \sum_{j=2}^{m} c_j \frac{|B'|_{ij}}{|B'|} \tag{A102}$$

where $|B'|$ is the determinant of degree $(m-1)$ formed from the elements B_{ij} with $i, j \neq 1$.

60.

$$\frac{\mathrm{d}\mu}{\mathrm{d}T} = -\frac{(S^{(\alpha)} - S^{(\beta)})/V}{\rho^{(\alpha)} - \rho^{(\beta)}} = -\frac{(H^{(\alpha)} - H^{(\beta)})/V}{T(\rho^{(\alpha)} - \rho^{(\beta)})}. \tag{A103}$$

*61.

(a) From

$$\frac{\mathrm{d}(\mu/T)}{\mathrm{d}(1/T)} = -\left(\frac{\partial(U/V)}{\partial \rho}\right)_{T,\mu/T}, \tag{A104}$$

$$\frac{\mathrm{d}\mu}{\mathrm{d}T} = -\left(\frac{\partial(S/V)}{\partial \rho}\right)_{T,\mu} \tag{A105}$$

we obtain

$$\left.\begin{aligned} U/V &= \text{const.} \\ S/V &= \text{const.} \end{aligned}\right\} \text{at } T = \text{const.} \tag{A106}$$

(b) For the co-existence of neighbouring phases an infinite number of equations of the form

$$\left.\begin{aligned}(S/V)'\mathrm{d}T-\mathrm{d}P-\rho'\mathrm{d}\mu &= 0\\ (S/V)''\mathrm{d}T-\mathrm{d}P-\rho''\mathrm{d}\mu &= 0\end{aligned}\right. \tag{A107}$$

is valid. The last term on the right-hand side vanishes since $\mu = 0$ at all temperatures in the region of co-existence. Because of (A106) the set of equations (A107) reduces to the one equation

$$\frac{\mathrm{d}P}{\mathrm{d}T} = \frac{S}{V}. \tag{A108}$$

The co-existence of an infinite number of phases therefore does not contradict the phase rule. Equation (A108) determines the temperature dependence of the 'saturation pressure' and therefore takes the place of the Clausius–Clapeyron equation.

62. $$\frac{\mathrm{d}Z}{\mathrm{d}T} = \left(\frac{\partial Z}{\partial T}\right)_P + \left(\frac{\partial Z}{\partial P}\right)_T \frac{\mathrm{d}P}{\mathrm{d}T} \tag{A109}$$

where $\mathrm{d}P/\mathrm{d}T$ is determined by the Clausius–Clapeyron equation.

63. $$\frac{\mathrm{d}(\Delta V^*)}{\mathrm{d}T} = \Delta(\alpha V^*) - \Delta(\kappa V^*)\frac{L}{T\Delta V^*}. \tag{A110}$$

64. If the quantities referring to the gas phase are distinguished by $'$ we find the approximate expression

$$\frac{\mathrm{d}V^{*\prime}}{\mathrm{d}T} = \frac{V^{*\prime}}{T}\left(1 - \frac{L}{RT}\right). \tag{A111}$$

65. $$\frac{\mathrm{d}L}{\mathrm{d}T} = \Delta C_P + \frac{L}{T} - \frac{L}{\Delta V^*}\Delta(\alpha V^*). \tag{A112}$$

66. $$\left.\begin{aligned}T\frac{\mathrm{d}S^{*\prime}}{\mathrm{d}T} &= C'_P - \frac{L}{\Delta V^*}\left(\frac{\partial V^{*\prime}}{\partial T}\right)_P\\ T\frac{\mathrm{d}S^{*\prime\prime}}{\mathrm{d}T} &= C''_P - \frac{L}{\Delta V^*}\left(\frac{\partial V^{*\prime\prime}}{\partial T}\right)_P\end{aligned}\right\}. \tag{A113}$$

67. $$T\frac{\mathrm{d}S^{*\prime}}{\mathrm{d}T} = C'_s, \quad T\frac{\mathrm{d}S^{*\prime\prime}}{\mathrm{d}T} = C''_s. \tag{A114}$$

68. Gas phase:
$$C'_s = C'_P - \frac{L}{T}, \tag{A115}$$

condensed phase:
$$C''_s = C''_P - \alpha'' \frac{PV^{*''}}{RT} L \approx C''_P. \tag{A116}$$

69.
$$C'_s = C'_P - T\left(\frac{\partial V^{*'}}{\partial T}\right)_P \left(\frac{\partial P}{\partial T}\right)_{n'} = C'_P - T\left(\frac{\partial V^{*'}}{\partial T}\right)_P \frac{\mathrm{d}P}{\mathrm{d}T} \tag{A117}$$

and an analogous expression for the condensed phase.

71.
$$\ln\frac{P}{P^+} = -\frac{L_0}{RT} + \frac{\Delta S^{*0}}{R} + \frac{\Delta C_P}{R}\left(\ln\frac{T}{T_0} + \frac{T_0}{T} - 1\right). \tag{A118}$$

P^+ is a fixed standard pressure (usually 1 atm), L_0 the molar heat of transition at the temperature T_0, ΔS^{*0} the change in molar entropy associated with the transition condensed phase→gas phase under standard conditions (pressure P^+, temperature T_0),† and

$$\Delta C_P \equiv C'_P - C''_P,$$

the difference between the molar heat capacities at constant pressure.

For $T_0/T \approx 1$ we find from (A118) that

$$\ln\frac{P}{P^+} = -\frac{L_0}{RT} + \frac{\Delta S^{*0}}{R}. \tag{A118a}$$

*72.

$$[x''_1(s'_1 - s''_1) + (1 - x''_1)(s'_2 - s''_2)]\,\mathrm{d}T - [x''_1(v'_1 - v''_1) + (1 - x''_1)(v'_2 - v''_2)]\,\mathrm{d}P = \frac{x''_1 - x'_1}{1 - x'_1}\left(\frac{\partial\mu_1}{\partial x_1}\right)'_{T,P} \mathrm{d}x'_1, \tag{A119}$$

$$[x'_1(s'_1 - s''_1) + (1 - x'_1)(s'_2 - s''_2)]\,\mathrm{d}T - [x'_1(v'_1 - v''_1) + (1 - x'_1)(v'_2 - v''_2)]\,\mathrm{d}P = \frac{x''_1 - x'_1}{1 - x''_1}\left(\frac{\partial\mu_1}{\partial x_1}\right)''_{T,P} \mathrm{d}x''_1. \tag{A120}$$

73.
$$\frac{\mathrm{d}P}{\mathrm{d}x'_1} = \frac{(x'_1 - x''_1)(\partial\mu_1/\partial x_1)'_{T,P}}{(1 - x'_1)[x''_1(v'_1 - v''_1) + (1 - x''_1)(v'_2 - v''_2)]}, \tag{A121}$$

$$\frac{\mathrm{d}P}{\mathrm{d}x''_1} = \frac{(x'_1 - x''_1)(\partial\mu_1/\partial x_1)''_{T,P}}{(1 - x''_1)[x'_1(v'_1 - v''_1) + (1 - x'_1)(v'_2 - v''_2)]}. \tag{A122}$$

† The quantity ΔS^{*0} has been tabulated for $P^+ = 1$ atm and $T_0 = 298{\cdot}15$ K.

We know that $v_1' > v_1''$, $v_2' > v_2''$. Using (28.10) we thus obtain Konowalow's rules for $T = \text{const.}$:

I. At constant temperature the vapour pressure as a function of composition has a stationary point when, and only when, liquid and vapour have the same composition.

II. At constant temperature the vapour pressure is increased by the addition of the component whose concentration is greater in the vapour than in the liquid.

III. If the pressure is changed at constant temperature, the composition of the liquid changes in the same direction as that of the vapour.

74. If we put

$$\left(\frac{\partial \mu_1}{\partial x_1}\right)'_{P,T} - \left(\frac{\partial \mu_1}{\partial x_1}\right)''_{P,T} \equiv \Delta, \tag{A123}$$

the azeotropic point on the isobaric boiling point curve is

a minimum for $\Delta > 0$,

a maximum for $\Delta < 0$.

If $x_1^{(a)}$ is the mole fraction in the azeotropic mixture we have further that

$$\frac{\mathrm{d}x_1^{(a)}}{\mathrm{d}T} = \frac{(1 - x_1^{(a)})(L_1 - L_2)}{T\Delta} \tag{A124}$$

with

$$L_i = h_i' - h_i'' > 0. \tag{A125}$$

Wrewsky's theorem follows from (A123) and (A124).

75.
$$\frac{\mathrm{d}T}{\mathrm{d}x_i'} = \frac{T(x_1'' - x_1')(\partial \mu_1/\partial x_1)'_{T,P}}{(1 - x_1')[x_1''(h_1' - h_1'') + (1 - x_1'')(h_2' - h_2'')]}, \tag{A126}$$

$$\frac{\mathrm{d}T}{\mathrm{d}x_1''} = \frac{T(x_1'' - x_1')(\partial \mu_1/\partial x_1)''_{T,P}}{(1 - x_1'')[x_1'(h_1' - h_1'') + (1 - x_1')(h_2' - h_2'')]}. \tag{A127}$$

76. The following equations are valid for the binary liquid:

(a) Co-existence with pure solid component 1

$$\frac{\mathrm{d}T}{\mathrm{d}x_1} = \frac{T}{L_{f1}}\left(\frac{\partial \mu_1}{\partial x_1}\right)_{T,P}. \tag{A128}$$

(b) Co-existence with pure solid component 2

$$\frac{\mathrm{d}T}{\mathrm{d}x_1} = -\frac{T}{L_{f2}}\frac{x_1}{1 - x_1}\left(\frac{\partial \mu_1}{\partial x_1}\right)_{T,P}. \tag{A129}$$

In case (a) $\mathrm{d}T/\mathrm{d}x$ has the same sign as L_{f_1}, in case (b) $\mathrm{d}T/\mathrm{d}x$ has the opposite sign to L_{f_2}. The point of intersection of the two co-existence curves (eutectic point) represents a univariant equilibrium.

77. If the value of x_1 for the solid chemical compound is denoted by a, the co-existing melt obeys the equation

$$\frac{\mathrm{d}T}{\mathrm{d}x_1} = \frac{a-x_1}{(1-x_1)\Lambda_{12}} T\left(\frac{\partial \mu_1}{\partial x_1}\right)_{T,P} \tag{A130}$$

with

$$\Lambda_{12} = a(h_1'' - h_1') + (1-a)(h_2'' - h_2'). \tag{A131}$$

For $\Lambda_{12} > 0$, we have

$$\begin{aligned} \frac{\mathrm{d}T}{\mathrm{d}x_1} &> 0 \quad \text{for} \quad a > x_1 \\ \frac{\mathrm{d}T}{\mathrm{d}x_1} &< 0 \quad \text{for} \quad a < x_1 \\ \frac{\mathrm{d}T}{\mathrm{d}x_1} &= 0 \quad \text{for} \quad a = x_1. \end{aligned} \tag{A132}$$

The isobaric melting point curve therefore has a maximum at this point (dystectic point).

78. The length ED′ of the curve obeys (A130) together with the first of the conditions (A132). The length $\mathrm{D'M_B}$ obeys eq. (A128) together with (P27). The gradient $\mathrm{d}T/\mathrm{d}x$ of the curve is therefore positive in both cases. (A130) and (A128) generally give different values of $\mathrm{d}T/\mathrm{d}x$ at the point D′. The curve therefore has a kink. Pure solid substance B, the solid compound of composition a, and the melt of composition a' co-exist at the point D′. The solid compound cannot, therefore, be in equilibrium with a melt of the same composition (unlike the state of affairs at a dystectic point). On melting, the compound decomposes into the melt of composition a' and a corresponding amount of solid component B. The point D′ is therefore called the melting point of an incongruently melting compound (incongruent melting point) and represents a univariant equilibrium.

79.

(a) If the composition of the phase $CaCO_3$ is characterized by the mole fraction of CaO, i.e. we put $x_1 = \frac{1}{2}$, eq. (31.23) gives

$$\frac{\mathrm{d}\ln P}{\mathrm{d}T} = \frac{2L}{RT^2} \tag{A133}$$

with

$$L = \tfrac{1}{2}H^{*\prime} + \tfrac{1}{2}H^{*\prime\prime} - H^{*\prime\prime\prime} \tag{A134}$$

where $'$ denotes the phase CaO, $''$ the gaseous phase CO_2, and $'''$ the phase $CaCO_3$.

(b) If we remember that the molar enthalpy $H^*_{CaCO_3}$ of the chemical compound $CaCO_3$ is related to the mean molar enthalpy $H^{*\prime\prime\prime}$ of the 'mixed phase' $CaO + CO_2$ by the expression

$$H^*_{CaCO_3} = 2H^{*\prime\prime\prime}, \tag{A135}$$

eqs. (A133) and (A134) are again obtained.

80.

(a) Let the reaction under discussion by I→II. Let

$$\text{I} \to 1 \to 2 \to \ldots \to n \to \text{II}$$

be a reaction path *via* an arbitrarily chosen number of intermediate reactions. The First Law then gives

$$\begin{aligned} \sum u_{\text{II}} - \sum u_{\text{I}} \equiv \Delta U &= \sum [\sum u_i - \sum u_{i-1}] \\ &\equiv \sum_{i=1}^{n+1} \Delta U_i. \end{aligned} \tag{A136}$$

Since V is likewise a function of state we find the heat of reaction at constant pressure

$$\Delta H = \sum_{i=1}^{n+1} \Delta H_i. \tag{A137}$$

(b)

$$6C + 6O_2 = 6CO_2;\ 6\Delta H_{\text{I}}, \tag{I}$$

$$3H_2 + \tfrac{3}{2}O_2 = 3H_2O;\ 3\Delta H_{\text{II}}, \tag{II}$$

$$C_6H_6 + \tfrac{15}{2}O_2 = 6CO_2 + 3H_2O;\ \Delta H_{\text{III}}. \tag{III}$$

If (I) and (II) are added together and (III) is subtracted from the sum, the result is

$$6C + 3H_2 = 3C_6H_6;\quad (6\Delta H_{\text{I}} + 3\Delta H_{\text{II}} - \Delta H_{\text{III}}). \tag{A138}$$

*81.

$$\left(\frac{\partial \Delta U}{\partial T}\right)_V = \sum_i \nu_i C_{V_i}, \tag{A139}$$

$$\left(\frac{\partial \Delta H}{\partial T}\right)_P = \sum_i \nu_i C_{P_i}, \tag{A140}$$

$$\frac{dQ}{dT} = \sum \nu_i C_i + \left(\frac{\partial P}{\partial T}\right)_\xi \Delta V - \left(\frac{\partial P}{\partial \xi}\right)_T \left(\frac{\partial V}{\partial T}\right)_\xi \tag{A141}$$

where the summations are taken over the participants in the reaction with the sign conventions of eq. (33.15).

83.
$$\frac{4\alpha^2}{1-\alpha^2}P = K_p, \tag{A142}$$

$$\frac{4\alpha^2}{1-\alpha}c_0 = K_c \tag{A143}$$

where $c_0 = n_0/V$ is the volume concentration of the gas without any dissociation having taken place.

84. If $C_{V\mathrm{u}}$ is the molar heat capacity at constant volume for the undissociated gas and $C_{V\mathrm{d}}$ is the corresponding quantity for the dissociation product, we have

$$C_V = (1-\alpha)\,C_{V\mathrm{u}} + 2\alpha\,C_{V\mathrm{d}} + \Delta U\left(\frac{\partial\alpha}{\partial T}\right)_v \tag{A144}$$

where ΔU is the heat of dissociation at constant volume. From (A143) we obtain

$$\left(\frac{\partial\alpha}{\partial T}\right)_v = \frac{\alpha(1-\alpha)}{2-\alpha}\,\frac{\Delta U}{RT^2} \tag{A145}$$

which, with (A144), gives

$$C_V = C_{V\mathrm{u}} + \alpha(2C_{V\mathrm{d}} - C_{V\mathrm{u}}) + \frac{(\Delta U)^2}{RT^2}\,\frac{\alpha(1-\alpha)}{2-\alpha}. \tag{A146}$$

85.

(a) We have

$$\left(\frac{\partial u}{\partial v}\right)_T = \Delta U\left(\frac{\partial\alpha}{\partial v}\right)_T, \tag{A147}$$

$$\left(\frac{\partial\alpha}{\partial v}\right)_T = \frac{\alpha(1-\alpha)}{v(2-\alpha)}, \tag{A148}$$

$$\left(\frac{\partial v}{\partial T}\right)_P = \frac{R}{P}(1+\alpha) + \frac{RT}{P}\,\frac{\alpha(1-\alpha^2)}{2RT^2}\,\Delta H. \tag{A149}$$

From these we get

$$C_P - C_V = \frac{[2RT + \alpha(1-\alpha)\,(\Delta U + RT)]^2}{2RT^2(2-\alpha)} \tag{A150}$$

(b)
$$\frac{C_P}{C_V} = \frac{C_{V\mathrm{u}} + \alpha(2C_{V\mathrm{d}} - C_{V\mathrm{u}}) + R(1+\alpha) + [(\Delta H)^2/RT^2]\,[\alpha(1-\alpha^2)/2]}{C_{V\mathrm{u}} + \alpha(2C_{V\mathrm{d}} - C_{V\mathrm{u}}) + [(\Delta U)^2/RT^2]\,[\alpha(1-\alpha)/(2-\alpha)]}. \tag{A151}$$

86.

(a) For $\alpha \approx 0{\cdot}57$, the function $\alpha(T)$ shows a point of inflexion and the function $\partial\alpha/\partial T$ a maximum.

87. y is a maximum for $a = b$, i.e. for an initial reaction mixture of stoichiometric composition.

88.

(a)
$$\ln K_p = -\frac{\Delta H}{RT} + \text{const.} \tag{A152}$$

(b) for $y \ll 1$
$$y \approx P\left(\frac{K_p}{2{\cdot}37}\right)^{\frac{1}{2}} \tag{A153}$$

is approximately true. The degree of formation therefore increases proportionally to the pressure whereas it decreases with increasing temperature. The reaction must be carried out at relatively high temperatures to get an adequate reaction rate. The high temperature causes a decrease in yield which must be compensated by the application of high pressures. These thermodynamic arguments form the basis of the manufacture of ammonia. Departures from ideal gas behaviour must be taken into account in quantitative calculations (because of the high pressures).

89. $$\ln P = -\frac{L_0}{RT} + \frac{C''_{P0}}{R}\ln T + \frac{1}{R}\int_0^T \frac{\mathrm{d}T'}{T'^2}\int_0^T (C''_{Ps} - C'_P)\,\mathrm{d}T'' + i \tag{A154}$$

where L_0 is the limiting value of the heat of vaporization at $T \to 0$.

90.

(a) $$\ln K_p = -\frac{\Delta H_0}{RT} + \frac{\Delta C_{P0}}{R}\ln T + \frac{1}{R}\int_0^T \frac{\mathrm{d}T'}{T'^{12}}\int_0^T \Delta C_{Ps}\,\mathrm{d}T'' + I \tag{A155}$$

where ΔH_0 is the limiting value of the enthalpy of reaction at $T \to 0$.

(b) The expression (P35) is obtained from (34.3), (34.5), and (34.24).

91.

(a)
$$\Delta G = \Delta H_0 - T\int_0^T \frac{\mathrm{d}T'}{T'^2}\int_0^{T'} \Delta C_P\,\mathrm{d}T'' - RT\Delta(j - i) \tag{A156}$$

where ΔH_0 refers to the crystalline state.

(b) According to (38.10), (38.17), and (A156) the Nernst heat theorem can be expressed as

$$i = j \tag{A156a}$$

for all pure condensed phases.

92. According to the Nernst heat theorem (Planck's formulation) we have

$$\left(\frac{\partial^2 S}{\partial P^2}\right)_{T\to 0} = -\frac{\partial}{\partial P}\left(\frac{\partial V}{\partial T}\right)_T = -\frac{\partial}{\partial T}\left(\frac{\partial V}{\partial P}\right)_T \to 0. \tag{A157}$$

93.

(a) We find from (41.16) or (41.19) that the quantity $(\partial V/\partial T)_P$ only appears as its square in the stability conditions. No statement about its sign can, therefore, be made.

(b) The inequality (P37) has nothing to do with the stability conditions but follows from the Maxwell relation (24.24).

94. The heat introduced during a quasi-static isothermal expansion is

$$\mathrm{d}Q = T\left(\frac{\partial P}{\partial T}\right)_V \mathrm{d}V. \tag{A158}$$

Since, for gases, the coefficient of thermal expansion $\alpha > 0$, we find with (41.18) and (41.23) that $(\partial P/\partial T)_V > 0$.

95. We have

$$\left(\frac{\partial P}{\partial V}\right)_S = \frac{C_P}{C_V}\left(\frac{\partial P}{\partial V}\right)_T. \tag{A159}$$

According to (41.16) and (41.21) we have

$$\left(\frac{\partial P}{\partial V}\right)_T < 0, \quad \frac{C_P}{C_V} > 0 \tag{A160}$$

from which the required statement is obtained.

96. Problem 95 is equivalent to the example cited in the text for the purpose of illustrating the Le Châtelier–Braun principle. We see from (43.8) that the stability conditions used here are (41.18) and (41.20). These are equivalent to (A160).

97.

(a) From $|\mathrm{d}_1 S| < |\mathrm{d}_2 S|$ we find that

$$C_P > C_V \tag{A161}$$

which agrees with (41.21).

(b) The stability conditions used are

$$\left(\frac{\partial T}{\partial S}\right)_P > 0, \quad \left(\frac{\partial P}{\partial V}\right)_S < 0. \tag{A162}$$

The simplest way of obtaining them is by means of the enthalpy.

(c) If the validity of (43.2) is assumed, (43.7) only shows that the derivatives $(\partial P_j/\partial X_j)_{P_i}$ and $(\partial X_i/\partial P_i)_{X_j}$ must have the same sign. No stability conditions are therefore obtainable from (43.2). The Le Châtelier–Braun principle is a weaker statement than the stability conditions.

98. For V, T = const. we have

$$\mathrm{d}\sum_i \nu_i \mu_i = \sum_i \frac{RT}{n_i} \nu_i \mathrm{d}n_i \tag{A163}$$

and, therefore,

$$\left[\frac{\mathrm{d}}{\mathrm{d}\xi}\sum_i \nu_i \mu_i\right]_{T,V} = \sum_i \frac{RT}{n_i} > 0. \tag{A164}$$

The transformation theory of § 41 shows that this condition is also fulfilled for P = const.

*99.

(a) If $\delta^2 n$ is transformed into a sum of squares, the result is

$$\delta^2 u = \tfrac{1}{2}[\lambda_1(\delta y_1)^2 + \lambda_2(\delta y_2)^2] \tag{A165}$$

with

$$\lambda_1 = \frac{T}{C_V}, \quad \lambda_2 = -\left(\frac{\partial P}{\partial v}\right)_T. \tag{A166}$$

The definition of the critical point is that (A165) becomes positive semi-definite. If $C_V < 0$, $\delta^2 u$ would be indefinite or negative semi-definite. If $C_V = 0$, $\delta^2 u$ would not exist and there would be no critical point in the ordinary sense.

(b) $\lambda_2 = 0$ at the critical point. In general, therefore, C_V must necessarily be less than $+\infty$ since, otherwise, $\delta^2 u$ would generally be zero which contradicts the definition in (a).

*100. We have the general expression

$$\delta^2 u = \frac{1}{2}\sum_i \lambda_i(\delta y_i)^2. \tag{A167}$$

If s is chosen as the first independent variable, we always have

$$\lambda_1 = \frac{\partial^2 u}{\partial s^2} = \frac{T}{C_V}. \tag{A168}$$

The statement $C_V > 0$ is obtained by a method analogous to that of problem 99. If it were generally true that $C_V \to \infty$, this would lead to the definition of an additional equation for a critical phase which is impossible according to § 45.

101. Variations of the mole fractions vanish for a phase of constant composition. The conditions for a critical phase therefore become identical with those for a one-component system for which the mechanical stability condition is not fulfilled.

102. At the critical solution point we have

$$\begin{vmatrix} \dfrac{\partial^2 F^*}{\partial V^{*2}} & \dfrac{\partial^2 F^*}{\partial V^* \partial x_1} \\ \dfrac{\partial^2 F^*}{\partial x_1 \partial V^*} & \dfrac{\partial^2 F^*}{\partial x_1^2} \end{vmatrix} = \frac{\partial^2 F^*}{\partial V^{*2}} \frac{\partial^2 F^*}{\partial x_1^2} - \left(\frac{\partial^2 F^*}{\partial V^* \partial x_1}\right)^2 = 0. \quad \text{(A169)}$$

$\partial^2 F^*/\partial V^{*2} = 0$ is therefore possible only for particular cases for which $\partial^2 F^*/\partial V^* \partial x_1 = 0$ is also true.

103.
$$\frac{\mathrm{d}P}{\mathrm{d}T} = \frac{1}{vT} \frac{(\partial C_P/\partial T)''_P - (\partial C_P/\partial T)'_P}{(\partial \alpha/\partial T)''_P - (\partial \alpha/\partial T)'_P}, \quad \text{(A170)}$$

$$\frac{\mathrm{d}P}{\mathrm{d}T} = \frac{(\partial \alpha/\partial T)''_P - (\partial \alpha/\partial T)'_P}{(\partial \kappa/\partial T)''_P - (\partial \kappa/\partial T)'_P}, \quad \text{(A171)}$$

$$\frac{\mathrm{d}P}{\mathrm{d}T} = \frac{(\partial \alpha/\partial P)''_T - (\partial \alpha/\partial P)'_T}{(\partial \kappa/\partial P)''_T - (\partial \kappa/\partial P)'_T}. \quad \text{(A172)}$$

104.
$$\alpha_i = -\sum_j \kappa_{ij} \beta_j, \quad \text{(A173)}$$

$$\beta_i = -\sum_j c_{ij} \alpha_j. \quad \text{(A174)}$$

105. According to the Nernst heat theorem we have

$$\left(\frac{\partial S}{\partial T_i}\right)_{T, T_{k \neq i}} \to 0 \quad \text{for} \quad T \to 0 \quad \text{(A175)}$$

and, therefore, because of the Maxwell relation obtained from (55.5),

$$\alpha_i \equiv \left(\frac{\partial S_i}{\partial T}\right)_{\{T_j\}} \to 0 \quad \text{for} \quad T \to 0. \quad \text{(A176)}$$

Correspondingly, we have

$$\left(\frac{\partial S}{\partial S_i}\right)_{T, S_{k \neq i}} \to 0 \quad \text{for} \quad T \to 0 \quad \text{(A177)}$$

and, therefore, using (55.2),

$$\beta_i \equiv \left(\frac{\partial T_i}{\partial T}\right)_{\{S_j\}} \to 0 \quad \text{for} \quad T \to 0. \quad \text{(A178)}$$

106.
$$C_{\{T_j\}} - C_{\{S_j\}} = Tv \sum_{i,j} c_{ij} \alpha_i \alpha_j = Tv \sum_{i,j} \kappa_{ij} \beta_i \beta_j. \tag{A179}$$

107. From
$$\left(\frac{\partial^2 S}{\partial S_i \partial S_j}\right)_{T, S_{k \neq i,j}} \to 0 \quad \text{for} \quad T \to 0 \tag{A180}$$

we find that
$$\frac{\partial}{\partial S_j}\left(\frac{\partial T_i}{\partial T}\right)_{\{S_k\}} = \frac{\partial}{\partial T}\left(\frac{\partial T_i}{\partial S_j}\right)_{T, S_{k \neq j}} \equiv \frac{\partial c_{ij}}{\partial T} \to 0 \quad \text{for} \quad T \to 0. \tag{A181}$$

108.
$$c_{ij}^{(S)} = c_{ij} + \frac{T v_0 \beta_i \beta_j}{C_{\{S_j\}}}. \tag{A182}$$

*109.
$$\frac{C_{\{T_j\}}}{C_{\{S_j\}}} = \frac{|c_{ij}^{(S)}|}{|c_{ij}|}. \tag{A183}$$

*110.
$$\begin{bmatrix} c_{11} & c_{12} & c_{13} & 0 & 0 & 0 \\ c_{12} & c_{11} & c_{13} & 0 & 0 & 0 \\ c_{13} & c_{13} & c_{33} & 0 & 0 & 0 \\ 0 & 0 & 0 & c_{44} & 0 & 0 \\ 0 & 0 & 0 & 0 & c_{44} & 0 \\ 0 & 0 & 0 & 0 & 0 & c_{66} \end{bmatrix}. \tag{A184}$$

*111. The solution is given in the text by eq. (56.21).

*112. The matrix of the elastic compliance coefficients has the same form as (56.21).

We have
$$\kappa_{11} = \frac{c_{11} + c_{12}}{(c_{11} - c_{12})(c_{11} + 2c_{12})}, \tag{A185}$$

$$\kappa_{12} = -\frac{c_{12}}{(c_{11} - c_{12})(c_{11} + 2c_{12})}, \tag{A186}$$

$$\kappa_{44} = \frac{1}{c_{44}}. \tag{A187}$$

*113. For the elastic compliance coefficients, the Cauchy relation is
$$(\kappa_{11} - \kappa_{12})(\kappa_{11} + 2\kappa_{12}) = -\kappa_{12} \kappa_{44}. \tag{A188}$$

The isotropy condition becomes

$$2(\kappa_{11}-\kappa_{12}) = \kappa_{44}. \tag{A189}$$

If the Cauchy relation is obeyed by an isotropic substance, we have

$$\kappa_{11} = -4\kappa_{12} = \tfrac{4}{10}\kappa_{44}. \tag{A190}$$

114.

(a) Triclinic crystal:

$$\kappa = \kappa_{11}+\kappa_{22}+\kappa_{33}+2\kappa_{12}+2\kappa_{13}+2\kappa_{23}. \tag{A191}$$

(b) Cubic crystal:

$$\kappa = \frac{3}{c_{11}+2c_{12}}. \tag{A192}$$

(c) Isotropic substance:

$$\kappa = \frac{3}{2\mu_{\mathrm{L}}+3\lambda_{\mathrm{L}}} = \frac{3(1-2\nu)}{E_{\mathrm{Y}}}. \tag{A193}$$

115.

(a) Triclinic crystal:

$$|c_{ij}| \text{ with all principal minors } > 0. \tag{A194}$$

(b) Tetragonal crystal:

$$\begin{aligned} &c_{11}>0, \quad c_{33}>0, \quad c_{44}>0, \quad c_{66}>0 \\ &c_{11}^2-c_{12}^2>0 \\ &c_{33}(c_{11}+c_{12})-2c_{13}^2>0. \end{aligned} \tag{A195}$$

(c) Cubic crystal:

$$\begin{aligned} &c_{11}>0, \quad c_{44}>0 \\ &c_{11}^2-c_{12}^2>0 \\ &c_{11}(c_{11}+c_{12})-2c_{12}^2>0. \end{aligned} \tag{A196}$$

(d) Isotropic substance:

$$\mu_{\mathrm{L}}>0, \quad \lambda_{\mathrm{L}}+\mu_{\mathrm{L}}>0. \tag{A197}$$

116.

(a) Tetragonal crystal:

$$\begin{aligned} &\kappa_{11}>0, \quad \kappa_{33}>0, \quad \kappa_{44}>0, \quad \kappa_{66}>0 \\ &\kappa_{11}^2-\kappa_{12}^2>0 \\ &\kappa_{33}(\kappa_{11}+\kappa_{12})-2\kappa_{13}^2>0. \end{aligned} \tag{A198}$$

(b) Cubic crystal:

$$\kappa_{11} > 0, \quad \kappa_{44} > 0$$

$$\kappa_{11}^2 - \kappa_{12}^2 > 0 \tag{A199}$$

$$\kappa_{11}(\kappa_{11} + \kappa_{12}) - 2\kappa_{12}^2 > 0.$$

(c)
$$E_Y < 0, \quad 1 - \nu > 0. \tag{A200}$$

117.
$$\left(\frac{\partial S}{\partial S_1}\right)_T = -\left(\frac{\partial T_1}{\partial T}\right)_{S_1}. \tag{A201}$$

According to this, an increase in temperature at constant extension must increase the stress if an extension is associated with an entropy decrease, i.e. with evolution of heat.

*118.

(a) No information about χ can be obtained from the stability conditions since the sign of the term $\boldsymbol{E}.\mathrm{d}\boldsymbol{P}$ in eq. (58.9) is determined by the sign of χ.

(b)
$$\varepsilon > 0. \tag{A202}$$

119.

(a)
$$s_i = s_{0i} + \frac{1}{2}\left(\frac{\partial \chi_i}{\partial T}\right)_{P,E,n} \boldsymbol{E}^2. \tag{A203}$$

(b)
$$h_i = h_{i0} + \frac{1}{2}\left[\chi_i + T\left(\frac{\partial \chi_i}{\partial T}\right)_{P,E,n}\right] \boldsymbol{E}^2. \tag{A204}$$

(c)
$$v_i = v_{0i} - \frac{1}{2}\left(\frac{\partial \chi_i}{\partial P}\right)_{T,E,n} \boldsymbol{E}^2. \tag{A205}$$

120.

(a)
$$\frac{\mathrm{d}T}{\mathrm{d}\boldsymbol{E}} = -\frac{T\Delta(v\chi)\,\boldsymbol{E}}{\Delta H^*}, \quad (P = \text{const.}) \tag{A206}$$

(b)
$$\frac{T - T_0}{T} = \frac{1}{2}\frac{\Delta(v\chi)}{\Delta H_0^*}\boldsymbol{E}^2. \tag{A207}$$

121.

(a)
$$\frac{\mathrm{d}x_1'}{\mathrm{d}\boldsymbol{E}} = \frac{(1-x_1')\,[x_1''(\chi_1' - \chi_1'') + (1-x_1'')\,(\chi_2' - \chi_2'')]\,\boldsymbol{E}}{(x_1'' - x_1')\,(\partial\mu_1/\partial x_1)'_{T,P,E}}, \tag{A208}$$

$$\frac{\mathrm{d}x_1''}{\mathrm{d}\boldsymbol{E}} = \frac{(1-x_1'')\,[x_1'(\chi_1' - \chi_1'') + (1-x_1')\,(\chi_2' - \chi_2'')]\,\boldsymbol{E}}{(x_1'' - x_1')\,(\partial\mu_1/\partial x_1)''_{T,P,E}}. \tag{A209}$$

(b)

$$\lim_{x_1'-x_1''\to 0} \frac{\mathrm{d}x_1''/\mathrm{d}\boldsymbol{E}}{\mathrm{d}x_1'/\mathrm{d}\boldsymbol{E}} = \frac{(\partial\mu_1/\partial x_1)'_{T,P,\boldsymbol{E}}}{(\partial\mu_1/\partial x_1)''_{T,P,\boldsymbol{E}}} \tag{A210}$$

For $x_1' = x_1''$ we therefore have, in general, that $\mathrm{d}x_1''/\mathrm{d}\boldsymbol{E} \neq \mathrm{d}x_1'/\mathrm{d}\boldsymbol{E}$. The application of an electric field will, therefore, destroy the azeotropic character.

122.
$$C_{P,\boldsymbol{E}}(T,\boldsymbol{E}) = C_{P,\boldsymbol{E}}(T,0) + \frac{[\mu]^2}{3RT^2}\boldsymbol{E}^2. \tag{A211}$$

123.
$$K_p(T,\boldsymbol{E}) = K_p(T,0)\exp\left(\frac{\Delta\bar{\chi}\boldsymbol{E}^2}{2RT}\right) \tag{A212}$$

with

$$\Delta\bar{\chi} = \sum \nu_i \chi_i. \tag{A213}$$

124.
$$C_{P,\boldsymbol{H}}(T,\boldsymbol{H}) = C_{P,\boldsymbol{H}}(T,0) + \frac{CT}{(T-\Theta)^3}\boldsymbol{H}^2. \tag{A214}$$

125. If the initial state is denoted by ′ and the final state by ″, we have

$$\ln\frac{T''}{T'} = \tfrac{1}{2}\ln\frac{B+C\boldsymbol{H}''^2}{B+C\boldsymbol{H}'^2}. \tag{A215}$$

Bibliography

A. LARGE GENERAL WORKS AND TEXTBOOKS

J. W. Gibbs. *Collected Works*, vol. I, Yale University Press, New Haven, 1957.

J. D. van der Waals and Ph. Kohnstamm. *Lehrbuch der Thermostatik*, Barth, Leipzig, 1927.

W. Schottky, H. Ulich and C. Wagner. *Thermodynamik*, Springer, Berlin, 1929.

M. Planck. *Vorlesungen über Thermodynamik*, 11th ed., De Gruyter, Berlin, 1964.

G. N. Lewis and M. Randall. *Thermodynamics*, 2nd ed., McGraw-Hill, New York, 1961.

M. W. Zemanski. *Heat and Thermodynamics*, 4th ed., McGraw-Hill, New York, 1957.

I. Prigogine and R. Defay. *Chemical Thermodynamics*, Longmans, Green, London, 1954.

A. B. Pippard. *The Elements of Classical Thermodynamics*, Cambridge University Press, Cambridge, 1957.

R. Haase. *Thermodynamik der Mischphasen*, Springer, Berlin, 1956.

H. Callen. *Thermodynamics*, Wiley, New York, 1960.

J. G. Kirkwood and I. Oppenheim. *Chemical Thermodynamics*, McGraw-Hill, New York, 1961.

E. A. Guggenheim. *Thermodynamics*, 5th ed., North-Holland, Amsterdam, 1967.

B. AXIOMATICS

A. Landé. *Die Carathéodory'sche Axiomatik* (*Geiger-Scheel Handbuch der Physik*), vol. IX, Springer Verlag, Berlin, 1926.

G. Falk and H. Jung. *Axiomatik der Thermodynamik* (Handbuch der Physik, (S. Flügge Ed.) Vol. III. 2), Springer Verlag, Berlin, 1959.

P. T. Landsberg, *Thermodynamics*, Interscience, New York, 1961.

R. Giles. *Mathematical Foundations of Thermodynamics*, Pergamon, Oxford, 1964.

H. A. Buchdahl. *The Concepts of Classical Thermodynamics*, Cambridge University Press, Cambridge, 1966.

L. Tisza. *Generalized Thermodynamics*, M.I.T. Press, Cambridge (Mass.), 1966.

C. HETEROGENEOUS EQUILIBRIA

H. W. Bakhuis Roozeboom. *Die heterogenen Gleichgewichte*, Friedr. Viehweg, Braunschweig, 1901–1911.
M. Hansen. *The Constitution of Binary Alloys*, 2nd ed., McGraw-Hill, New York, 1958.

D. NERNST'S HEAT THEOREM

W. Nernst. *Theoretische und experimentelle Grundlagen des neuen Wärmesatzes*, 2nd ed., W. Knapp, Halle, 1924.
K. Bennewitz. *Der Nernst'sche Wärmesatz* (*Geiger-Scheel Handbuch der Physik*), vol. IX, Springer Verlag, Berlin, 1926.
F. Simon. *Ergebnisse der exakten Naturwissenschaften*, 9, 222, Springer, Berlin, 1930.
F. Simon. *Science Museum Handbook*, **3**, 61, 1937.
J. Wilks. *The Third Law of Thermodynamics*, Oxford University Press, Oxford, 1961.

E. HIGHER ORDER TRANSITIONS

L. Tisza. *On the General Theory of Phase Transitions* (Phase Transformations in Solids, R. Smoluchowski, J. E. Mayer and W. A. Weyl, Eds.), p. 1, Wiley, New York, 1951.

F. SOLIDS

W. Voigt. *Lehrbuch der Kristallphysik*, Teubner, Leipzig, 1928.
A. Sommerfeld. *Mechanik der deformierbaren Medien* (Vorlesungen über theoretische Physik), vol. II, 4th ed., Akad. Verlagsges., Geest and Portig K.-G., Leipzig, 1957.
L. D. Landau and E. M. Lifshitz. *Theory of Elasticity* (Course of Theoretical Physics), vol. 7, Pergamon, Oxford, 1959.

G. SYSTEMS IN ELECTRIC AND MAGNETIC FIELDS

A. Sommerfeld. *Elektrodynamik* (Vorlesungen über Theoretische Physik), vol. III, 3rd ed., Akad. Verlagsges., Leipzig, 1961.
L. D. Landau and E. M. Lifshitz. *Electrodynamics of Continuous Media* (Course of Theoretical Physics), vol. 8, Pergamon, Oxford, 1960.
J. D. Jackson. *Classical Electrodynamics*, Wiley, New York, 1962.
W. Känzig. *Ferroelectrics and Antiferroelectrics* (Solid State Physics, F. Seitz and D. Turnbull, Eds.), vol. 4, p. 1, Academic, New York, 1957.
E. Fatuzzo and W. J. Merz. *Ferroelectricity*, North Holland, Amsterdam, 1967.
J. H. van Vleck. *Theory of Electric and Magnetic Susceptibilities*, Oxford University Press, Oxford, 1952.
H. M. Rosenberg. *Low Temperature Solid State Physics*, Oxford University Press, Oxford, 1963.
P. G. de Gennes. *Superconductivity of Metals and Alloys*, Benjamin, New York, 1966.

H. ELECTROCHEMICAL SYSTEMS

H. Falkenhagen. *Elektrolyte,* Hirzel, Leipzig, 1953.
H. S. Harned and B. Owen. *The Physical Chemistry of Electrolytic Solutions,* 3rd ed., Reinhold Publ. Corporation, New York, 1958.
R. A. Robinson and R. H. Stokes. *Electrolyte Solutions,* 2nd ed., Butterworth, London, 1959.

I. STATISTICAL THERMODYNAMICS

R. H. Fowler and E. A. Guggenheim. *Statistical Thermodynamics,* Cambridge University Press, Cambridge, 1949.
A. Münster. *Statistical Thermodynamics,* 2nd ed., Springer Verlag, Berlin–New York, 1969.

J. TABLES

F. Rossini. *Selected Values of Chemical Thermodynamic Properties,* Carnegie Press, Washington, 1952.
K. K. Kelley. *Entropies of Inorganic Substances,* Bur. of Mines Bull. 477, U.S.G.P.O., 1950.
K. K. Kelley. *High Temperature Heat Content,* Heat Capacity and Entropy Data, Bur. of Mines Bull. 476, U.S.G.P.O., 1949.
Landolt-Börnstein. *Zahlenwerte und Funktionen,* 6th ed., vol. II, 2nd–4th parts, Springer Verlag, Berlin, 1956–1964.

K. MATHEMATICAL AIDS

H. Margenau and G. M. Murphy. *The Mathematics of Physics and Chemistry,* 2nd ed., D. Van Nostrand Company Inc., New York, 1962.

Index

Absolute
  temperature 33
    Thomson's definition 12
  zero, unattainability of 150
Activities 137
Activity coefficients 136
  for solid solutions 305
  of electrolyte, mean 286
    determination 304
  of ions 285
  of solid mixed phases 137, 139
  of solvent 285
  standardization 137
Adiabatic
  curve 20
  demagnetization 154, 270
  process 12
  wall 20
Affinity 48
Amount of substance 316
Azeotropic
  curve 118
  mixtures 117
  point 125

Barometric height equation 312
Binodal curve 178
Bivariant equilibria 124–127
Boiling point of solvent, elevation 319–322
Boudouard equilibrium 138
Bridgeman's method 95–97

Carathéodory's principle 30
Carnot
  cycle 12
  theorem 11
Cells—*see* Galvanic cell *and* Half-cell
Celsius scale 22
Characteristic function 69
Chemical
  constants 338
  equilibrium,
    general conditions for 129, 131
    in gravitational field 310
    stability 176, 177
  potential 50, 76, 103
    for gases, standard values 133
    in presence of electric field 246
    of electrolyte 287
    of liquid mixtures, standard states 136
    of solvent 107
  reactions,
    phase rule in 140
    thermodynamic calculation 141–144
Clausius–Clapeyron equation 120
  generalized,
    for bivariant and multivariant equilibria 125
    for univariant equilibrium 123
Clausius'
  inequality 16
  principle 9
  relation 324
Coefficients,
  activity 136, 137, 139, 285, 286, 304, 305
  elastic compliance 231
  elastic stiffness 229, 341
  Joule–Thomson 101
  of thermal expansion 95
  osmotic 288
  stoichiometric 46
  thermal strain 340

Coefficients (*contd.*)
thermal stress 340
thermodynamic 205
work 46
Co-existence curves 119, 178
Compressibility,
isentropic 166, 170
isothermal 170, 237
Congruently melting compound 334
Conjugate parameters 76
Connodals 178
Consolute point,
lower 179
upper 179
Contact equilibrium 49
Corresponding states, principle of 158
Critical
opalescence 192
parameters 213
phase 178
degrees of freedom 182
general equations 188
transformation of equations 189–192
phenomena 191
points 178
and higher-order transitions, Tisza's theory 208–215
temperature—*see* Temperature
Cryoscopic constant 322
Crystalline solid, vapour pressure 337
Crystals,
ferroelectric 255
pyroelectric 254
Curie point, ferroelectric 255
Curie–Weiss law 271
Cyclic process 8
reversible, efficiency of 10

Dalton's law of partial pressures 131
Deformation,
infinitesimal, quasi-static work done in 226
small, mathematical description 217
Degree of
dissociation, dependence of molar heat capacity 336
long-range order 202
Degrees of freedom of a critical phase 182
Densities 73
Diamagnetic substances 269
Diathermic wall 20
Dielectric
constant 239
tensor 239
Diesel engine 325
Differential equations
for isothermal vapour pressure diagram 333
for temperature dependence of heat of reaction 335
Differentials—*see also* Pfaff differentials
complete 26
incomplete 25
Donnan equilibrium 293–295
Dystectic points 128, 334

Ebullioscopic constant 322
Ehrenfest equations for second order transitions 198, 277
Einstein condensation 332
Elastic
compliance coefficients, isothermal 231
stiffness coefficients, isothermal 229
Electrical
neutrality, condition of 280
susceptibility 242
Electric field,
chemical potentials in presence of 246
fundamental equation for body in 245
Gibbs free energy for body in 246
Helmholtz free energy for body in 246
quasi-static work done on system in 242, 244
Electrocaloric effect 251
Electrochemical
equilibrium,
general conditions for 282–284
in galvanic cell 300
potential 283
systems 279

Electrochemical systems (*contd.*)
fundamental equation 283
Electrode potentials, single 286, 303
Electrode, standard 302
Electrolytes 279
activity coefficients, mean 286
determination 304
binary 280
volatile, vapour pressure equilibrium of aqueous solution 290
chemical potential 287
mole fraction 287
mean 287
stoichiometric 287
solutions, membrane equilibria 292–295
strong, solutions of 284–292
Electromotive force (e.m.f.),
of galvanic cell 297
standard values,
for galvanic cell 301
for half-cell 303
temperature dependence 304
Electrostriction 247
Energy—*see also* Free energy
internal 9, 23
representation 68, 76
stability conditions 165
Enthalpy 78—*see also* Heat content
of body in magnetic field 268
standardization 155
standard state 155
Entropy 13, 33, 50
calorimetric 157
conventional zero point 156
of dilution 107
representation 68
spectroscopic 157
standardization 156
standard state 156
Equations,
Clausius–Clapeyron 120, 123, 125
Ehrenfest 198, 277
fundamental 50, 67, 68, 69, 74, 76, 83, 228, 245, 268, 283
Gibbs–Duhem 69, 75, 78, 104, 105, 107, 229
Gibbs–Helmholtz 91, 92, 134, 149, 304
Equations (*contd.*)
Kirchhoff 320
of state 69, 72
caloric 72
thermal 72
thermal 22, 229, 316
van der Waals 324
van't Hoff 135
vapour pressure 333
Equilibrium 6, 52
bivariant 124–127
Boudouard 138
chemical,
general conditions 129, 131
in gravitational field 310
stability 176, 177
conditions,
for constant entropy, volume and mole number 88
for constant temperature, pressure and mole number 88
for constant temperature, volume and mole number 88
for homogeneous fluid in gravitational field 310
general, for heterogeneous systems 109–111
transformation 85–90
constants 132
contact 49
Donnan 293–295
electrochemical,
general conditions 282–284
in galvanic cell 300
frozen 145
Gibbs', conditions 56
hydrostatic, condition 311
invariant 119
membrane 111–113
non-osmotic 293
of electrolyte solutions 292–295
osmotic 112, 293–295
metastable 62, 160
multivariant 124–127
neutral 62
sedimentation,
equations for 312
in centrifugal field 313
stable 61

Equilibrium (*contd.*)
  thermal 21
  thermodynamic 52
    for galvanic cell 297
  univariant 120
  unstable 62
  vapour pressure, of aqueous solution of volatile binary electrolyte 290
Equivalence principle 8
Euler's theorem 67
Eutectic mixtures 123
Extensive parameters 49, 68

Faraday 283
Ferroelectric
  crystals 255
  Curie point 255
First Law of thermodynamics—*see* Thermodynamics
First-order transitions—*see* Transitions
Force, generalized 46
Free energy
  Gibbs 81
    for body in electric field 246
    for solids 230
    mean molar 102
    of body in magnetic field 269
  Helmholtz 80
    for body in electric field 246
    for solids 230
    of body in magnetic field 268
  of dilution 107
Freezing point, of solvent, lowering 322
Fugacities 132
Fundamental equation 50, 67, 68, 76, 83
  for body in electric field 245
  for electrochemical systems 283
  for solids 228
  of body in magnetic field 268
  per mole 74
  properties 69

Galvanic cell—*see also* Half-cell
  definition 296
Galvanic cell (*contd.*)
  electrochemical equilibrium, general condition for 300
  electromotive force (e.m.f.) 297
    standard value 301
  irreversible 296, 304
  reversible 296
  thermodynamic equilibrium 297
Gas
  ideal 22
  one-component, thermal equation of state 316
  reactions, homogeneous 131–136, 143
  thermometer scale 22
Gases,
  chemical potentials 133
  ideal mixture 131
Gay–Lussac's streaming experiment 36, 38, 326
Generalized force 46
Gibbs–Dalton law 133
Gibbs–Duhem equation 69, 75, 78
  for solids 229
  generalized 104
  isothermal–isobaric form 105, 107
Gibbs' equilibrium conditions 56
Gibbs free energy—*see* Free energy
Gibbs–Helmholtz equation 91, 304
  integration 92, 143, 149
Gibbs–Konowalow rule, generalized 125
Gibbs' phase rule 114
Gibbs' stability criterion 162
Gough–Joule effect 342
Grand potential 82

Half-cell 302—*see also* Galvanic cell
  standard e.m.f. values 303
Heat 24
  capacity,
    at constant volume 71
    molar,
      at constant pressure 80, 95, 169
      at constant strain 340
      at constant stress 340
      at constant volume 25, 99, 165, 169

Heat (*contd.*)
capacity, molar (*contd.*)
at saturation vapour pressure 333
dependence on degree of dissociation 336
conducting wall 20
content 78—*see also* Enthalpy
of dilution 107
of melting 120
of phase change 120
of reaction, differential equations for temperature dependence 335
of vaporization 120
pump 18
'uncompensated' 16
Helium, λ-transition 194, 199
Helmholtz free energy—*see* Free energy
Helmholtz' fundamental theorem of kinematics 218
Hess' Law 335
Heterogeneous
reactions 138–140
systems, stability conditions 171–173
Homogeneous
fluid, equilibrium conditions in gravitational field 310
functions 67
gas reactions 143
solution reactions, in liquid state 136, 137
system 7, 308
Hydrostatic equilibrium, condition for 311
Hyperstructure transitions—*see* Transitions

Incongruently melting compound 335
Ideal
gas 22
solution, dilute 137
Indifference conditions 117
Indifferent curve 118
Integrating factor, condition for existence 26
Intensive parameters 49, 68, 76
Internal
energy—*see* Energy
parameters 53
Invariant equilibria 119
Inversion curve 329
Ions, activity coefficients of 285
Irreversible processes 15
Isentropic
compressibility 166, 170
elastic stiffness coefficients 341
processes 13
Isothermal
compressibility 95, 170, 237
elastic compliance coefficients 231
elastic stiffness coefficients 229
process 12
vapour pressure diagram, differential equations 333
Isotherms 23, 178, 184

Jacobi determinants, method using 97–100
Joule–Thomson
coefficient 101
effect 100–102
integral (total) 331

Kelvin scale 12
Kinematics, Helmholtz' fundamental theorem 218
Kirchhoff's equation 320
Konowalow's rules 126

Lagrange method of undetermined multipliers 110, 289, 291, 310
Lamé constants 234
Law,
Curie–Weiss 271
Dalton's law of partial pressures 131
Gibbs–Dalton 133
Hess 335
of mass action 132
Le Châtelier–Braun principle 175
Le Châtelier's principle 173
Legendre transformations 63–67
Linear dilatations 221
Liquid
mixtures, standard states of chemical potentials of 136

Liquid (*contd.*)
state, homogeneous solution reactions 136, 137

Magnetic
field,
enthalpy of body in 268
fundamental equation of body in 268
Gibbs free energy of body in 269
Helmholtz free energy of body in 268
quasi-static work done on system in 268
permeability 265
substances—*see* Diamagnetic substances *and* Paramagnetic substances
susceptibility 267
Magnetization,
adiabatic demagnetization 270
density 267
total 268
Magnetocaloric effect 270
Magnetostatic work 265–267
Mass action,
law of 132
equilibrium constants of 132
Massieu–Planck functions 83
Mathematical description of small deformations 217
Maxwell relations 71, 93, 230
Mechanical stability condition 166
Membrane
equilibria 111–113
non-osmotic 293
of electrolyte solutions 292–295
osmotic 293–295
potential 293, 295, 303
Metastable equilibrium 160
Miscibility gap, closed 180
Molality 136
Molar heat
capacity—*see* Heat capacity
Molarity 136
Molar quantities 73
mean, of state 103
partial 103
Molecular weight 316
determination 315–322
thermodynamic definition 317
Mole fractions 102
of electrolyte 287
mean 287
stoichiometric 287
Motion, perpetual,
first kind 8
second kind 9
Multivariant equilibria 124–127

Nernst heat theorem 148, 304, 338, 340
Planck's formulation 156
Non-static
change of state 20
processes, Second law applied to 41–45

Onsager type of transition 198
Order, degree of long-range 202
Order–disorder transitions—*see* Transitions
Osmotic
coefficient 288
equilibrium 112, 293–295
pressure 113, 295
of binary solution 317

Paramagnetic substances 269
Parameters,
conjugate 76
extensive 49, 68
intensive 49, 68, 76
internal 53
Partial
derivatives, conversion 94–102
pressures 131
Dalton's law of 131
Pfaff
differential equation 26
differentials 25–30
Phase 7, 308—*see also* Critical phase
solid mixed, activity coefficients 137, 139
Phase
reaction 115
rule in chemical reactions 140

Piezo-electricity 252
Poisson's ratio 234
Polarization,
  density 241
  total 244
Potential—*see also* Chemical potential
  electrochemical 283
  grand 82
  membrane 293, 295, 303
  single electrode 286, 303
Pressure—*see* Osmotic pressure, Partial pressure *and* Vapour pressure
Progress variable 47, 53, 141, 299
Pyroelectric
  crystals 254
  effect 254

Quadruple points 119
Quasi-static
  change of state 20
  processes, Second Law applied to 30–35
  work—*see* Work

Reaction isochore 135
Refrigerator 17
Retrograde
  condensation 191
  vaporization 191
Rotation transitions—*see* Transitions

Second Law of thermodynamics—*see* Thermodynamics
Second-order transitions—*see* Transitions
Sedimentation equilibrium,
  equations for 312
  in centrifugal field 313
Single electrode potential 286
Solids,
  crystalline,
    ferroelectric 255
    pyroelectric 254
    vapour pressure 337
  free energy,
    Gibbs 230
Solids (*contd.*)
  free energy (*contd.*)
    Helmholtz 230
  fundamental equation 228
  Gibbs–Duhem equation 229
  stability conditions 236
  symmetry properties 231–237
  thermal equation of state 229
Solid solutions
  activity coefficients 305
Solution,
  binary, osmotic pressure 317
  ideal dilute 137
Solution point, upper critical 179
Solvent,
  activity coefficient 285
  elevation of the boiling point 319–322
  lowering,
    of the freezing point 322
    of the vapour pressure 318, 319
Spinodal curve 185
Stability
  conditions 61, 62, 89, 90, 207, 208
    for heterogeneous systems 171–173
    for solids 236
    in energy representation 165
    mechanical 166
    thermal 165
    transformation 85–90, 166–171
  of chemical equilibria 176, 177
Standard
  electrode—*see* Electrode
  state,
    for enthalpy 155
    for entropy 156
    of chemical potentials of liquid mixtures 136
State 5—*see also* Corresponding state, Equations of state *and* Standard state
  change of 20
    non-static 20
    quasi-static 20
  thermal equation 22, 229, 316
  variable of 6
Stoichiometric coefficient 46

Strain
  coefficients, thermal 340
  ellipsoid 220
  tensor 220
Stress
  coefficients, thermal 340
  ellipsoid 224
  tensor 224
Stresses,
  normal 224
  rectangular 224
  tangential 224
Superconduction 272–278
Susceptibility,
  electrical 242
  magnetic 267
Svedberg's formula 314
System 5, 6
  binary 7
  closed 6
  heterogeneous 7
  homogeneous 7
  independent components 7
  isolated 7
  open 7
  simple 8
  ternary 7

Tangent method 106
Temperature,
  absolute 33
    Thomson's definition 12
  critical 178
  dependence,
    of e.m.f. 304
    of heat of reaction, differential equations for 335
  empirical 22
  very low, measurement of 39–41
Thermal
  equation of state 22
    for solids 229
    of one-component gas 316
  equilibrium 21
  expansion coefficient 95
  processes 19
  stability condition 165
  strain coefficients 340
  stress coefficients 340
Thermodynamic
  calculation of chemical reactions 141–144
  coefficients 205
  definition of molecular weight 317
  degrees of freedom 114
  equilibrium 52
    for galvanic cell 297
  moduli 204
  potentials 76
Thermodynamics,
  First Law 8, 9, 23–25
  Second Law 9–17
    applied to non-static processes 41–45
    applied to quasi-static processes 30–35
    general formulation 50
Thomson's principle 9
Tisza's theory
  of critical points and higher-order transitions 208–215
Transition points
  electric 195
  magnetic 195
λ-Transition of helium 194, 199
Transitions,
  displacive 214
  first-order 195
  higher order 194
    and critical points, Tisza's theory 208–215
  hyperstructure 194, 201, 215
  Onsager type 198
  order-disorder 213
  rotation 195
  second order 197, 277
    Ehrenfest equations 198, 277
Triple points 119

Ultracentrifuge 315
Univariant equilibrium 120

van der Waals' equation of state 324
van't Hoff equation 135
Vapour pressure
  constant 337

Vapour pressure (*contd.*)
  diagram, isothermal, differential equations 333
  equation 333
  equilibrium, of aqueous solution of binary electrolyte 290
  of crystalline solid 337
  of solvent, lowering of 318, 319
  saturation, molar heat capacities 333
Virial
  coefficients 317
  equations 317
Virtual displacement 56

Wall,
  adiabatic 20
  diathermic 20
  heat conducting 20
Work
  coefficient 46
  co-ordinates 46, 216
Work,
  electrostatic 238–242
  magnetostatic 265–267
  quasi-static
    done on, deformable body 228
      material system in electric field 242
      system in magnetic field 268
        in electric field 244
    in infinitesimal deformation 226
Wrewsky's theorem 334

Young's modulus 234

Zero point entropy, conventional 156
Zero, unattainability of absolute 150